Advances in Pattern Recognition

T0142868

Advances in Pattern Recognition is a series of books which brings together current developments in all areas of this multi-disciplinary topic. It covers both theoretical and applied aspects of pattern recognition, and provides texts for students and senior researchers.

Springer also publishes a related journal, **Pattern Analysis and Applications**. For more details see: http://springlink.com

The book series and journal are both edited by Professor Sameer Singh of Loughborough University, UK.

Also in this series:

Principles of Visual Information Retrieval
Michael S. Lew (Ed.)
1-85233-381-2

Statistical and Neural Classifiers: An Integrated Approach to Design
Šarūnas Raudys
1-85233-297-2

Advanced Algorithmic Approaches to Medical Image Segmentation
Jasjit Suri, Kamaledin Setarehdan and Sameer Singh (Eds)
1-85233-389-8

NETLAB: Algorithms for Pattern Recognition
Ian T. Nabney
1-85233-440-1

Object Recognition: Fundamentals and Case Studies
M. Bennamoun and G.J. Mamic
1-85233-398-7

Computer Vision Beyond the Visible Spectrum
Bir Bhanu and Ioannis Pavlidis (Eds)
1-85233-604-8

Hexagonal Image Processing: A Practical Approach
Lee Middleton and Jayanthi Sivaswamy
1-85233-914-4

Support Vector Machines for Pattern Classification
Shigeo Abe
1-85233-929-2

Digital Document Processing
Bidyut B. Chaudhuri (Ed.)
978-1-84628-501-1

Recent Advances in Pattern Recognition
Sameer Singh and Maneesha Singh (Eds)
978-1-84628-944-6

Human Ear Recognition by Computer
Bir Bhanu and Hui Chen
978-1-84800-128-2

Zheng-Hua Tan and Børge Lindberg

Automatic Speech Recognition on Mobile Devices and over Communication Networks

 Springer

Zheng-Hua Tan, BSc, MSc, PhD
Børge Lindberg, MSc
Department of Electronic Systems, Aalborg University, Aalborg, Denmark

Series editor
Professor Sameer Singh, PhD
Research School of Informatics, Loughborough University, Loughborough, UK

Advances in Pattern Recognition Series ISSN 1617-7916
ISBN-13: 978-1-84996-736-5   e-ISBN-13: 978-1-84800-143-5
DOI 10.1007/978-1-84800-143-5

British Library Cataloguing in Publication Data
A catalogue record for this book is available from the British Library

Printed on acid-free paper

9 8 7 6 5 4 3 2 1

Springer Science+Business Media
springer.com

Preface

The remarkable advances in computing and networking have sparked an enormous interest in deploying *Automatic Speech Recognition on Mobile Devices and Over Communication Networks*, and the trend is accelerating. This yields an abundance of practical systems, operational algorithms and scientific publications. There is, however, no integrated book available that portrays the whole picture of this area. Our primary impetus for editing this book is to fill this gap by providing a comprehensive and unified introduction to the field.

The prevalence of mobile devices, coupled with the proliferation of wireless networks, creates new opportunities for speech recognition technology. Mobile devices are small in size and are used while on the move, both of which make speech-enabled user interfaces attractive in comparison with other interaction modes like keypad and stylus. The opportunities come along with challenges as well. For instance, it is not an easy task to port state-of-the-art speech recognition systems onto computationally limited devices such as mobile phones, PDAs and automobiles where they are highly desirable. Fortunately, the barriers are being removed because of increasingly powerful embedded platforms and pervasive network connections. Still, however, the accompanying research and engineering issues are many: computational constraints and power limitations on the devices, speech coding and transmission deteriorations over the networks, diverse operating systems and hardware configurations, to name just a few. To address these issues requires a wide scope of knowledge and experience.

This book brings together leading researchers and practitioners from academia and industry to provide an in-depth review of methods and standards, share working knowledge, and present state-of-the-art systems and applications. We cover network speech recognition, distributed speech recognition and embedded speech recognition, which are expected to co-exist in the coming years.

Organization and Features

The book begins with an overview chapter and is then divided into four parts: network speech recognition, distributed speech recognition, embedded speech recognition, and systems and applications.

Chapter 1 gives a comprehensive overview of network, distributed and embedded speech recognition and discusses the pros and cons of the presented approaches. This chapter sets the scene for the entire book.

Part I, **Network Speech Recognition**, focuses on remote speech recognition that uses conventional speech coders for the transmission of speech from a client device to a recognition server where feature extraction and recognition decoding take place. This part consists of three chapters.

Chapter 2 first describes the commonly used speech coding standards for mobile and IP networks, and then investigates the effect of speech codecs on speech and speaker recognition performance, with or without packet loss. Chapter 3 addresses issues related to speech recognition over mobile networks, and presents solutions to the performance degradation caused by speech coding algorithms, transmission errors and environmental noise. Chapter 4 reviews robustness techniques against packet loss in the context of voice over IP-based network speech recognition, and introduces a CELP-type speech coder optimized for speech recognition over IP networks.

Part II, **Distributed Speech Recognition**, makes a thorough presentation of speech recognition that adopts the client-server architecture by placing feature extraction in the client and recognition decoding in the server. It begins with a review of distributed speech recognition standards. The subsequent four chapters cover the major blocks of distributed speech recognition.

Chapter 5 provides a comprehensive overview of the industry standards for distributed speech recognition developed in ETSI, 3GPP and IETF in addition to a summary of substantial performance testing and comparisons to AMR coded speech. Chapter 6 presents techniques for feature extraction and back-end speech reconstruction from the MFCC features on the basis of voicing and fundamental frequency information either transmitted from the client device or predicted from the received features. Chapter 7 describes a series of schemes for quantizing the MFCC features, including scalar quantization, vector quantization and block quantization, where the optimization objective is to maximize recognition accuracy. Chapter 8 presents a survey of error recovery methods for transmitting the quantized features over error-prone channels, including both forward error control coding that adds redundancy to the feature stream and interleaving that creates spread in it. Client-side error recovery cannot completely prevent the occurrence of residual bit errors or packet loss. Chapter 9 therefore concentrates on sever-side error concealment to reduce the detrimental effect induced by transmission errors.

Part III, **Embedded Speech Recognition**, addresses the main problems in realizing a speech recognition system fully on a mobile device. The problems are approached from both algorithm and arithmetic sides through three dedicated chapters.

Chapter 10 presents an overview of algorithm implementations and optimizations aimed at a speech recognition system with a low computational complexity and thus suitable for deployment on embedded platforms. To complement this, Chap. 11 primarily targets a low memory footprint and emphasizes on techniques for compressing HMMs by removing redundancies from HMMs through parameter tying and state- or density-clustering and by quantizing HMMs. Chapter 12 reviews problems concerning the fixed-point arithmetic implementation of speech recognition algorithms and presents fixed-point methods that give the same recognition accuracy as that of floating-point algorithms.

Part IV, **Systems and Applications**, introduces practical work and knowledge. It starts with the introduction to architecture considerations in a network environment. The succeeding three chapters present speech recognition systems and applications tailored for mobile phones, PDAs and automobiles, respectively. The last chapter presents energy-aware speech recognition for mobile devices.

Chapter 13 examines software architectures for mobile speech applications from an industrial viewpoint with a thorough comparison between embedded and distributed speech engines and a highlight on supporting multimodal user interaction. Chapter 14 presents applications of speech recognition for mobile phones and puts the focuses on multilinguality, noise robustness, and footprint and complexity reduction. Chapter 15 presents a two-way free-form speech-to-speech translation system that includes a large vocabulary continuous speech recognizer, a translation module and a multi-language speech synthesis system and is completely hosted on a PDA. Chapter 16 describes the development of speech technology components for various automotive applications and reviews issues and challenges related to automotive platforms. With a concern that battery technology significantly lags behind semiconductor technology, Chap. 17 investigates the system-level energy consumption from both computation and communication of distributed speech recognition on a wireless device and presents a set of optimization algorithms that can increase the battery lifetime by an order of magnitude.

A comprehensive **index** is provided at the end of this book. Index words are highlighted in the text by using italic font.

While chapters are complemented to each other and are presented in a unified manner with a clear flow from chapter to chapter, each chapter is written to be self-contained and can be read and understood independently. As such, certain redundancy is kept in the book. The book contains chapters of a tutorial nature as well as chapters on research advances and practical applications.

Target Audiences

The book is primarily intended for students, engineers and scientists working in speech processing and recognition. This book can also be a reference for practitioners and researchers involved in user interface and application design for mobile devices, speech communication over networks, Internet and wireless communications, and data compression.

Supplementary Materials

For more information about software, databases, literature and related links, please refer to the book's Web site, http://asr.es.aau.dk.

Acknowledgements

We warmly thank the authors, our friends and colleagues, for their outstanding contributions and for their hard work in making a timely, high-quality publication of this book a reality. We are grateful to Wayne Wheeler, Catherine Brett, Beverley Ford and Helen Callaghan at Springer for their great support and assistance.

Zheng-Hua Tan
Børge Lindberg

Aalborg, Denmark

Contents

Part III Embedded Speech Recognition

Contributors

Abeer Alwan
University of California, Los Angeles,
Department of Electrical Engineering,
Los Angeles, CA, USA
alwan@ee.ucla.edu

Alexis Bernard
University of California, Los Angeles,
Department of Electrical Engineering,
Los Angeles, CA, USA
Bernard@UCLAlumni.net

Laurent Besacier
University J. Fourier,
LIG Laboratory,
Grenoble, France
Laurent.Besacier@imag.fr

Enrico Bocchieri
AT&T Labs Research, Florham Park,
New Jersey, USA
enrico@research.att.com

Bengt J. Borgström
University of California, Los Angeles,
Department of Electrical Engineering,
Los Angeles, CA, USA
jonas@ee.ucla.edu

Brian Delaney
Massachusetts Institute of Technology,
Lincoln Laboratory, Information Systems
Technology Group,
Lexington, MA, USA
bdelaney@ll.mit.edu

Jonathan Engelsma
Motorola Labs, USA
jonathan.engelsma@motorola.com

James C. Ferrans
Motorola Labs,
USA
james.ferrans@motorola.com

Yuqing Gao
IBM T. J. Watson Research Center,
USA
yuqing@us.ibm.com

Reinhold Haeb-Umbach
University of Paderborn,
Department of Communications
Engineering,
33095 Pad erborn, Germany
haeb@nt.uni-paderborn.de

Harald Höge
Siemens AG, Corporate Technology,
81739 München, Germany
harald.hoege@siemens.com

Sascha Hohenner
Siemens AG, Corporate Technology,
81739 München, Germany
sascha.hohenner@siemens.com

Valentin Ion
University of Paderborn,
Department of Communications
Engineering,
33095 Paderborn, Germany
ion@nt.uni-paderborn.de

Bernhard Kämmerer
Siemens AG, Corporate Technology,
81739 München, Germany
bernhard.kaemmerer@siemens.com

Hong Kook Kim
Gwangju Institute of Science
and Technology, Department
of Information and Communications,
Gwangju, Korea
hongkook@gist.ac.kr

Imre Kiss
Nokia, Finland
imre.kiss@nokia.com

Niels Kunstmann
Siemens AG, Corporate Technology,
81739 München, Germany
niels.kunstmann@siemens.com

Børge Lindberg
Aalborg University,
Department of Electronic Systems,
9220 Aalborg, Denmark
bli@es.aau.dk

Ben Milner
University of East Anglia,
School of Computing Sciences,
Norwich, NR4 7TJ, United Kingdom
b.milner@uea.ac.uk

Miroslav Novak
IBM T.J Watson Research Center,
Speech and Language Technologies,
USA
miroslav@us.ibm.com

Kuldip K. Paliwal
Griffith University,
Griffith School of Engineering,
Signal Processing Laboratory,
QLD 4222 Australia
k.paliwal@griffith.edu.au

David Pearce
Motorola Labs,
Applications Research Centre,
Basingstoke, UK
david.pearce@motorola.com

Richard C. Rose
McGill University, Department
of Electrical and Computer Engineering,
Montreal, Quebec, Canada
rose@ece.mcgill.ca

Stefanie Schachtl
Siemens AG, Corporate Technology,
81739 München, Germany
stefanie.schachtl@siemens.com

Martin Schönle
Siemens AG, Corporate Technology,
81739 München, Germany
martin.schoenle@siemens.com

Panji Setiawan
Universität der Bundeswehr München,
München, Germany
panji.setiawan@yahoo.de

Stephen So
Griffith University,
Griffith School of Engineering,
Signal Processing Laboratory,
QLD 4222 Australia
s.so@griffith.edu.au

Zheng-Hua Tan
Aalborg University,
Department of Electronic Systems,
9220 Aalborg, Denmark
zt@es.aau.dk

Imre Varga
Siemens AG,
Corporate Technology, Germany
imre.varga@nsn.com

Marcel Vasilache
Nokia, 33100 Tampere, Finland
marcel.vasilache@nokia.com

Wei Zhang
IBM T. J. Watson Research Center,
USA
zhangwei@us.ibm.com

Bowen Zhou
IBM T. J. Watson Research Center,
USA
zhou@us.ibm.com

Weizhong Zhu
IBM T. J. Watson Research Center,
USA
zhuwe@us.ibm.com

1

Network, Distributed and Embedded Speech Recognition: An Overview

Zheng-Hua Tan and Imre Varga

Abstract. As mobile devices become pervasive and small, the design of efficient user interfaces is rapidly developing into a major issue. The expectation for speech-centric interfaces has stimulated a great interest in deploying automatic speech recognition (ASR) on devices like mobile phones, PDAs and automobiles. Mobile devices are characterised as having limited computational power, memory size and battery life, whereas state-of-the-art ASR systems are computationally intensive. To circumvent these restrictions, a great deal of effort has therefore been spent on enabling efficient ASR implementation on embedded platforms, primarily through fixed-point arithmetic and algorithm optimisation for low computational complexity and memory footprint. The restrictions can also be largely bypassed from the architecture side: Distributed speech recognition (DSR) splits ASR processing into the client based feature extraction and the server based recognition. The relief of computational burden on mobile devices, however, comes at the cost of network deteriorations and additional components such as feature quantisation, error recovery and concealment. An alternative to DSR is network speech recognition that uses a conventional speech coder for speech transmission from client to server. Over the past decade, these areas have undergone substantial development. This chapter gives a comprehensive overview of the areas and discusses the pros and cons of different approaches. The optimal choice is made according to the complexity of ASR components, the resources available on the device and in the network and the location of associated applications.

1.1 Introduction

Computing is penetrating every corner of our life: Mobile devices bring computers all over the place and networks connect everywhere to computing resources. Today masses of *mobile devices* are being used as digital assistants, for communication or simply for fun. Examples are PDAs, mobile phones, MP3 players, GPS devices, digital cameras and the like. With mobile phones alone, the number of subscriptions exceeded 2.7 billion by the end of 2006 according to Informa's report, Mobile Market Status 2007 (http://www.informatm.com). The number is expected to hit 3.5 billion by 2010. On the networking side, the goal has long been to achieve network access anywhere, anytime and from any devices. Besides the fast development of various *network* forms such as 3G, wireless LAN, Bluetooth and IP networks, the concept of free wireless connection for the public is widely accepted and in many places, has been implemented or is under serious considerations.

In this ubiquitous computing environment, the use of keypad, stylus and small screen is inconvenient and speech-centric user interface is foreseen to be a desirable interaction paradigm where automatic speech recognition (ASR) is the enabling technology. This has led to the growing interest in deploying speech recognition on mobile devices.

As ASR technology has been optimised primarily for general computers in a centralised architecture, specific care is required when incorporating the technology into mobile devices and communication networks, both of which place significant constraints on the use of ASR to its full potential. In comparison with contemporary desktop computers, mobile devices are inherently featured with compromised computing power, reduced CPU (central processing unit) clock, limited-speed memory access, small memory size and limited battery life. Fortunately, the 'always-on' network connectivity for mobile devices opens up new opportunities to circumvent these constraints by delivering some of the ASR computing tasks into remote servers. The price to pay, however, is the effect of limitations enforced by networks themselves, which for instance are not always reliable or even not available for some periods or locations. 'Always-on' usually means connectivity with some drop-outs, hence over less than 100% of time. In fact, placing ASR in the remote server is an efficient option for network based applications which can tolerate natural drop-outs in radio network connectivity. In other cases, placing ASR in the mobile devices represents the only possibility.

Due to the existence of means of interaction, the user expects perfection from speech interfaces, presenting a significant challenge for both academia and industry. While efforts have been put in all aspects of ASR technology to meet the expectation, in the attempt of utilising the resources available from devices and networks and addressing the accompanying hindrances, three approaches have been devised: *network speech recognition* (NSR), *distributed speech recognition* (DSR) and *embedded speech recognition* (ESR).

In *NSR*, speech signal, in most cases encoded by a conventional speech coder, is transmitted to the server where feature extraction and recognition decoding are conducted (Kim and Cox 2001). The apparent and major advantage of the NSR approach is that numerous commercial applications are developed on the basis of speech coding. This enables a plug and play of ASR systems at the server side while no changes are required for the existing devices and networks. It further shares all the advantages of server based solutions in terms of system maintenance and update and device requirements. In addition to network dependency, the downside of NSR is that speech coding and transmission may degrade the recognition performance due to such factors as data compression, transmission errors, training-test mismatch, production model oriented parameterisation and transcoding (Euler and Zinke 1994; Lilly and Paliwal 1996; Peinado and Segura 2006). Among the factors, effect of information loss over transmission channels has shown to be the most significant.

The curse of dimensionality is a well-known problem in pattern recognition. In ASR, feature extraction process is applied to the speech signal to obtain a representation with a low dimension and less redundant information. The generated features are therefore well suitable for compression and transmission. *DSR* directly quantises these features and transmits them through networks (Pearce 2004; Tan et al.

2005). In the server the features are decoded and used for recognition. With recent advances in source coding, channel coding and error concealment, this approach both achieves a low bit rate and avoids the distortion introduced by speech coding. To provide the possibility for human listening, effort has also been put into the re-construction of speech from ASR features with or without supplementary speech features such as pitch information and the results are quite encouraging (Milner and Shao 2007). The key barrier for deploying DSR is that it lacks foundation in the existing devices and networks that NSR has. Stronger motivation and more effort will be needed to make DSR grow in visibility and importance.

In *ESR*, all ASR processing is conducted in the target device (Varga et al. 2002). Such fully embedded ASR is independent of network connectivity and has the advantage of not introducing extra distortion to speech signals. However, the re-quirements to the client are high in terms of computing, memory and power con-sumption. Also, when the ASR involves large databases residing in networks, e.g. for compiling application specific grammars, bandwidth requirement and security concern turn out to be nontrivial. Update of the ASR engine is also inconvenient due to the widespread, numerous devices. In many cases ASR is merely an integrated part of user interfaces, so ASR is not supposed to consume a large proportion of computational resources and scarce battery. Fixed-point arithmetic and algorithm optimisation are therefore required to realise ASR in embedded platforms (Lam et al. 2003). The hope lies in the continuous advance in semiconductor technology implying a rapid evolution of computing speed and memory size so the complexity of ASR is expected to become less and less of a bottleneck in the future.

This chapter presents an overview of the various ASR areas and discusses the pros and cons of different approaches. The remainder of this chapter is organised as follows. Section 1.2 presents the basics of ASR and limitations of mobile devices and networks. Sections 1.3, 1.4 and 1.5 sequentially present network, distributed and embedded speech recognition. This chapter is ended with discussions.

1.2 ASR and Its Deployment in Devices and Networks

1.2.1 Automatic Speech Recognition

Automatic speech recognition converts a speech signal to a word sequence (Deller et al. 1999). Modern ASR systems are firmly based on the principles of statistical pattern recognition, in particular the use of *hidden Markov models* (HMMs). Given the observation data Y, which are feature vectors extracted from the speech signal, the most likely sequence of words $\hat{W}$ is found through the following *Bayesian decision rule*:

$$\hat{W} = \arg\max_{W} P(W \mid Y) = \arg\max_{W} P(W)P(Y \mid W) \qquad (1.1)$$

where $P(W)$ is the a priori probability of observing some specified word sequence W and is given by a language model, and $P(Y \mid W)$ is the probability of observing speech data Y given word sequence W and is determined by an acoustic model, often an HMM.

The architecture of a typical ASR system, depicted in Fig. 1.1, shows a sequential structure of ASR including such components as speech signal capturing, front-end feature extraction and back-end recognition decoding. Feature vectors are first extracted from the captured speech signal and then delivered to the ASR decoder. The decoder searches for the most likely word sequence that matches the feature vectors on the basis of the *acoustic model, lexicon* and *language model* (LM). The output word sequence is then forwarded to a specific application.

The partition between the ASR components is sharp, enabling flexible architectures when deploying it on the device and in the network. Speech is always captured in the client and the application can reside either in the client or in the server. The decision on where to place the remaining ASR components distinguishes three approaches: NSR, DSR and ESR, as shown in the bottom panel of Fig. 1.1. The choice of approaches is driven by a number of factors including complexity of components, resources available on the device and in the network, and location of the application.

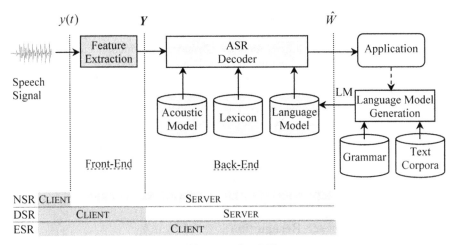

Fig. 1.1 Architecture of an ASR system

Although the acoustic model is as well related to and may adapt to application, the language model has much stronger dependency on it, especially when the model is constructed from rule-based context-free grammars. Grammar based LM often dynamically changes along the application dialogue flow and necessitates data from application and databases. Stochastic language models, such as data-driven *n*-gram trained from text corpora, however, are less dependent on individual applications and can be generated offline. The data location, the size of grammar and the frequency of change in the grammar are among the decisive factors in choosing embedded or remote ASR, see Chap. 13.

The next factor to consider is the complexity of various ASR components. In general, the front-end processing is less resource demanding. Nevertheless, the HMM based back-end is much more computationally intensive than the front-end and has a high demand for memory and CPU resources. First, the acoustic model normally consists of several millions of parameters and the system usually has a large lexicon and language model to store and access. Secondly, decoding word sequences takes a substantial amount of CPU resources due to the needs for calculating observation likelihood and for searching over a huge space. The storage of intermediate results brings in further demand for memory. During the decoding, memory is frequently accessed making memory access speed an important factor. Finally, it consumes a significant amount of energy. When implemented in embedded platforms, these demands for resources appear to be a considerable obstacle and optimisation is therefore necessary to pursue.

In the following we discuss the constraints of mobile devices and communication networks.

1.2.2 Resources and Constraints of Mobile Devices

Key concerns with *mobile devices* are computing power, fixed-point arithmetic, memory size, memory access speed and power consumption (or battery lifetime). These factors are common for all low-cost consumer electronic devices including PDAs, mobile phones, car kits and game devices. Although resources are generally scarce on consumer devices, we have to carefully distinguish between various scenarios. The basic aspect is the targeted speech recognition application in relationship to the available resources: The needs of e.g. digit dialling, keyword spotting or continuous dictation are largely different and a specific device will be able to run speech recognition up to a certain complexity level.

From an ASR implementation point of view, mobile devices and car kits may be classified into at least two classes: high-end and low-resourced platforms. It is important to mention that as of today, computing power, memory size and speed in a consumer product are usually chosen according to the requirements of the main functionality of the device. Speech recognition software is part of the handset software infrastructure hence ASR based applications are considered as well although no driving forces when determining the actual resource level of a platform. The consequence is that we have to choose the actual speech recognition solution according to the capabilities of the given platform. Examples for high-end devices are PDAs, featured car kits and smart phones, and plain mobile phones for low resourced platforms. For discussion purpose, it is as well interesting to somehow touch one more class of consumer devices—any other unit with a microphone including telephone and home electronic appliances.

Typically, users of high-end devices expect the support of advanced features, for example, video telephony, audio-video streaming or mobile TV, messaging service, interactive content delivery—all these applications already require a (relatively) high-resourced platform. Speech recognition based applications may make benefit of the availability of those resources. Command-and-control by speech assists the user in a more comfortable user interface. Furthermore, some advanced features like keyword spotting may be offered as well, in addition to name and digit dialling. The resources on mobile phones or game devices are still limited today to support a large vocabulary continuous speech-to-text dictation application. However, resources of high-end devices, such as PDAs, smart phones and eventually car kits, have reached the level to support full-featured dictation useful for SMS and email. As smart phones and PDAs are more and more enriched by new features, we may expect a positive effect on speech recognition based applications as well.

On the other hand, plain mobile phones basically used for telephony are no ideal platform for sophisticated speech recognition yet. They still may be equipped by speech recognition based applications: Isolated-word digit dialling is the best example although name dialling using a combination of speaker independent and speaker dependent training fits simple phones well. Good progress has been demonstrated in this area over the last years as continuous digit dialling becomes available as well. Nevertheless, the wish of having enriched ASR applications in plain phones presents opportunities for DSR and NSR, which require a thin client only.

Battery lifetime (around 3–5 h in a mobile phone when talking) represents a major constraint in addition to limited computing power and memory size since robust signal processing algorithm computing, large storage with fast access and increased CPU speed imply increased power consumption. In addition, power consumption further increases when a video screen is present e.g. for video telephony, video streaming and mobile TV applications. The impact is even less power for ASR applications. Although high power drain of video applications urges manufacturers to improve the battery situation, this circumstance does not imply necessarily more power for speech recognition applications. Chapter 17 is dedicated to managing and optimising battery lifetime for mobile devices through techniques like energy aware speech recognition.

After elaborating the impacts of scarce resources onto the feasibility of speech recognition applications in mobile devices, let us take a closer look at the platform constraints themselves. A major constraint is the available memory: In a consumer device like a mobile phone, game device, car kit, the typical size is 4–16 MB for RAM memory with slow access and up to 32 kB for cache. So the amount of signal processing algorithms that can run simultaneously is limited and they also limit the size of language and acoustic models. The result is a compromised performance. Computing power of the CPU is limited which implies the use of suboptimal methods in speech recognition and hence performance degradation. In addition, the CPU runs on *fixed-point* arithmetic, which implies the need for fixed-point algorithm code, or a floating-point arithmetic that is emulated on the CPU's fixed-point hardware. The second approach allows the implementation of floating-point code but at a reduced speed, further decreasing the available computing power. Moreover, there is no low-level access to the operating system by the programmer of signal processing algorithms; high-level programming is more comfortable but results in a less efficient code. Resource scarcity is even worse when using the device in adverse acoustic environments, which is usually the case for mobile phones, PDAs or car kits. Car noise, street noise, office noise and reverberant speech all represent major impairment factors to the input speech commonly referred as adverse acoustic conditions. Sophisticated signal processing algorithms are needed to cope with the negative effect of the adverse acoustic environment—their implementation is not always possible in highest quality due to memory and speed constraints.

Besides physical resource situation, it is worth drawing our attention to further aspects of properties of mobile device platforms with respect to speech recognition applications. Speech input has to compete with existing and well-accepted *user interface* methods, for all of command-and-control, dialling or text input. The existing methods, like typing on a keypad or pushing buttons on a phone, pointing with stylus, use of touch screen, are all well established. New users seem to learn typing of buttons on a phone quickly and especially young people are fast when typing SMS text. However, there are some limiting factors of conventional user interface methods in consumer devices. One of them is that due to potentially increased risk of accidents, law prohibits the use of hand-held devices while driving in a number of countries. Furthermore, the size of consumer device keypads is becoming smaller and smaller in the course of miniaturisation. Use of tiny keypads is neither comfortable for some people nor reliable enough. Finally, as high-end devices are enriched by more and more features, their handling becomes increasingly sophisticated. Indeed, handling of phones is computer-like today already and it does not resemble that of conventional phones in any respect. Navigation in complex menu structures seems inevitable although not manageable for everyone. All these factors strengthen the need for an alternative user interface—the most natural solution is the use of speech recognition.

1.2.3 Resources and Constraints of Communication Networks

Networking facility is becoming a standard component on mobile devices; wired and wireless *network* accesses are broadly available, though not ubiquitous yet. Furthermore, network service is gradually moving towards a flat-rate subscription-based business model in which the user pays a certain fee for unlimited connection. Variants usually differ in service grades like basic-enhanced-premium services. All these factors together assure an '*always-on*' networking and the quality of connections in relationship with costs, rather than network connectivity, becomes the major concern. From this viewpoint, we may distinguish between circuit-switched and packet-switched types of networks as detailed in the following.

Circuit-switched networks set up a dedicated circuit (or channel) between the two parties for the duration of a communication and this gives a constant delay and a constant throughput. In contrast, packet-switched networks break data into small packets and based on the destination address in each packet, route them through nodes and data links that are shared with other traffic. Note that the previously mentioned data may refer to any type of information, such as text of an email or segments of digitised speech signal in telephony service. Once all the packets constituting a message arrive at the destination, they are reassembled in the proper order to restore the original message.

Circuit-switched networks are ideal for communications that require data to be delivered to its destination in real-time and in its original order. Example communications are speech conversation (telephony) and video telephony. Packet-switched networks are rather oriented to non-real time data transfer, and they are more efficient and robust if some amount of delay is tolerable. Nowadays, packet-switched networks are also used for speech conversation (named VoIP) although this service lacks the quality common for circuit-switched telephony and suffers from large call latency. Extensive efforts are made on the QoS area and on speech coding so that quality of VoIP based service improves steadily. Due to overall advantages in terms of flexibility and costs, packet-switched IP networks are the development trend and will be the dominating network form in the future.

Landline telephone networks are circuit-switched and are considered reliable, whereas radio channels cannot be considered as always reliable because fading and interference introduce errors into transmitted data. Specifically, in circuit-switched wireless channels, transmission impairments arise in the form of bit errors. In packet-switched networks, the impairment is in packet errors: Packets are queued or buffered in each network node, and due to congestion at the nodes, packets can be lost or get delayed and thus have to be declared as lost by real-time applications. Packet-switched networks implement packet loss concealment mechanisms to improve the subjective quality of the speech signal in the presence of packet losses. Bit error and packet loss are two different types of channel noises, but one thing in common is that both tend to be burst-like, making error recovery and concealment a challenging task.

Lossless transmission schemes are applied for data transmission, so that channel noise is reflected as delays rather than deterioration of data quality. For real-time services such as speech conversation and remote speech recognition, delay above a certain threshold is not acceptable. As a result, transmission errors inevitably remain in the data and degrade ASR performance. Techniques for error recovery and concealment must be applied and take effect within certain range of time for both NSR and DSR.

Although network capacity has been expanded dramatically, more and more new applications are constantly deployed. Thus, bandwidth is obviously a concern and data compression is always welcomed for transmission of speech information. Low-bit-rate compression in NSR is a source of performance degradation, though not as severe a source as transmission errors. In contrast, the effect of data compression on DSR is often negligible.

1.2.4 Architectural Solutions for ASR in Devices and Networks

Through the discussions above, we get a picture about ASR and its deployment environments. From the system architecture point of view, ESR may be considered as the simplest approach since all recognition related processing is performed in the client and no signal or data is sent from the client device to a remote server based engine. This simplicity is conditioned on that the ASR related application is embedded on the device, or the communication between the ASR and the application (if network based) is restricted to merely the recognition results. Otherwise security concern and data dependence may favour a remote ASR solution. Furthermore, due to the limitations of embedded system platforms, the implementation of ESR requires customised fixed-point conversion and algorithm optimisation to reduce its consumption of memory, computation and power (Novak 2004). Finally, porting and update of ESR systems are up to the user.

The downsides of ESR exactly represent the benefits of a remote ASR, and vice versa. The rule of thumb for data-intensive computing is to place computation where the data is, instead of moving the data to the point of computation (Bryant 2007). When the ASR acquires more data from the network than from the microphone, a network based ASR may be preferable. Another favourable scenario for network-based ASR is when the ASR computation is a big burden for the device. The network based approaches also offer some opportunities that ESR cannot offer. For example, humans can assist the ASR in the background to provide semi-automatic speech transcription service.

In remote ASR, speech signals are transmitted from the device to the server as either coded speech (NSR) or as ASR features (DSR), both of which can be efficiently compressed to a bit rate of several kbps. Speech signal quality and (noise and channel) robustness are important parameters for choosing DSR while the wide deployment of high-quality speech coders makes NSR a favourite.

Due to the pros and cons of the three different approaches, they are expected to co-exist in the years to come.

1.3 Network Speech Recognition

In NSR, speech encoded by conventional speech coders normally used for telephony voice conversation is transmitted to the server in which ASR is conducted. At the server side, there are two ways to extract ASR features from the bitstream of the coded speech. One is to reconstruct speech signal first and extract features subsequently; in this case, NSR is essentially the concatenation of a conventional speech coding and decoding (codec) system and a speech recognition system. The other way is to estimate features directly from the bitstream without decoding (reconstructing) the speech; this method has demonstrated a superior performance to the former in terms of both computational complexity and recognition accuracy (Kim et al. 2001; Peláez-Moreno et al. 2001).

The ubiquitous presence of *speech coding* on mobile devices largely leverages the deployment of NSR as this enables a plug and play of ASR systems at the server side without touching the massive clients. For some devices such as for a telephone which have no computing power for basic front-end processing, NSR represents the only possibility to have an ASR-driven interface.

The disadvantages of NSR are network dependency and distortion introduced by speech transmission specifically by low bit-rate coding and error-prone channels. Coding distortion occurs mainly since speech coders are optimised for receiver-side reconstruction and human listening rather than for computer recognition. For instance, parameterisation of speech coding is mainly based on a speech production model and thus the use of linear prediction coding (LPC) coefficients while speech recognition widely employs Mel-frequency cepstral coefficients (MFCCs) that are extracted on the basis of human perception. This difference can be overcome by directly estimating features from the bitstream of coded speech without re-constructing the speech, see Chap. 3. In Kim et al. (2001), for a connected digit recognition task, the word error rate (WER) for wireline speech is 3.83% and it is 5.25% for IS-641 coder at 7.4 kbps. Their proposed bitstream-based front-end achieved a WER of 3.76%. However, techniques of this kind are tailored for each specific coder. Recently Kim further proposed a CELP-type speech coder that uses MFCCs to represent the spectral envelop, see Chap. 4.

As end-users of an NSR system may use various speech coders, the resulting mismatch between training and test is a source of degradation as well. In Euler et al. (1994), it is found that with matched training and test conditions, WERs for a speaker independent isolated word recognition task for 64 kbps A-law speech and for 4.8 kbps CELP speech are 1.48% and 2.57%, respectively. When acoustic models trained on 64 kbps A-law data are used for testing the 4.8 kbps CELP speech, the WER for it increases to 3.96%. Nevertheless, this observation is in contrast with that in Hirsch (2002) where training using PCM speech (no coding) generally gives better performance. For example, the weighted WER for PCM Aurora-2 speech is 26.77% and it is 29.84% for AMR (*Adaptive Multi-Rate*) 4.75 mode when training and testing the recogniser by using the same coder. When using PCM speech for training and AMR 4.75 for testing, the WER is 28.17%, which is better than the matched coding condition. In contrast with the above moderate drops in ASR performance, certain audio codecs, such as the MPEG layer-2 8 kbps codec, can substantially degrade the ASR performance, or even result in almost random ASR output, see Chap. 2. The degradation becomes gradually less significant with better speech coding quality. That is achievable by using more sophisticated coding algorithms or increasing the bit rate and enlarging the audio bandwidth (wideband speech at 16 kHz sampling frequency). For example, it was shown (Fingscheidt et al. 2002) when using the EFR or AMR codec in GSM at 12.2 kbps, the impact of speech coding itself is negligible on ASR performance while radio channel errors were found to be the main source of impairment. Overall, one firm conclusion is that low-bit-rate speech coding and transmission decreases ASR performance while transmission of >10 kbps coded speech over good channels has a negligible effect.

The effect of *packet loss* on NSR has been extensively investigated in (Mayorga et al. 2003). The authors reveal that packet loss may imply substantial degradation of recognition performance. In contrast, speech coding is a less severe problem, but when coupled with packet losses, it can make ASR out of function. One of the reasons is that speech coders usually exploit inter-frame correlation to achieve high compression ratio so that one frame loss affects subsequent frames—the phenomenon of error propagation (Pearce 2004). Lately some frame-independent coders have been developed. Furthermore, in a low-bit-rate coder, one packet contains a large amount of information making the effect of packet loss even more severe.

Various applications have been developed on the basis of NSR. For instance, it is used in interactive voice response (IVR) systems to accomplish complex transactions that are difficult for touch tone based interaction to handle if a complicated application menu structure is to be avoided. A significant move in this direction is the introduction of the W3C's standard *VoiceXML*, which enables voice applications to be developed in a similar way to HTML based web applications. It aims 'to bring the advantages of web-based development and content delivery to interactive voice response applications' (http://www.voicexml.org).

1.4 Distributed Speech Recognition

The high complexity of an ASR decoder makes it tempting to adopt a client-server architecture: placing the front-end in the client and the computation-intensive back-end in the server. Since feature extraction is located in the client, the process of speech coding and decoding is eliminated. Instead, the feature vectors are directly compressed and sent to the server for recognition decoding. As data transmission may take place via heterogeneous networks, the use of a DSR codec further avoids the problem of transcoding.

To optimise DSR performance over adverse transmission channels, considerable efforts have been made ranging from front-end processing, source coding/decoding, channel coding/decoding, packetisation to error concealment (EC) (Tan et al. 2005). A diagram of a typical DSR system is shown in Fig. 1.2. The major building blocks are introduced briefly in this section and are extensively covered by Chaps. 6, 7, 8 and 9 in addition to a review of DSR standards in Chap. 5.

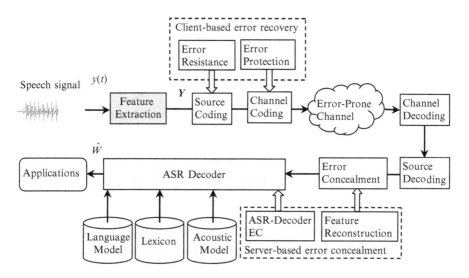

Fig. 1.2 Diagram of a DSR system

1.4.1 Feature Extraction

The extraction of discriminative and reliable features is a key issue in speech recognition. Over the years, *MFCCs* (Davis and Mermelstein 1980) have become the de facto standard features and are therefore used in DSR systems as the primary choice.

For human listening purpose an interesting exploration consists in speech reconstruction from MFCC features (Milner et al. 2007). This effort together with the attempt of using MFCC features for speech coding (Chap. 4) imply a convergence of speech coding and DSR feature extraction where the DSR codec becomes one member of the speech coding family. The difference lies in that the optimisation criterion for speech coding is primarily perceptual quality whereas it is recognition performance for DSR feature extraction.

The other features that have been investigated are the *PLP* (perceptual linear predictive) features, which have the advantage of being efficiently coded into as low as 0.3 kbps while providing recognition accuracy comparable to the unquantised system (Bernard and Alwan 2002). It, however, lacks the possibility for speech reconstruction.

Acoustic environments in which mobile devices operate are typically noisy. In hands-free mode a far field microphone further decreases the signal-to-noise ratio of speech. Noise robustness in adverse conditions is therefore a key issue to deal with. Since many robustness techniques are applied in the time and frequency domains and features sent to the DSR server are in the Mel-frequency domain, those robustness (speech enhancement) techniques must be implemented in the front-end at the client side.

1.4.2 Source Coding

Source coding is applied to compress speech features for transmission over bandwidth-limited channels. Techniques include scalar quantisation, vector quantisation (VQ) and transform coding (So and Paliwal 2006). In general lossy coding is applied for DSR.

The widely used Split VQ partitions each feature vector into sub-vectors and quantises each sub-vector independently by using its own codebook. Digalakis et al. (1999) have extensively evaluated the use of split VQ and scalar quantisation for compressing MFCC features. As compared with full VQ and scalar quantisation, split VQ has a better trade-off between storage and computation requirements and quantisation performance. It was found that 2 kbps is sufficient for 13-dimentional MFCCs.

Speech features contain a substantial amount of redundant information. In transform coding, the redundant information or correlation in the features are removed by transforming them, and thereafter quantisation is applied in the transformed domain. This is also known as block coding. An example is the two dimensional discrete cosine transform (2D-DCT) (Hsu and Lee 2004; Zhu and Alwan 2001).

Tan and Lindberg (2007) presented a scalable coding scheme based on a variable frame rate analysis where the target bit rate is met by adjusting frame rate. Prior to recognition in the server, frames are repeated so that the original frame rate is restored to fit the frame rate with the applied HMM models.

The ETSI-DSR front-end compresses speech source into a *bit rate* of 4.4 kbps and gives a WER of 0.95% on the Aurora 2 database (Hirsch and Pearce 2000). The 2D-DCT achieves a bit rate of 1.45 kbps and a WER of 1.58% (Hsu et al. 2004). The run-length coding method obtains a WER of 0.89% at a bit rate of 1.40 kbps and a WER of 1.15% at a bit rate of 1.06 kbps. The performance of run-length coding is dependent on the amount of steady regions in the signal, so does transform coding.

Source coding also plays a role in robustness against transmission errors. Error-resilient source coding techniques should be effective to prevent error propagation and minimise distortions. When inter-frame correlation is exploited, the resulting inter-frame dependency will make the coder more sensitive to transmission errors. Some coders are considered joint source-channel coding such as layered coding (Srinivasamurthy et al. 2006) and multiple description coding (Tan et al. 2007a) which are used for DSR as well.

Histogram-based quantisation (HQ) was recently proposed for feature compression (Wan and Lee 2006), which performs the quantisation of a feature parameter based on the histogram or order statistics of that feature parameter within a moving segment. This method needs no fixed codebook and eliminates the mismatch between the corrupted feature vectors and the fixed codebook. Another recent scheme applies the group of pictures concept (GoP) from video coding to DSR to achieve a variable bit rate compression scheme (Borgstrom and Alwan 2007).

1.4.3 Channel Coding and Packetisation

Channel coding aims at protecting information from channel related errors through adding redundancy to the data (Bossert 2000). Channel coding techniques are measured by, among others, error detection capability and error correction capability. In applying techniques known as backward error correction (BEC), errors are detected but not corrected; upon detection of errors, a retransmission is requested. Retransmission is not deployed for DSR since speech interaction is considered a real time application so that Real-time Transport Protocol (RTP) is used. Retransmission mechanism further relies on duplex communication. Instead, server-side EC can be used in combination with BEC. The other type of techniques, known as forward error correction (FEC), aims at not only detecting errors but also recovering the message from errors without referring back to the client. For instance, Boulis et al. (2002) applied Reed-Solomon codes to DSR to cope with packet erasure. In general, channel coding techniques trade bandwidth for redundancy and thus error resilience.

FEC and EC techniques are efficient in handling randomly distributed errors, but inefficient when errors are burst-like. Therefore, they are better used in connection with appropriate packetisation, which redistributes errors or erasures. For example, interleaving is such a technique that is able to randomise transmission errors though at the cost of delay (James and Milner 2004).

Though being efficient, client-driven techniques have drawbacks like increased bandwidth, additional delay, computational overhead and weak compatibility.

In applying channel coding to DSR, error detection is more important than error correction (Bernard et al. 2002) as error detection in combination with EC is quite effective for speech recognition. This is further supported by a frame based CRC (cyclic redundancy check) for error detection, which shows a significant performance improvement with a marginal bandwidth increase (Tan et al. 2005).

1.4.4 Error Concealment

The final stronghold for error robustness is error concealment. The prerequisite for EC is error detection, which can be done in two ways. One is to apply channel coding including simple BEC techniques (e.g. parity check, checksum, CRC) and more sophisticated FEC ones where the amount of errors may extend beyond their capability of error correction. The other is to exploit the redundancy in the speech signal. Due to the real-time constraint, detection of errors does not result in a request for retransmission to the client, but in a server-based EC.

EC first aims at feature reconstruction through repetition, interpolation, splicing, or substitution among which repetition usually gives a superior performance. Sub-vector based EC is a repetition at the sub-vector level which uses speech correlation to identify consistent, thus potentially correct, features within erroneous vectors (Tan et al. 2007a). This is proved to be quite effective and well suitable for combining with ASR-decoder EC such as weighted Viterbi decoding.

To benefit from a priori information about speech features, statistical techniques exploit the statistical information about speech for feature reconstruction (Gomez et al. 2003). Reliability information from channel decoding can also be used either for feature reconstruction or for ASR decoding, resulting in a class of soft-feature decoding based techniques (Peinado et al. 2003).

Since we have a computer (speech recogniser) as destination rather than a person, the quality of feature reconstruction can be deployed in the ASR, resulting in ASR-decoder based EC. At the ASR decoding stage, the reliability of the channel decoded features is integrated into the recognition process by using modified Viterbi decoding algorithm such that contributions made by observation probability associated with features estimated from erroneous features are decreased. The concept of uncertainty decoding has also been applied for EC in DSR (Ion and Haeb-Umbach 2006; Wan et al. 2006).

Server-based EC has a good compatibility e.g. with the ETSI-DSR standards.

1.4.5 DSR Standards

A number of DSR standards have been produced by the STQ Aurora DSR working group in ETSI. The first standard was published in 2000 that defines a feature-extraction processing and a source and channel coding scheme (FE) (ETSI ES 201 108 2000). It aims to handle the degradations of ASR over mobile channels due to lossy speech coding and transmission errors.

As mobile devices often operate in adverse acoustic environment and denoising techniques are applied in the front-end, ETSI upgraded the basic front-end by including a noise robustness component to the advanced front-end (AFE) in 2002 (ETSI ES 202 050 2002). The bit rate for both FE and AFE is 4.8 kbps of which 4.4 kbps is used for source coding and 0.4 kbps for channel coding.

A further update is to respond to the needs for server-side speech reconstruction and for tone language ASR. This is done by including fundamental frequency information in the feature stream and has led to the extended versions of the two issued DSR standards: XFE and XAFE (ETSI ES 202 211 2003; ETSI ES 202 212 2003). The bit rate for extended front-ends is 5.6 kbps where 5.1 kbps is for source coding.

Extensive industrial tests have been organised by the 3rd Generation Partnership Project (*3GPP*) and the results justified the superior performance of the DSR XAFE to adaptive multi-rate codecs (3GPP TR 26.943 2004). As compared to AMR 4.75 mode, XAFE obtained a 36% reduction in WER. The gain with using XAFE is even more significant in the presence of transmission errors due to the frame independency in DSR codecs. Consequently 3GPP chose the XAFE as the codec for speech enabled services and published a new specification that provides a *fixed-point* implementation of XAFE (3GPP TS 26.243 2004). The significance of the selection by 3GPP is that we can look forward to the widespread deployment of DSR in future GSM and 3G mobile devices (Pearce 2004).

In the Internet Engineering Task Force (IETF), the RTP payload formats have also been defined for these DSR codecs (Xie and Pearce 2004).

The introduction of front-end standards enables interoperability over networks and gets rid of transcoding, which often is needed for speech transmission over heterogeneous networks.

1.4.6 A Configurable DSR System

Based on the ETSI XAFE (3GPP TS 26.243 2004) and the SPHINX IV speech recogniser (Walker et al. 2004), a configurable DSR system is implemented in (Xu et al. 2006). The system supports simultaneous access from a number of clients each with its own requirements to the recognition task. The recogniser allows multiple recognition modes including isolated word recognition, grammar based recognition and large vocabulary continuous speech recognition (LVCSR). The client part of the system is realised on a H5550 IPAQ with a 400 MHz Intel® XScale CPU and 128 MB memory. Evaluation shows that conversion from floating-point AFE to fixed-point AFE reduces the computation time by a factor of 5 and most of the computation comes from the noise reduction algorithm deployed in the front-end and the MFCC calculation itself is computation light. With regard to memory consumption in the client, the size of the client DLL library file is only around 74 kB, and the maximal memory consumption at run-time is below 29 kB.

1.5 Embedded Speech Recognition

Commonly, embedded speech recognition (ESR) refers to a technique in which all speech recognition processing is located in the target mobile or handheld consumer device. That is the case if no network connection is available and also for certain speech recognition applications even when a communication link is available, while others may use NSR and DSR methods. Example consumer devices are PDAs, mobile phones, car kits, game devices.

1.5.1 ESR Scenario

From a system architecture point of view, embedded speech recognition may be considered as the simplest approach when implementing speech recognition. In contrast to network or distributed speech recognition, there is no signal or data sent from the client device to a remote server based engine. Hence the application is always ready to use, irrespective of radio link existence and conditions.

Given that, it becomes immediately clear there is a price to be paid for the architecture simplicity: The complex speech recognition algorithm has to run on a generically low-resourced consumer device. In fact, we are forced to develop special techniques to cope with limited resources in terms of computing speed (MIPS) and memory on the platform. The result of the efforts is that consumer platforms are generally able to accommodate some kind of ASR based applications. The limits today are best demonstrated by the availability of LVCSR recognisers (dictation) only on the most powerful consumer platforms, on the latest PDAs (Zhou et al. 2004). Also, all maintenance and upgrading activity falls on the user or service of the consumer device.

Fortunately, continuous advance in semiconductor technology implies a rapid evolution of computing speed of microprocessors and improvement of power consumption of memory devices. So the complexity of speech recognition algorithms is expected to become less and less of a bottleneck in the future when implemented in an embedded manner. Nevertheless, server-based speech recognition will always have an advantage in terms of available resources. The result of increasing computing resources and at the same time, more sophisticated methods to cope with low resources may be expected to be a convergence of embedded and remote recognition in terms of application: The border between applications realised by these techniques will disappear which allows for advanced features like the use of natural language understanding instead of simple command-and-control system.

Resource scarcity limits the available applications; on the other hand it forces the algorithm designer to optimise techniques in order to guarantee sufficient speech recognition performance even in adverse conditions and on limited platforms, and to optimise memory usage.

1.5.2 Applications and Platforms

Mobile phones, PDAs, game devices, car kits are all attractive target products for the application of speech recognition. Typical applications in car environment are continuous digit dialling and name dialling with hands-free car kits, and command-and-control for menus and navigation systems. Mobile phones implement speaker-dependent (trained) name dialling and digit dialling, also command-and-control functionality. Games benefit from command-and-control feature. Next, mobile phones will offer speaker-independent dialling and simple dictation features for SMS. Command-and-control applications will extend to interactive man-machine interfaces. Chapters 14, 15 and 16 are dedicated to speech recognition in mobile phones, PDAs and car kits, respectively.

Embedded speech recognition may be implemented on a general purpose processor available in the consumer device already, or on a specialised IC in the device designed to run speech recognition only. While the former approach allows a higher degree of customisation, the latter one is of benefit in terms of cost reduction if a very large quantity can be produced. An example for a general purpose processor in mobile phones is the ARM family: ARM7/ARM9/ARM11 offer 50–600 MHz processing speed usable on proprietary or common (Windows CE, Linux, Symbian) OS. Car kits often apply a DSP of 50–200 MIPS (TriCore, OMAP, Blackfin, C55) or a RISC processor.

1.5.3 Fixed-Point Arithmetic

Use of fixed-point processors is the key for low cost and for low power consumption, which are important aspects for consumer devices. Moreover, the higher computational power of fixed-point devices as compared to floating-point processors may make the integration of complex speech recognition, for example of LVCSR, possible at all on consumer devices.

Whether a general purpose hardware platform or a specific one (custom IC) is used for ESR influences the applied optimisation criteria and techniques which include software level optimisation and custom hardware architecture design. In the following discussion, we address both cases.

A convenient way to develop the ASR software is using C or C++ language in floating-point in order to have a reference code. The next step is to convert it to fixed-point. The fixed-point C code serves then as the basis for assembler implementation on the target CPU. A basic requirement is that the numeric precision of the fixed-point code should not be worse than that of the floating-point reference code otherwise the performance may suffer.

Fixed-point data types must be used in the fixed-point version and the corresponding fixed-point operations have to be defined. A convenient approach has been introduced in ITU-T and ETSI for speech codec specification with the use of basic operators which model the instruction set of a hypothetical but characteristic 16 bit fixed-point DSP. The basic operators are defined as ANSI-C functions for typically used arithmetic (addition, subtraction, multiplication, division and shift) and other operations (logarithm, square root etc.). All speech codecs of the last decade are specified in ITU-T and ETSI using the set of 16 bit basic operators. Following this practice, the DSR extended advanced front-end was defined using the ETSI 16 bit fixed-point basic operators in 3GPP TS 26.243 (2004). This method may be suitable for ESR implementations but apparently this approach has not been followed yet.

The Very Smart Recogniser (VSR) presented in Varga et al. (2002) addresses a method to imitate the mantissa-and-exponent representation of a floating-point data type by fixed-point one. This is realised by shifting the value to a range where the data is optimally used and storing the shift level in a second variable. Exact imitation is not possible unfortunately because data types are CPU and implementation dependent. In addition, often a code of complex modules (division, FFT) are provided by the DSP manufacturer specially optimised for the given DSP but use of these modules cannot be recommended for general fixed-point reference C code purposes. In VSR feature extraction due to MFCC logarithm, the feature values can easily be compressed into a signed 8 bit type which is used in both floating-point and fixed-point versions. As shown, more than 96% of the features are identical in the fixed-point and floating-point versions and less than 0.01% have a numerical difference of more than ± 1. This high accuracy ensures no degradation of the recognition performance of fixed-point software using floating-point trained HMMs. The compressing nature of the logarithm and the smoothing nature of linear discriminant analysis (LDA) help to reduce numeric differences. In addition, the complexity is not high.

For the development of a customised VLSI IC for an embedded isolated word recognition system, a purely software level optimisation method was proposed in (Lam et al. 2003) in a way to optimise for chip area. All floating-point operations are replaced by fixed-point routine calls (for arithmetic operations) or look-up table implementations (for cosine and logarithm functions). For that, a C++ class named Fixed was developed. They first find a minimum word length implementation for each operand and then they optimise for a minimum circuit area of arithmetic operations by further fraction size optimisation. After optimisation of fraction size in the whole isolated word recogniser, the fraction size of LPC processor, VQ and HMM decoder is optimised subsequently. Minimisation of fraction size for LPC processor showed the most significant effect. Not just the same accuracy can be achieved by fixed-point. For the same speech recognition accuracy as with floating-point, they even show a circuit area improvement of 29.7% with fixed-point arithmetic, with training in floating-point.

Direct hardware level optimisation is achieved by the introduction of a low complexity custom arithmetic architecture based on high-speed lookup tables (Li et al. 2006). At the price of a small additional 59 kB of lookup table memory, a speed improvement of at least three times is expected.

Chapter 12 reviews methods for fixed-point implementation of ASR systems, focusing on introduction of a practical approach to the implementation of the frame-synchronous beam search Viterbi decoder, N-grams language models, HMM likelihood computation and Melcepstrum front-end. The fixed-point recogniser is shown as accurate as the floating-point recogniser in several experiments with different types of acoustic front-ends and HMM's. This allows highly accurate LVCSR algorithms with the same performance on the device as on the server.

1.5.4 Optimisation

The complexity constraint in consumer devices is in fact a major challenge for signal processing algorithm design. Section 1.2.2 presented the resources and constraints on mobile devices. As pointed out, signal processing design has to be such to cope with the effects of reduced computing power (CPU speed) and limited amount of memory. Next we address some optimisation techniques, which aim to overcome these difficulties in order to get satisfactory performance of embedded speech recognition and optimise memory usage.

For small vocabulary ASR applications most resources concerning memory and computing power are needed for HMM parameter storage and for calculating the emission probabilities. VSR (Varga et al. 2002) has to be able to run on a platform with 50 MHz processing power and a memory of less than 64 kB. VSR uses the properties of the LDA, discriminative training and HMM parameter coding. Discriminative training is used to achieve high recognition rate with a moderate amount of Gaussians. Here a performance measure like the minimum word error (MWE) is applied for training. After Viterbi based maximum likelihood training, 10 iterations of MWE based training were performed. HMM parameters are coded using Subspace Distribution Clustering HMMs (SDCHMMs) where the Gaussians are represented by pointers to a codebook. The VSR uses Continuous Densities HMMs (CDHMMs). The WERs show that discriminative training is most effective for small model sizes: In case of single density modelling the error rate on the test set is almost reduced by 50%. The experiments show that the use of discriminative training allows high performance HMMs with limited costs in terms of memory. The emission computation is highly processing power consuming. The SDCHMM allows computing emission probabilities very effectively. For each frame and every codeword the stream likelihoods can be pre-calculated once. The log likelihood is then computed as the sum of the pre-calculated stream log likelihoods. The results have shown that it is possible to reduce the memory requirement of HMM-parameters by a factor of three.

In Chap. 10, speech recognition optimisation techniques are presented that are especially suitable for ESR. Focus is on front-end, feature extraction and search. Specific algorithmic improvements are discussed while the best solution can be achieved by a dedicated combination of particular improvements depending on platform and speech recognition task.

The treatment of Chap. 11 focuses on long-term memory requirements and on acoustic model compression in which redundancy in data and parameter representation accuracy limits are exploited. Considering data redundancies specific to HMM based acoustic models, parameter tying and state or density clustering algorithms are presented with cases like semi-continuous HMMs (SCHMMs) and SDCHMMs. Regarding parameter representation a simple scalar quantised representation is shown for the case of quantised HMMs (qHMMs).

1.5.5 Robustness

Noise robustness is an important requirement since the acoustic environment in mobile usage is quite different from laboratory: Adverse acoustic environment is common when using the device in a car or on the street. Although enrichment of application portfolio would require so, direct transfer of speech recognition solutions designed for high-resourced platforms like PCs to handheld consumer products is usually not possible—dictation is still too complex even for relatively high-powered consumer appliances and the acoustic environment in mobile usage, especially hands-free, is much more difficult.

Experience with early speaker dependent digit dialling shows a big difference between the attractiveness of say keyword spotting in the lab as compared to using speaker dependent name dialling on the phone in a car or on the street. That is true both from handling point of view (need for training in speaker dependent case, comfort with keyword spotting) and from an accuracy point of view (ideal in lab, impaired in real mobile environment).

Robustness means a set of multiple requirements: robustness against adverse acoustic conditions, background noise, Lombard reflex, gender, different pronunciations, non-native talker, spontaneous speech. The front-end has to adapt to these conditions and also so-called robust HMM models are of advantage. In VSR (Varga et al. 2002), a maximum likelihood channel adaptation is used in feature extraction and a suitable database representing mobile usage is applied for training resulting in robust HMM models. VSR includes a spectral attenuation and a frame dropping algorithm. The spectral attenuation algorithms regard noise as an additive noise superimposed on undisturbed speech where the noise is regarded as statistically independent of the undisturbed speech. The goal of the algorithms is to create a time-varying filter function based on estimates of the short-term power spectrum of noise to attenuate the noisy spectrum. A Wiener filter is calculated for every spectral bin as the attenuation function in the first stage called short-time spectral attenuation. In the second stage of this basic spectral subtraction scheme, the noise power spectrum is estimated by the minima of the smoothed power spectrum within a moving interval. The advantage is that no explicit detection of non-speech segments is needed. For every frequency bin the noise estimate is subtracted from the noisy speech signal where flooring is employed.

1.6 Discussion

This chapter presented an extensive overview on speech recognition on mobile devices and over communication networks.

We analyzed the system architecture and requirements of speech recognition, the resource situation and constraints on various targets like mobile devices and networks, and presented the characteristics of three main solutions in detail: network speech recognition, distributed speech recognition and embedded speech recognition. These are different solutions addressing how to provide speech recognition based applications when using them on a mobile device.

Improved noise robustness and recognition accuracy in conjunction with algorithm complexity reduction for low-resourced consumer platforms represent the major challenge of embedding speech recognition in mobile devices. Increasing resources and optimisation techniques will certainly facilitate the deployment of embedded systems although resources will remain scarce for all consumer devices in near future for high-complexity applications like dictation systems. For such applications, use of distributed architecture is promising since this structure efficiently divides the system into two parts with a robust data link between them. Moreover, use of network based speech recognition is an excellent solution as well for sophisticated applications like large-vocabulary continuous dictation because high-quality speech transmission can be achieved from mobile phone to server due to high-quality speech codecs that are being used in the network for speech transmission. That is especially true if wideband (16 kHz sampling) speech will become widely deployed. Still the drawback of effect of packet losses remains, which will imply the need for implementation of effective packet loss concealment algorithms.

References

3GPP TS 26.243 (2004) ANSI C Code for the fixed-point distributed speech recognition extended advanced front-end.

3GPP TR 26.943 (2004) Recognition performance evaluations of codecs for Speech Enabled Services (SES).

Bernard, A., and Alwan, A. (2002) Low-bitrate distributed speech recognition for packet-based and wireless communication. *IEEE Transanctions on Speech and Audio Processing*, vol. 10, no. 8, pp. 570–579.

Borgstrom, B.J., and Alwan, A. (2007) A packetization and variable bitrate interframe compression scheme for vector quantizer-based distributed speech recognition. In *Proceedings of Interspeech*, Antwerp, Belgium.

Bossert, M. (2000) *Channel Coding for Telecommunications*. John Wiley & Sons.

Boulis, C., Ostendorf, M., Riskin, E. A., and Otterson, S. (2002) Graceful degradation of speech recognition performance over packet-erasure networks. *IEEE Transanctions on Speech and Audio Processing*, vol. 10, no. 8, pp. 580–590.

Davis, S., and Mermelstein, P. (1980) Comparison of parametric representations for monosyllabic word recognition in continuously spoken sentences. *IEEE Transactions on. Acoustics, Speech, and Signal Processing*, vol. 28, no. 4, pp. 357–366.

Deller, J., Hansen, J., and Proakis, J. (1999) *Discrete-Time Processing of Speech Signals*, 2nd Edition. Wiley-IEEE Press.

Digalakis, V., Neumeyer, L., and Perakakis, M. (1999) Quantization of cepstral parameters for speech recognition over the World Wide Web. *IEEE Journal on Selected Areas in Communications*, vol. 17, no. 1, pp. 82–90.

ETSI Standard ES 201 108 (2000) Distributed speech recognition; front-end feature extraction algorithm; compression algorithm, v1.1.2.

ETSI Standard ES 202 050 (2002) Distributed speech recognition; advanced front-end feature extraction algorithm; compression algorithm.

ETSI Standard ES 202 211 (2003) Distributed speech recognition; extended front-end feature extraction algorithm; compression algorithm, back-end speech reconstruction algorithm.

ETSI Standard ES 202 212 (2003) Distributed speech recognition; extended advanced front-end feature extraction algorithm; compression algorithm, back-end speech reconstruction algorithm.

Euler, S., and Zinke, J. (1994) The influence of speech coding algorithms on automatic speech recognition. In *Proceedings of ICASSP*, Adelaide, Australia.

Fingscheidt, T., Aalburg, S., Stan, S., and Beaugeant, C. (2002) Network based vs. distributed speech recognition in adaptive multi-rate wireless systems. In *Proceedings of ICSLP*, Denver, USA.

Gomez, A.M., Peinado, A.M., Sanchez, V., and Rubio, A.J. (2003) A source model mitigation technique for distributed speech recognition over lossy packet channels. In *Proceedings of Eurospeech*, Geneva, Switzerland.

Hirsch, H.G., and Pearce D. (2000) The Aurora experimental framework for the performance evaluation of speech recognition systems under noisy conditions. In *Proceedings of ISCA ITRW ASR*, Paris, France.

Hsu, W.-H., and Lee, L.-S. (2004) Efficient and robust distributed speech recognition (DSR) over wireless fading channels: 2D-DCT compression, iterative bit allocation, short BCH code and interleaving. In *Proceedings of ICASSP*, Montreal, Canada.

Ion, V., and Haeb-Umbach, R. (2006) Uncertainty decoding for distributed speech recognition over error-prone networks, *Speech Communication*, vol. 48, pp. 1435–1446.

James, A.B., and Milner, B.P. (2004) An analysis of interleavers for robust speech recognition in burst-like packet loss. In *Proceedings of ICASSP*, Montreal, Canada.

Kim, H.K., and Cox, R.V. (2001) A bitstream-based front-end for wireless speech recognition on IS-136 communications system. *IEEE Transanctions on Speech and Audio Processing*, vol. 9, no. 5, pp. 558–568.

Lam, Y.-M., Mak, M.-W., and Leong, Ph. H.-W. (2003) Fixed-point implementations of speech recognition systems. In *Proceedings of ISPC*, Dallas, USA.

Lilly, B.T., and Paliwal, K.K. (1996) Effect of speech coders on speech recognition performance. In *Proceedings of ICSLP*, pp. 2344–2347, Philadelphia, PA, USA.

Li, X., Malkin, J., and Bilmes, J. (2006) A high-speed, low-resource ASR back-end based on custom arithmetic. *IEEE Transanctions on Speech and Audio Processing*, vol. 14, no. 5, pp. 1683–1693.

Mayorga, P., Besacier, L., Lamy, R., and Serignat, J.-F. (2003) Audio packet loss over IP and speech recognition. In *Proceedings of Automatic Speech Recognition and Understanding*, Virgin Islands, USA.

Milner, B., and Shao, X. (2007) Prediction of fundamental frequency and voicing from Mel-frequency cepstral coefficients for unconstrained speech reconstruction. *IEEE Transactions on Audio, Speech and Language Processing*, vol. 15, no. 1, pp. 24–33.

Novak, M. (2004) Towards large vocabulary ASR on embedded platforms. In *Proceedings of ICSLP*, Jeju Island, Korea.

Pearce, D. (2000) Enabling new speech driven services for mobile devices: An overview of the ETSI standards activities for distributed speech recognition front-ends. In *Proceedings of Applied Voice Input/Output Society Conference*, San Jose, CA, USA.

Pearce, D. (2004) Robustness to transmission channel—The DSR approach. In Proceedings of COST278 & ISCA Research Workshop on Robustness Issues in Conversational Interaction, Norwich, UK.

Peinado, A., and Segura, J.C. (2006) *Speech Recognition Over Digital Channels*. Wiley.

Peinado, A., Sanchez, V., Perez-Cordoba, J., and de la Torre, A. (2003) HMM-based channel error mitigation and its application to distributed speech recognition. *Speech Communication*, vol. 41, pp. 549–561.

Peláez-Moreno, C., Gallardo-Antolín, A., and Díaz-de-María, F. (2001) Recognizing voice over IP: A robust front-end for speech recognition on the World Wide Web. *IEEE Transanctions on Multimedia*, vol. 3, no. 2, pp. 209–218.

So, S., and Paliwal, K.K. (2006) Scalable distributed speech recognition using Gaussian mixture model-based block quantisation. *Speech Communication*, vol. 48, pp. 746–758.

Srinivasamurthy, N., Ortega, A., and Narayanan, S. (2006) Efficient scalable encoding for distributed speech recognition. *Speech Communication*, vol. 48, no. 8, pp. 888–902.

Tan, Z.-H., Dalsgaard, P., and Lindberg, B. (2005) Automatic speech recognition over error-prone wireless networks. *Speech Communication*, vol. 47, no. 1–2, pp. 220–242.

Tan, Z.-H., Dalsgaard, P., and Lindberg, B. (2007a) Exploiting temporal correlation of speech for error-robust and bandwidth-flexible distributed speech recognition. *IEEE Transactions on Audio, Speech and Language Processing*, vol. 15, no. 4, pp. 1391–1403.

Tan, Z.-H., and Lindberg, B. (2007b) A variable frame rate method for distributed speech recognition over wireless networks. In *Proceedings of the 10th International Symposium on Wireless Personal Multimedia Communications*, Jaipur, India.

Varga, I., Aalburg, S., Andrassy, B., Astrov, S., Bauer, J.G., Beaugeant, Ch., Geissler, Ch., and Höge, H. (2002) ASR in mobile phones—An industrial approach. *IEEE Transactions on Speech and Audio Processing*, vol. 10, no. 8, pp. 562–569.

Walker, W., Lamere, P., Kwok, P., Raj, B., Singh, R., Gouvea, E., Wolf, P., and Woelfel, J. (2004) Sphinx-4: A Flexible Open Source Framework for Speech Recognition. Technical report TR-2004-139, Sun corporation, USA.

Wan, C.-Y., and Lee, L.-S. (2006). Joint uncertainty decoding (JUD) with histogram-based quantization (HQ) for robust and/or distributed Speech Recognition. In *Proceedings of ICASSP*, Toulouse, France.

Xie, Q., and Pearce, D. (2004) RTP Payload Formats for ETSI ES 202 050, ES 202 211, and ES 202 212 Distributed Speech Recognition Encoding.

Xu, H., Tan, Z.-H., Dalsgaard, P., Mattethat, R., and Lindberg, B. (2006) A configurable distributed speech recognition system. In H. Abut, J.H.L. Hansen, and K. Takeda (eds.), *Digital Signal Processing for In-Vehicle and Mobile Systems 2*. Springer Science, New York.

Zhou, B., Dechelotte, D., and Gao, Y. (2004) Two-way speech-to-speech translation on handheld devices. In *Proceedings of ICSLP* Jeju Island, Korea.

Zhu, Q., and Alwan, A. (2001) An efficient and scalable 2D DCT-based feature coding scheme for remote speech recognition. In *Proceedings of ICASSP*, Salt Lake City, USA.

Part I

Network Speech Recognition

2

Speech Coding and Packet Loss Effects on Speech and Speaker Recognition

Laurent Besacier

Abstract. This chapter is related to the speech coding and packet loss problems that occur in network speech recognition where speech is transmitted (and most of the time coded) from a client terminal to a recognition server. The first part describes some commonly used speech coding standards and presents a packet loss model useful to evaluate different channel degradation conditions in a controlled fashion. The second part evaluates the influence of different speech and audio codecs on the performance of a continuous speech recognition engine. It is shown that MPEG transcoding degrades the speech recognition performance for low bit rates whereas performance remains acceptable for specialized speech coders like G723. The same system is also evaluated for different simulated and real packet loss conditions; in that case, the significant degradation of the automatic speech recognition (ASR) performance is analyzed. The third part presents an overview of joint compression and packet loss effects on speech biometrics. Conversely to the ASR task, it is experimentally demonstrated that the adverse effects of packet loss alone are negligible, while the encoding of speech, particularly at a low bit rate, coupled with packet loss, can reduce the speaker recognition accuracy considerably. The fourth part discusses these experimental observations and refers to robustness approaches.

2.1 Introduction

Today in the context of industry and telecommunication, speech technologies are ever increasingly used for several tasks, including speech and *speaker recognition*. In this framework, a widely used architecture is client-server based where a distant speech or speaker recognition server is remotely accessed by a client. Compression of the speech signal is then generally necessary to reduce transmission delays and to respect bandwidth constraints. Many problems can appear with this kind of architecture, particularly when the transmission is made via the internet or wireless networks:

- First, transcoding (the process of coding and decoding) modifies the spectral characteristics of the speech signal, and thereby can adversely affect the system performance;

- Secondly, transmission errors can occur on the transmission line: thus, data packets can be lost (for example with UDP transport protocols over the Internet which do not implement any error recovery);
- Finally, the time response of the system is increased by coding, transmission and possible error recovery processes. This delay (termed "jitter" as used in the domain of computer networks) can be potentially very disturbing. For example, in some applications (e.g. man-machine dialogue), speech recognition is only one subsystem amongst a number of other subsystems. In such cases, the effective operation of the whole system depends heavily on the response time of the individual subsystems.

This chapter presents an overview of the speech coding and packet loss problems that occur in network speech recognition. Section 2.2 describes some commonly used speech coding standards and presents a packet loss model useful to evaluate different channel degradation conditions in a controlled fashion. Section 2.3 evaluates the influence of different speech and audio codecs on the performance of a continuous speech recognition engine. A common ASR system is also evaluated for different simulated and real packet loss conditions. Section 2.4 presents an overview of joint compression and packet loss effects on speech biometrics. Section 2.5 discusses these experimental observations and concludes this chapter.

This chapter is not dedicated to the proposal of robust methods to speech compression and packet loss. While these issues have been addressed by the author of this chapter, for instance, in (Mayorga et al. 2003), they will be deeply discussed in other chapters of this book (notably Chaps. 3 and 4).

2.2 Sources of Degradation in Network Speech Recognition

2.2.1 Speech and Audio Coding Standards

Different human-machine interfaces use speech recognition technology. For instance, voice servers (used to obtain information via the telephone) are more and more developed. Nowadays, access to a voice server is not only made through the conventional telephone network, but voice can also be transmitted through wireless networks (with mobile phones or mobile devices) or through *IP* networks (through H323 videoconferencing standard for instance). Nowadays, the number of standard and proprietary coders developed to compress speech and audio data has been quickly increased. It is thus impossible to present a detailed view of all of them in this chapter. For more details on *speech coding* standards and algorithms, the interested reader may refer to (Goldberg and Riek 2000) or to international organizations websites like ITU (www.itu.int) or ETSI (www.etsi.org).

As a consequence, we decided to present, in this section, only the coders that are used in the experiments further described in this chapter. Theses coders are nevertheless widely used in different applications: GSM (used in European mobile wireless communication), G711 and G723 (used in some VoIP protocols) and MPEG (used for audio compression).

2.2.1.1 GSM (Global System for   Mobile Communications)Coders

There exist different GSM speech coders; among them, we find the full rate, half rate and enhanced full rate coders. Their corresponding European telecommunications standards are the GSM 06.10, GSM 06.20 and GSM 06.60, respectively. These coders work on a 13-bit uniform PCM speech input signal, sampled at 8 kHz. The input is processed on a frame-by-frame basis, with a frame size of 20 ms (160 samples). A brief description of these coders follows.

Full Rate (FR) Speech Coder

The FR coder was standardized in 1987. This coder belongs to the class of Regular Pulse Excitation-Long Term Prediction—linear predictive (RPE-LTP) coders. In the encoder part, a frame of 160 speech samples is encoded as a block of 260 bits, leading to a *bit rate* of 13 kbps. The decoder maps the encoded blocks of 260 bits to output blocks of 160 reconstructed speech samples. The GSM full rate channel supports 22.8 kbps. Thus, the remaining 9.8 kbps are used for error protection. The FR coder is described in GSM 06.10 down to the bit level, enabling its verification by means of a set of digital test sequences which are also given in GSM 06.10. A public domain bit exact C-code implementation of this coder is available (http://kbs.cs.tu-berlin.de/~jutta/toast.html).

Half Rate (HR) Speech Coder

The HR coder standard was established to cope with the increasing number of subscribers. This coder is a 5.6 kbps VSELP (Vector Sum Excited Linear Prediction) coder from Motorola (Gerson and Jasiuk 1993). In order to double the capacity of the GSM cellular system, the half rate channel supports 11.4 kbps. Therefore, 5.8 kbps are used for error protection. The measured output speech quality for the HR coder is comparable to the quality of the FR coder in all tested conditions, except for tandem and background noise conditions. The normative GSM 06.06 gives the bit exact ANSI-C code for this algorithm, while GSM 06.07 gives a set of digital test sequences for compliance verification.

Enhanced Full Rate (EFR) Speech Coder

The EFR coder was standardized later. This coder is intended for utilization in the full rate channel, and it provides a substantial improvement in quality compared to the FR coder (Järvinen 1997). The EFR coder uses 12.2 kbps for speech coding and 10.6 kbps for error protection. The speech coding scheme is based on Algebraic Code Excited Linear Prediction (ACELP). The bit exact ANSI-C code for the EFR coder is given in GSM 06.53 and the verification test sequences are given in GSM 06.54.

2.2.1.2 G711 and G723.1 Coders

Nowadays some popular speech coders in voice transmission over *IP* (VoIP) are: G.723.1, G.729, G.728, G.726/7 and G.711. This set of coders is also used in video transmission and is part of the standard H323. There are several software packages for videoconferencing which can also be used for voice transmission on the Internet, for example Microsoft's *NetMeeting* uses H323. Recently, some VoIP softwares like *Skype,* for instance, use private standards. We will use in our experiments the H323 audio codec which has the lowest bit rate: G723.1 (6.4 and 5.3 kbits/s), and the one with the highest bitrate: G711 (64 kbits/s: 8 kHz, 8 bits) while we also transmitted PCM speech without any compression.

While G711 coder is very low complexity (it basically corresponds to a speech stream downsampled to 8 kHz with 8 bits per sample only), G723 is from the ACELP family (ETSI Consortium 1998). The Mean Opinion Score (MOS) which measures the perceptual quality of a coder is 3.9 for G723.1 whereas it is above 4 for G711.

2.2.1.3 MPEG Audio Coders

Unlike GSM and G7XX which are specific speech coders, MPEG coders can compress any audio signal. In fact, MPEG audio coding is generally not used for transmission of speech data but for compression of audiovisual data (TV programs for instance). Another application of speech recognition is the transcription of broadcast news and TV programs or films for archiving and retrieval. It is thus interesting to test the influence of MPEG audio coding algorithms on speech recognition performance. Moreover, MPEGI audio coding supports a variable bit rate (from 8 to 64 kbits/s), which allows us to test speech recognition on more and more compressed speech. For the experiments on MPEG transcoded speech, we used a PCX11+ specialized board for layers 1, 2 and 3 of MPEG I and for different bit rates. The perceptual quality of these coders is similar to the one of ITU coders with similar bit rates. MPEG4 implements a specific speech coder that can operate below 2 kbits/s but it is not considered in our experiments.

2.2.2 Packet Loss

While "live transmission" of a complete database over the network seems to be the best approach to evaluate packet loss and ASR degradation in real conditions (Metze et al. 2001), it is most of the time difficult to obtain a large range of degradation conditions with this method, which also needs numerous and time consuming connections between distant sites. Another possibility is to simulate how the packets are lost on the network. In the experiments further reported in this chapter, we will use both real and simulated approaches which are more deeply described in the following sections.

2.2.2.1 Packet Loss Simulation: The Gilbert Model

If we suppose that the speech packets are transmitted over the Internet, the process of audio packet loss can be characterized with the Gilbert model (Yajnik et al. 1999) of two states, as we can see in Fig. 2.1. One of the states (state 1) represents a packet loss; the other state (state 0) represents the case where packets are correctly transmitted. In this model p is the probability of going from state 0 to state 1, and q the probability of going from state 1 to state 0. This model is then characterized by two parameters, p and q, which indicate the probability of transition from either state. The different values of p and q define different packet loss conditions that may occur on the Internet. The probability that at least n consecutive packets are lost is $p (1 - q)^{n-1}$. If $(1 - q) > p$, the probability of losing a packet is greater after having already lost one packet than after having successfully received a packet; which is generally the case on Internet data transmission where packet losses occur in bursts. Note that $p + q$ is not necessarily equal to 1. When p and q parameters are fixed, the mean number of consecutive packets lost depends on p/q. The higher the quantity is, the stronger the degradation should be. For our experiments, this model was applied to obtain five different degraded versions of an existing database (Table 2.1).

Table 2.1 Different packet loss conditions

Condition	1	2	3	4	5
p	0.10	0.05	0.07	0.20	0.25
q	0.70	0.85	0.67	0.50	0.40
p/q	0.14	0.06	0.10	0.4	0.62

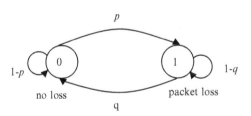

Fig. 2.1 Gilbert Model

2.2.2.2 Packet Loss in Real Transmission (Per   IP)

In order to observe what happens in real transmissions, the speech signals of the same database can be passed through different coders and different network conditions on the Internet as well. For our experiments, we decided to play and record our test database at both points of an *IP* connection with *NetMeeting*™ software. We did this by playing our speech database into a computer setup for videoconferencing. We initiated a transatlantic connection with videoconferencing software but we replaced the microphone (on the emitting site) by a computer playing the test database. These connections were established at different times of the day and at different days of the week in order to investigate a large variety of real-life network conditions. Finally, the packet loss rate (always found to be very high for the transatlantic connections) was measured for each codec and each connection, and the speech or speaker recognition performance was evaluated (results will be presented in Sects. 3 and 4 of this chapter).

2.3 Effects on the Automatic Speech Recognition Task

2.3.1 Experimental Setup

Our continuous French speech recognition system uses the Janus-III toolkit from CMU (Finke et al. 1997). The context dependent acoustic model (750 CD codebooks, 16 Gaussians each) was learned on a corpus, which contains 12 h of continuous speech of 72 speakers extracted from Bref 80 database (Lamel et al. 1991). The system uses 24-dimensional LDA features obtained from 43-dimensional acoustic vectors (13 MFCC, 13 ΔMFCC, 13 $\Delta\Delta$MFCC, E, ΔE, $\Delta\Delta$E, zero-crossing parameter) and extracted every 10 ms. The vocabulary contains nearly 5,500 phonetic variants of 2,900 distinct words; it is specific to the tourist reservation and information domain. The trigram language model that we used for our experimentation was computed using an interpolation between two LMs trained on task specific documents and on more general documents gathered from the Internet, as described in (Vaufreydaz et al. 1999).

We conducted a series of recognition experiments with 120-recorded sentences focused on reservation and tourist information task. The database was duplicated into several versions, according to the degradation methodology described in Sect. 2 (database either transcoded or passed through a packet loss process).

2.3.2 Degradation Due to Simulated Packet Loss

The first experiment was performed to show the influence of the degradation conditions (described in Table 2.1) of the Gilbert model, on speech recognition performance for different audio packet sizes (10, 20, 30 or 60 ms). In Fig. 2.2, the results for each packet size and for each condition are shown (for the degradation, we assumed that PCM wave signals were transmitted on the simulated network, without any codec applied). These word error rate (WER) measurements were done without applying any reconstruction. It can be observed that the WER tends to be relatively independent of the packet size (it only increases very slightly when the packet size increases). From this figure, we can observe that the most severe condition is the condition 5, followed by the condition 4, then condition 3 and 1, and the least severe one is the condition 2. As expected, the performance is correlated with the p/q ratio (Table 2.1). This figure also shows that the ASR degradation can be very significant in strong adverse conditions (high packet loss rate).

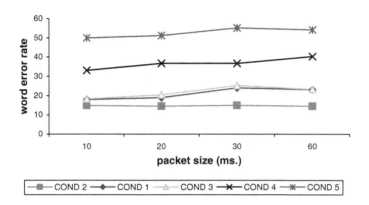

Fig. 2.2 Degradation by packet size and by condition (baseline performance without degradation is 14.4% WER) (From Mayorga and Besacier 2003, © 2003 IEEE)

2.3.3 Degradation with Real Transmissions

A second experiment was performed to show the influence of the degradation due to real transmissions over *IP*, transmitting different audio bitstreams: G723 (low bit rate codec), G711 (high bit rate codec) and PCM (no codec). The speech recognition performance was assessed and we show a summary of the results in Table 2.2. For each series of experiments (several connections were performed for each type of audio bitstream), the mean packet loss rate (PLR), the mean word error rate (WER) and the correlation coefficients between both series of PLR and WER were measured.

In real VoIP conditions, there are three additional problems: (1) noise due to our experimental transmission protocol (we noticed that playing our speech database into a computer setup for videoconferencing, as explained in Sect. 2.2.2.2, sometimes introduced signal degradation which is not quantified here), degradation due to (2)

Table 2.2 Results of WER and Packet Loss Rate (PLR) in real VoIP conditions (10 transmissions/audio bitstream); baseline = 14.4% WER

Audio bitstream	G723	G711	PCM
Mean PLR	31.8	29.8	30.5
Mean WER	81.8	62.9	53.5
Correlationcoeff. (WER,PLR)	0.28	0.49	0.64

speech compression, and (3) lost packets. Comparable average PLR's are found for the three bitstreams: 31.8% for *G723*, 29.8% for *G711* and 30.5% for PCM, which means that the same quantity of signal is lost on average. But, as we can see in Table 2.2, the highest WER is for G723, with an average of 81.8%, then 62.9% for G711, and 53.5% for PCM. Thus, for a same packet loss rate, the higher the compression level is, the higher the value of WER will be. This difference may be due to the effect of the compression itself, but also to the fact that in the case of real transmissions with G723 (the highest compression degree), one packet lost represents a bigger quantity of consecutive speech information lost, compared to the case where G711 codec or no codec (PCM) is used. In other words, lost information occurs dramatically as long bursts for G723 whereas it is more spread for G711 and PCM data transmitted. If we compare this with results in Fig. 2.2, it may be found surprising that packet size does not matter much in the simulated case: one explanation of this might be that, in this case, the packet size only varied from 10 to 60 ms (factor 6 maximum) whereas the ratio between PCM (256 kbits/s) and G723 (5.3 kbits/s) packets is much more important (50). The correlation between WER and PLR was also measured and the results show that the real conditions do not really lead to the same ideal and predictable results obtained in simulated conditions. In the simulated case, a correlation value of 0.98 was obtained whereas in the real conditions, the correlation between WER and PLR is smaller (0.64 for PCM instead of 0.98) and tends to decrease with additional factors like speech compression (0.28 and 0.49 for G711 and G723 respectively).

2.3.4 Degradation Due to Speech and Audio Codecs

The results are presented in Table 2.3 where the *MPEG* codecs were all applied on 16 kHz speech signals while the test database was downsampled to 8 kHz before the use of G711 and G723 codecs (which are generally applied on telephonic signals). Consequently, the acoustic model used in the last two lines of this table, was also trained on a downsampled version of our training database.

Results in Table 2.3 show that above 32 kbits/s bit rate, no significant degradation of speech recognition performance is observed, whereas below this threshold, performance starts to decrease dramatically. Moreover, performance is better for MPEG layer 3 than for MPEG layer 2 which is again better than MPEG layer 1.

These results are in correspondence with the known perceptual speech quality of the different *MPEG* layers. The results of this table also show that G711 and *G723* transcoding alone do not significantly degrade the speech recognition performance. Moreover, we do not see much difference between G711 and *G723* performance whereas G723 is a very low bit rate coder (5.3 kbits/s) compared to G711 coder (64 kbits/s). This result and the result from the previous section lead us to think that packet loss is certainly the biggest source of degradation for ASR whereas speech compression, if not too drastic, does not have such a big influence on the performance.

Table 2.3 Effect of different audio and speech codecs on speech recognition performance (same test database[a] transcoded with different codecs) (From Besacier 2001 © 2001 IEEE)

Coder for test	Word error rate
None (16 kHz sig.)	7.7%
MPEG Lay3 64 kbits/s	7.8%
MPEG Lay3 32 kbits/s	7.9%
MPEG Lay3 24 kbits/s	8.4%
MPEG Lay3 16 kbits/s	14.6%
MPEG Lay3 8 kbits/s	66.2%
MPEG Lay2 64 kbits/s	7.5%
MPEG Lay2 32 kbits/s	7.7%
MPEG Lay2 24 kbits/s	29.4%
MPEG Lay2 16 kbits/s	41.7%
MPEG Lay2 8 kbits/s	93.8%
MPEG Lay1 32 kbits/s	27.0%
G711 (model 8 kHz)	8.1%
G723 (model 8 kHz)	8.8%

[a]But different LM used compared to previous experiments which explains the different baseline performance.

2.4 Effect for the Automatic Speaker Verification Task

This part presents the same methodology for evaluating the speaker verification performance over compressed speech and packet loss. The idea is to duplicate an existing and well-known database used for speaker verification by passing its speech signals through different coders and different network conditions representative of what can occur over the Internet or wireless networks. First section is dedicated to the effect of joint speech compression and packet loss over *IP* networks on speaker verification while the second section evaluates the effect of GSM speech coding on speaker verification (SV) performance.

2.4.1 Speaker Verification Experiments Over Compressed Speech and Packet Loss

2.4.1.1 Experimental Setup

In acquiring the XM2VTS database (Messer et al. 1999), 295 volunteers from the University of Surrey visited a recording studio four times at approximately one month intervals. On each visit (session) two recordings (shots) were made. The first shot consisted of speech while the second consisted of rotating head movements. The experiments described in this chapter were made on the speech part of this database where the subjects were asked to read three sentences twice. The three sentences remained the same throughout all four recording sessions and a total of 7,080 speech files were made available on 4 CD-ROMs. The audio, which had originally been stored in mono, 16 bit, 32 kHz, PCM wave files, was down-sampled to 8 kHz. This is the input sampling frequency required in the speech codecs considered in this study. As previously, we used in our experiments the codec which has the lowest bit rate: *G723.1* (6.4 and 5.3 kbps), and the one with the highest bit rate: *G711* (64 kbps).

The speaker verification system used here is based on the ELISA framework (The ELISA Consortium 2000; Magrin-Chagnolleau et al. 2001). It is a GMM-based system including audio parameterization as well as score normalization techniques for speaker verification.

For the purpose of this investigation, the Lausanne protocol (configuration 2) is adopted. This has already been defined for the XM2VTS database (Messer et al. 1999). There are 199 clients in the XM2VTS DB. The training of the client models is carried out using full session1 and full session2 of the clients part of XM2VTS. 398 client test accesses are obtained using full session4 (×2 shots) of the clients part. 111,440 impostor accesses are obtained using the impostor part of the database (70 impostors × 4 sessions × 2 shots × 199 clients = 111,440 impostor accesses). The 25 evaluation impostors of XM2VTS are used to develop a World Model. The text independent speaker verification experiments are conducted in matched conditions (same training/test conditions).

The speaker verification system on XM2VTS is similar to the one presented in (Meignier et al. 2002). The speaker verification system uses 32 parameters (16 LFCC + 16 DeltaLFCC). Silence frame removal is applied as well as Cepstral Mean Subtraction. For the world model, 128 Gaussian component GMM was trained using Switchboard II phase II data (8 kHz landline telephone) and then adapted [MAP (Gauvain and Lee 1994), mean only] on XM2VTS data (25 evaluation impostors set). The client models are 128 Gaussian component GMM developed by adapting (MAP, mean only) the previous world model. Decision logic is based on using the conventional log likelihood ratio (LLR). No LLR normalization is applied here before the decision process.

2.4.1.2 Results

The speaker verification performance with the simulated degraded versions of XM2VTS is presented in Table 2.4. Based on these results, it can be concluded that the degradation due to packet loss is negligible regarding the one due to compression for text-independent speaker verification, even with bad network conditions. Comparing these results with those for speech recognition detailed in Sect. 3, it can be said that the speaker verification performance is far less sensitive to packet loss. It is probably due to the fact that the modeling is GMM which considers every frame as an independent entity. Then GMMs are not sensitive to temporal breakdown induced by packet loss and the only consequence is a reduction of the amount of signal data available for taking a decision. To our feeling, conclusions would be very different in a text-dependent mode where temporal information is important.

Table 2.4 EER (Equal Error Rate) of speaker verification results using degraded XM2VTS

No packet loss	Clean	G711	G723
	(128kbits/s)	(64kbits/s)	(5.3kbits/s)
	0.25%	0.25%	2.68%
Average	Clean	G711	G723
Network cond.	(128kbits/s)	(64kbits/s)	(5.3kbits/s)
p=0.1; q=0.7	0.25%	0.25%	6.28%
Bad	Clean	G711	G723
Network cond.	(128kbits/s)	(64kbits/s)	(5.3kbits/s)
p=0.25; q=0.4	0.50%	0.75%	9%

On the other hand, Table 2.4 shows that the speaker verification performance is adversely affected when the speech material is encoded at low bit rates (e.g. using *G723.1*).

2.4.2 Speaker Verification Experiments Over GSM Compressed Speech

Table 2.5 shows speaker verification experiments reported in (Besacier et al. 2003) where the used database (TIMIT in this paper) was downsampled from 16 kHz to 8 kHz and transcoded using the three *GSM* speech coders. All the experiments were carried out under matching conditions (i.e. training and testing are both made using the same database) and a GMM-based speaker verification system was used. For more details on this experiment see (Besacier et al. 2003).

The results of Table 2.5 show a significant performance degradation when using *GSM* transcoded databases, compared to the normal and downsampled versions of TIMIT. The results obtained are in correspondence with the perceptual *speech quality* of each coder. That is, the higher the speech quality is, the higher the measured recognition performance is.

Table 2.5 EER of speaker verification results for original and GSM transcoded speech

Original		GSM transcoded		
16 kHz	8 kHz	FR	HR	EFR
1.1%	5.1%	7.3%	7.8%	6.6%

2.5 Conclusion

This chapter presented an overview on the effect of speech coding and packet loss on two different tasks: automatic speech recognition and speaker verification. Concerning ASR, the effect of packet loss was first assessed. For this, two scenarios were considered: the simulation of lost audio packets, and the real audio transmission through *IP* networks. In the simulation case, a strong correlation between word error rate and packet loss ratio was obtained. This is less clear in real conditions where additional problems like speech compression may increase the degradation. In both cases, it was shown that packet loss can hurt the ASR performance very significantly. In a second experiment, it was shown, on the contrary, that the effect of transcoding alone is not a big issue for ASR since we have observed that the speech recognition performance remains acceptable for specialized speech coders like *G723* or reasonable bit rates of MPEG (above 24 kbits/s). To treat the critical degradations due to packet loss, packet recovering strategies can be used, like in (Mayorga et al. 2003). Some chapters of this book are more specifically dedicated to this issue: compensation for channel errors (Chaps. 3 and 4), distributed speech recognition architectures (Chap. 5), error recovery by channel coding (Chap. 8).

Concerning speech biometrics, the experiments have shown that the degradation due to packet loss is negligible regarding the one due to compression for text independent voice person authentication. It is probably due to the GMM models used which consider every frame as an independent entity. This is in contrast with the automatic speech recognition experiments where packet loss was found to reduce the accuracy significantly. However, a degradation of the speaker verification performance is observed when low bit-rate speech compression is applied to the speech signal (GSM and G723.1 codecs). In this case, packet loss can increase the degradation.

Acknowledgments

This paper is a compilation of different works made in collaboration with the following persons: P. Mayorga, R. Lamy, C. Fredouille, S. Meignier, J.-F. Bonastre and S. Grassi.

References

Besacier, L., Bergamini, C., Vaufreydaz, D., Castelli E. (2001). The effect of speech and audio compression on speech recognition performance. IEEE Multimedia Signal Processing Workshop, Cannes, France, October 2001.

Besacier, L., Bonastre, J.-F., Mayorga, P., Fredouille, C., and Meignier, S. (2003). Overview of compression and packet loss effects in speech biometrics. *IEE Proceedings Vision, Image & Signal Processing*—Special Issue on Biometrics on the Internet, vol. 150, no. 6.

ETSI Consortium. (1998). Telecommunication and internet protocol harmonization over networks: General aspects of quality of service. ETSI Technical Report.

Finke, M., Geutner, P., Hild, H., Kemp, T., Ries, K., and Westphal, M. (1997). The Karlsruhe-Verbmobil speech recognition engine. In *Proceedings of ICASSP*, Munich, Germany, vol. 1, pp. 83–86.

Gauvain, J.-L. and Lee, C.-H. (1994). Maximum a posteriori estimation for multivariate Gaussian mixture observations of Markov chains. *IEEE Transactions on Speech and Audio Processing*, vol. 2, pp. 291–298.

Gerson, I., and Jasiuk, M. (1993). A 5600 bps VSELP speech coder candidate for half rate GSM. In *Proceedings Eurospeech'93*, vol. 1, pp. 253–256.

Goldberg, R., and Riek, L. (2000). *A Practical Handbook of Speech Coders.* CRC Press, Boca Raton, FL.

Järvinen, K. (1997). GSM enhanced full rate codec. In *Proceedings of ICASSP*, vol. 2, pp. 771–774.

Lamel, L., Gauvain, J.-L., and Eskénazi, M. (1991). BREF, a large vocabulary spoken coprus for French. In *Proceedings of Eurospeech*, Gênes, Italy, vol. 2, pp. 505–508.

Magrin-Chagnolleau, I., Gravier, G., and Blouet, R. (2001). Overview of the ELISA consortium research activities. In *Proceedings. 2001: A Speaker Odyssey*, pp. 67–72.

Mayorga, P., Besacier, L., Lamy, R., and Serignat, J.-F. (2003). Audio packet loss over IP and speech recognition. In *Procedings ASRU 2003 (Automatic Speech Recognition & Understanding)*, Virgin Islands.

Meignier, S., Merlin, T., Blouet, R., and Bonastre, J.-F. (2002). NIST 2002 speaker recognition evaluation: LIA results. In *Proceedings NIST 2002 Speaker Recognition Workshop*, Vienna, Virginia.

Messer, K., Matas, J., Kittler, J., Luettin, J., and Maitre, G. (1999). XM2VTSbd: The extended M2VTS database. In *Proceedings of 2nd Conference on Audio and Video-Base Biometric Personal Verification (AVBPA99)*, Springer Verlag, New York.

Metze, F., McDonough, J., and Soltau, H. (2001). Speech recognition over netmeeting connection. In *Proceedings of Eurospeech*, Aalborg, Denmark.

The ELISA Consortium (2000). The ELISA systems for the NIST '99 evaluation in speaker detection and tracking. *Digital Signal Processing, a Review Journal*—Special Issue on NIST '99 Speaker Recognition Workshop, pp. 143–153.

Vaufreydaz, D., Akbar, M., Rouillard, J., and Caelen, J. (1999). Internet documents: A rich source for spoken language modeling. In *Proceedings ASRU Workshop*, Keystone, Colorado, pp. 277–280.

Yajnik, M., Moon, S., Kurose, J., and Towsley, D. (1999). Measurement and modelling of temporal dependence in packet loss. In *Proceedings IEEE Infocom'99*, New York.

3

Speech Recognition Over Mobile Networks

Hong Kook Kim and Richard C. Rose

Abstract. This chapter addresses issues associated with automatic speech recognition (ASR) over mobile networks, and introduces several techniques for improving speech recognition performance. One of these issues is the performance degradation of ASR over mobile networks that results from distortions produced by speech coding algorithms employed in mobile communication systems, transmission errors occurring over mobile telephone channels, and ambient background noise that can be particularly severe in mobile domains. In particular, speech coding algorithms have difficulty in modeling speech in ambient noise environments. To overcome this problem, noise reduction techniques can be integrated into speech coding algorithms to improve reconstructed speech quality under ambient noise conditions, or speech coding parameters can be made more robust with respect to ambient noise. As an alternative to mitigating the effects of speech coding distortions in the received speech signal, a bit-stream-based framework has been proposed. In this framework, the direct transformation of speech coding parameters to speech recognition parameters is performed as a means of improving ASR performance. Furthermore, it is suggested that the receiver-side enhancement of speech coding parameters can be performed using either an adaptation algorithm or model compensation. Finally, techniques for reducing the effects of channel errors are also discussed in this chapter. These techniques include frame erasure concealment for ASR, soft-decoding, and missing feature theory-based ASR decoding.

3.1 Introduction

Interest in voice-enabled services over mobile networks has created a demand for more natural human-machine interfaces (Rabiner 1997; Cox et al. 2000; Lee and Lee 2001; Nakano 2001), which has in turn placed increased demands on the performance of automatic speech recognition (ASR) technology. It is interesting that the evolution of mobile networks has fostered increased interest in ASR research (Chang 2000; Mohan 2001). This is because the performance of ASR systems over mobile networks is degraded by factors that are in general not important in more traditional ASR deployments (Euler and Zinke 1994; Lilly and Paliwal 1996; Milner and Semnani 2000). These factors can be classified as device-oriented noise and network-oriented noise.

Mobile communication technologies provide access to communications networks anytime, anywhere, and from any device. Under this framework, communications devices like cell phones and PDAs are becoming increasingly smaller to support various levels of mobility. Furthermore, different combinations of microphone technologies including close talking device mounted microphones, wired and wireless headsets, and device mounted far-field microphones may be used with a given device depending on the user's needs. All of these issues can result in a large variety of acoustic environments as compared to what might be expected in the case of a plain old telephony service (POTS) phone. For example, a handheld device can be considered as a distance-talking microphone, where the distance might be continually changing and thus background noise could be characterized as being time-varying and non-stationary. The issues of ASR under such a device-oriented noise condition have been discussed in the context of feature compensation and acoustic model combination under a background noise condition (Dufour et al. 1996; Rose et al. 2001) and acoustic echo cancellation (Barcaroli et al. 2005), distance speech recognition, and multiple-microphone speech recognition (Wang et al. 2005).

Network-oriented sources of ASR performance degradation include distortion from low-bit-rate speech coders employed in the networks and the distortions arising from transmission errors occurring over the associated communication channels. Even though a state-of-the-art speech coder can compress speech signals with near transparent quality from a perceptual point of view, the performance of an ASR system using the decoded speech can degrade relative to the performance obtained for the original speech (Euler and Zinke 1994; Lilly and Paliwal 1996). One of the major reasons is that the parameterization of speech for speech coding is different from that for speech recognition. For example, speech coding is mainly based on a speech production model, which represents the spectral envelope of speech signals using *linear predictive coding* (*LPC*) *coefficients*. However, feature representations used for speech recognition like, for example, *Mel-frequency cepstral coefficients* (MFCC), are usually extracted on the basis of human perception. In addition to speech coding distortion, mobile networks can introduce a range of transmission errors that impact speech quality at the speech decoder (Choi et al. 1999). Transmission errors are generally represented using measures like the carrier-to-interference (C/I) ratio or the frame erasure rate.

There are three general configurations used for extracting feature parameters for ASR over mobile networks; the decoded speech-based approach, the bitstream-based approach, and the distributed speech recognition (DSR) approach (Gallardo-Antolín et al. 1998; Milner and Semnani 2000; Kim and Cox 2001).

The decoded speech-based approach involves extracting speech recognition parameters from the decoded speech after transmission over the network. This corresponds to conventional ASR performed without explicitly accounting for the communication network.

The bitstream-based approach obtains speech recognition parameters for ASR directly from the transmitted bitstream of the speech coder. It exploits the decomposition of speech signals into spectral envelope and excitation components that is performed by the speech coder. The two components are quantized separately where the spectral envelope is represented as an all-pole model using LPC coefficients. Deriving ASR feature parameters directly from the bitstream is primarily motivated by the fact that ASR feature parameters are based on the speech spectral envelope and not on the excitation. Moreover, the distortion that is introduced by convolution of the spectral envelope with the quantized excitation signal while reconstructing speech in the decoder represents another source of performance degradation in ASR. Bitstream approaches avoid this source of degradation. It will be shown in this chapter that a bitstream based approach applies a feature transformation directly to the LPC-based spectral representation derived from the transmitted bitstream.

The DSR approach involves extracting, quantizing, and channel encoding the speech recognition parameters at the client before transmitting the channel encoded feature parameters over the mobile network. Thus, ASR is performed at the server using features that were quantized, encoded, and transmitted over a protected data channel. The general framework for DSR will be discussed in Chap. 5. For all three of the above configurations, approaches for compensating with respect to sources of spectral distortion and channel distortion can be applied both in the ASR feature space and in the acoustic model domain.

In this chapter, we focus on the techniques that can be applied to the bitstream-based approach and to overcoming network-oriented and device-oriented sources of ASR performance degradation. Following this introduction, Sect. 3.2 describes the techniques in more depth. Section 3.3 explains the bitstream-based approaches in detail. The transformation of spectral parameters obtained from the bitstream into MFCC-like parameters for the purpose of improving ASR performance is discussed in Sect. 3.4. We introduce compensation techniques for cellular channels, speech coding distortion, and channel errors in Sect. 3.5. Summary and conclusion are provided in Sect. 3.6.

3.2 Techniques for Improving ASR Performance Over Mobile Networks

This section addresses the general scenario of ASR over mobile networks. Figure 3.1 shows a series of processing blocks applicable to ASR over mobile networks. There are two processing paths: one is for the decoded speech-based approach and the other is for the bitstream-based approach. The processing blocks dedicated to the decoded speech-based approach include the speech decoding algorithm itself, enhancing the quality of the decoded speech in the signal domain, and extracting ASR features. Processing blocks such as spectral feature decoding and feature transformation from speech coding features to ASR features are used for the bitstream-based approach. In addition to these processing blocks, common processing blocks include: 1) frame loss concealment, 2) compensating for ASR features in communication channels, 3) adapting acoustic models to compensate for spectral distortion or channel errors, and 4) Viterbi decoding incorporating ASR decoder-based concealment.

It is assumed that all robust ASR techniques discussed in this chapter can be applied to the cases where ASR parameters are extracted either from speech reconstructed by

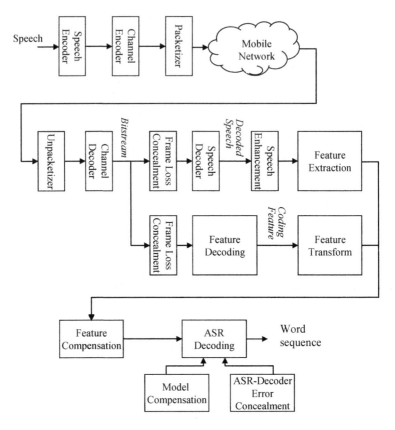

Fig. 3.1 Scenarios for the implementation of robust feature analysis and feature and model compensation for ASR over mobile networks

the decoder or directly from the transmitted bitstream. Previously-developed techniques for robust ASR in the conventional ASR framework can also be applied to the decoded speech-based approach by considering the effects of the mobile network to be similar in nature to the effects of an adverse environment. However, there are several techniques that will be presented here, which are strictly relevant to the bitstream-based approach. These include feature transformations from the feature representations used in the speech coding algorithm to the feature representations used in ASR and techniques for feature compensation in the bitstream domain. Moreover, the existence of network-oriented noise sources such as speech coding distortions and channel transmission errors has led to the development of compensation techniques in the signal space, feature space, and model space. A brief summary of the techniques developed for the bitstream-based approach and the network-oriented noise compensation is provided here.

In the *decoded speech-based approach*, the decoded speech is used directly for feature extraction on the receiver side of the network. There has been a great deal of work devoted to exploring the effect of speech coding on ASR performance and to training or improving ASR acoustic models to compensate for these effects. In (Euler and Zinke 1994; Lilly and Paliwal 1996; Nour-Eldin et al. 2004), it was shown that ASR performance degraded when the input speech was subjected to encoding/decoding from standard speech coding algorithms. One approach used to mitigate this problem was to train the model with the equivalence of the multi-style training of a hidden Markov model (HMM) by using utterances that are recorded over a range of communication channels and environmental conditions. This approach was followed successfully for ASR over cellular telephone networks in (Sukkar et al. 2002), who noted the severe impact of the acoustic environment in mobile applications on the ASR word error rate (WER). Another approach was to improve the average transmitted speech quality by adjusting the trade-off between the number of bits assigned to coded speech and the number of bits assigned to channel protection based on an estimate of the current network conditions. To this end, Fingscheidt et al. (Fingscheidt et al. 2002) investigated the effect of coding speech using the *adaptive multi-rate* (AMR) coder for ASR over noisy GSM channels. It was shown that the effects of communication channels on ASR WER could be significantly reduced with respect to WER obtained using standard fixed rate speech coders.

Bitstream-based techniques for robust ASR obtain speech recognition parameters directly from the bitstream transmitted to the receiver over digital mobile networks. The difference between bitstream-based techniques and techniques that operate on the decoded speech is that bitstream-based techniques avoid the step of reconstructing speech from the coded speech parameters. In this scenario, the transformation of speech coding parameters to speech recognition parameters is required to improve ASR performance (Peláez-Moreno et al. 2001). Since each mobile network relies on its own standardized speech coder, the bitstream-based approaches are dependent upon the characteristics of the mobile network. Moreover, each speech coder has a different spectral quantization scheme and different levels of resolution associated with its spectral quantizer. Therefore, dedicated feature extraction and transform techniques must be developed for each speech coder. Such techniques have been developed and published for GSM RPE-LTP (Huerta and Stern 1998; Gallardo-Antolín et al. 2005), the TIA standard IS-641 (Kim and Cox 2001), the ITU-T Recommendation G.723.1 (Peláez-Moreno et al. 2001), and the TIA standard IS-96 QCELP and IS-127 EVRC (Choi et al. 2000).

Frame loss concealment refers to a technique used to reconstruct ASR features even if the bitstream associated with a given transmitted frame is lost (Tan et al. 2005). In general, a frame loss concealment algorithm is embedded in the speech decoder. It allows the parameters of lost frames to be estimated by repeating those of the previous uncorrupted frame (ITU-T Recommendation G.729 1996). Consequently, the estimated parameters can be directly used for extracting ASR features in the bitstream-based approach. Otherwise, in the decoded speech approach, speech is reconstructed using the estimated parameters and ASR can be performed with this decoded speech. Furthermore, the frame erasure rate of the network or the indication of the lost frames can be used for ASR decoder-based error concealment (Bernard and Alwan 2001b).

Feature and model compensation techniques can be implemented without consideration of the speech coding algorithm used in the network. The approaches emphasize the development of robust algorithms for improving ASR performance without any explicit knowledge of the distortions introduced by the speech coder. These robust algorithms can be realized in the feature domain, the HMM model domain, and through modification of the ASR decoding algorithms. In the feature domain, *spectral distortion* is considered to be a nonlinear noise source, distorting the feature parameters in a variety of ways. Current methods compensate for these distortions by applying linear filtering, normalization techniques, or some other nonlinear processing applied to the feature parameters (Dufour et al. 1996; Kim 2004; Vicente-Pena et al. 2006). In addition, speech coding parameters can be directly enhanced in the coding parameter domain to compensate for speech coding distortion and environmental background noise (Kim et al. 2002). In the HMM model domain, model compensation or combination techniques can be applied by incorporating parametric models of the noisy environment (Gómez et al. 2006). In the ASR decoder, the effect of channel errors can be mitigated by incorporating probabilistic models that characterize the confidence associated with a given observation or spectral region. These techniques have been implemented under the headings of missing features and "soft" Viterbi decoding frameworks (Gómez et al. 2006; Siu and Chan 2006).

3.3 Bitstream-Based Approach

This section describes how ASR feature analysis can be performed directly from the bitstream of a code-excited linear predictive (CELP) *speech coder*, as produced by the channel decoder in a mobile cellular communications network. First, spectral analysis procedures performed in both CELP speech coders and ASR feature analysis procedures are compared. Then, techniques for taking coded representations of CELP parameters and producing ASR feature parameters are described.

It is important to understand the similarities and differences between the speech analysis performed for speech coding and that for speech recognition. In general, speech coding is based on the speech production model shown in Fig. 3.2. The reconstructed speech, $\hat{s}(n)$, is modeled as an excitation signal, $u(n)$, driving an all-pole system function, $1/A(z)$, which describes the spectral envelope of the vocal tract response. The spectral envelope can be represented by LPC or equivalent parameter sets, including *line spectral pairs* (LSP), immitance spectral pairs (ISP), and reflection coefficients. The model in Fig. 3.2 for CELP-type speech coders represents the excitation signal as a combination of: 1) periodic information, $g_p x_p(n)$, which includes parameters such as pitch or long-term predictor (adaptive codebook) lag and gain value; and 2) random source information, $g_c c(n)$, which is represented by the indices and gain coefficients associated with a fixed codebook containing random excitation sequences.

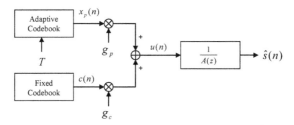

Fig. 3.2 General structure of code-excited linear prediction speech coding

The block diagram in Fig. 3.3 compares the steps that are typically involved in feature extraction for ASR and CELP speech coding. Figure 3.3(a) shows a typical example of frame-based ASR feature analysis. The speech signal is pre-emphasized using a first-order differentiator, (1-0.95z-1), and the signal is then windowed. In the case of *LPC-derived cepstral coefficient* (LPCC) analysis, a linear prediction polynomial is estimated using the autocorrelation method. Then, the shape and the duration of the analysis window are determined as a trade-off between time and frequency resolution. Typically, a Hamming window of length 30 ms is applied to the speech segment. The Levinson-Durbin recursion is subsequently applied to the autocorrelation coefficients to extract LPC coefficients. Finally, LPCCs are computed up to the 12 order, and a cepstral lifter can be applied to the cepstral coefficients. This analysis is repeated once every 10 ms, which results in a frame rate of 100 Hz.

Figure 3.3(b) shows the simplified block diagram of the LPC analysis performed in the IS-641 speech coder (Honkanen et al. 1997). In this analysis, undesired low frequency components are removed using a high-pass filter with a cutoff frequency of 80 Hz. Because of delay constraints that are imposed on the speech coder, an asymmetric analysis window is used, where one side of the window is half of a Hamming window and the other is a quarter period of the cosine function. Two additional processes are applied to the autocorrelation sequence; one is lag-windowing, and the other is white noise correction. The former helps smooth the LPC spectrum to remove sharp spectral peaks (Tohkura et al. 1978). The latter gives the effect of adding white noise to the speech signal and thus avoids modeling an anti-aliasing filter response at high frequencies with the LPC coefficients (Atal 1980). Finally, the Levinson-Durbin recursion is performed with this modified autocorrelation sequence, and LPC coefficients of order ten are converted into ten LSPs. The speech encoder quantizes the LSPs and then transmits them to the decoder. Of course, the LSPs recovered at the decoder differ from the unquantized LSPs by an amount that depends on the LSP spectral quantization algorithm.

The windowed spectral analysis procedures in Figs. 3.3(a) and 3.3(b) are similar in that they both extract the parameters of the spectral envelope filter $1/A(z)$, as shown in Fig. 3.2. However, there are two differences that are important when applying the procedure in Fig. 3.3(b) to obtain the ASR features. The first is that the frame

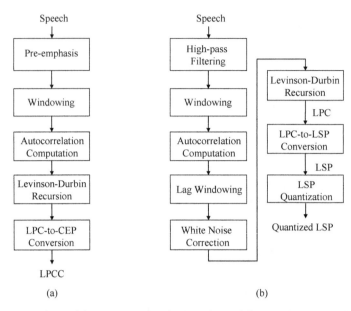

Fig. 3.3 Comparison of feature extraction for (a) ASR and (b) speech coding (After Kim and Cox 2001)

rate used for the LPC analysis in the speech coder is 50 Hz, as opposed to the 100 Hz frame rate used for ASR. This lack of resolution in time-frequency sampling can be mitigated by using an interpolation technique (Kim and Cox 2001), duplicating the frames under such a low frame rate condition, or reducing the number of HMM states (Tan et al. 2007). The second difference is the spectral quantization that is applied to the LSPs, where the distortion resulting from this LSP quantization cannot be recovered.

Figure 3.4 shows a procedure for extracting cepstral coefficients from the bit-stream of the IS-641 speech coder (Kim and Cox 2001). The figure displays the parameters that are packetized together for a single transmitted analysis frame. The bitstream for a frame is largely divided into two classes for vocal tract information and excitation information. 26 bits are allocated per frame for the spectral envelope which is represented using LSP quantization indices. 122 bits per frame are allocated for excitation information which includes pitch, algebraic codebook indices, and gains. The procedure shown in the block diagram begins with tenth order LSP coefficients being decoded from the LSP bitstream. In order to match the 50 Hz frame rate used for LPC analysis in the speech coder with the 100 Hz frame rate used in ASR feature analysis, the decoded LSPs are interpolated with the LSP coefficients decoded from the previous frame. This results in a frame rate of 100 Hz for the ASR front-end (Peláez-Moreno et al. 2001; Kim and Cox 2001). For the case of LPCC-based ASR feature analysis, cepstral coefficients of order 12 are obtained from the conversion of LSP to LPC followed by LPC-to-cepstrum conversion. The twelve liftered cepstral coefficients are obtained by applying a band-pass lifter to the cepstral coefficients. Lastly, an energy parameter is obtained by using the decoded excitation, $u(n)$, or the decoded speech signal, $\hat{s}(n)$, which is equivalent to a log energy parameter of the conventional MFCC feature (Davis and Mermelstein 1980).

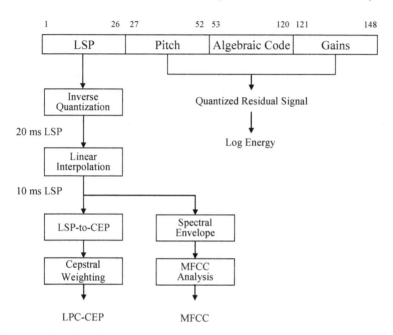

Fig. 3.4 A typical procedure of feature extraction in the bitstream-based approach (After Kim and Cox 2001)

Figure 3.4 also depicts the process of extracting MFCC-based ASR features from the IS-641 speech coder bitstream. As shown in Fig. 3.3(a), MFCC feature analysis can be performed for ASR by applying a 512-point fast Fourier transform (FFT) to compute the magnitude spectrum of the windowed speech signal. The magnitude spectrum is subsequently passed through a set of triangular weighting functions that simulate a filterbank defined over a Mel-warped frequency scale. For a 4 kHz bandwidth, 23 filters are used. The filterbank outputs are transformed to a logarithmic scale, and a discrete cosine transform (DCT) is applied to obtain 13 MFCCs. In order to obtain MFCC-based ASR features from the bitstream, the MFCCs can be obtained directly from the decoded LSPs. This LSP-to-MFCC conversion will be described in the next section.

There has been a large body of work on the bitstream-based approach in the context of a variety of standard speech coding algorithms. This work falls into two general categories. The first includes procedures for deriving ASR features from the bitstreams associated with standard speech coding algorithms. Peláez-Moreno et al. (2001) compared ASR WER using ASR features derived from the bitsream of the ITU–T G.723.1 speech coding standard with the WER obtained using reconstructed speech from the same coding standard and found that the bitstream derived parameters resulted in lower ASR WER. Additional work has been reported where ASR features were derived from the bitstream of the LPC-10E coder (Yu and Wang 1998), the Qualcomm CELP coder (Choi et al. 2000), and the continuously variable slope delta modulation (CVSD) waveform coder (Nour-Eldin et al. 2004). In all of these cases, the WER obtained by deriving ASR features from the bitstream was lower than that obtained by deriving features from the reconstructed speech. The second category of work on bitstream-based approaches includes techniques for compensating bitstream-based parameters to improve ASR robustness. Kim et al. (Kim and Cox 2002) proposed the enhancement of spectral parameters in the LSP domain at the decoder by estimating the background noise level. Yu and Wang (2003) proposed an iterative method for compensating channel distortion in the LSP domain, where the ITU-T G.723.1 coder was used for their experiments.

3.4 Feature Transform

Though bitstream-based ASR is known to be more robust than that using decoded speech, the spectral parameters used for speech coding are not adequate for ASR (Choi et al. 2000). Most speech coders operating at moderate bit-rates are based on a model of the type used in code-excited linear prediction as is illustrated in Fig. 3.2. In these coders, LPCs are further transformed into LSPs to exploit the coding efficiency, simple stability check for synthesis filters, and superior linear interpolation performance enjoyed by the LSP representation. There have been several research efforts focused on using LSP coefficients as feature representations for ASR (Paliwal 1988; Zheng et al. 1988). Signal processing steps that are thought to emulate aspects of speech perception including critical band theory and non-linear amplitude compression have been found to have a far greater impact in ASR. For example, LPC coefficients based on *perceptual linear prediction* (PLP) analysis are known to provide significantly better ASR performance than LPCs (Hermansky 1990). As such, one of the research issues associated with a bitstream-based ASR front-end is to obtain more robust parameters for ASR by transforming the spectral parameters that are used by the speech coder (Fabregas et al. 2005; Peláez-Moreno et al. 2006).

Figure 3.5 illustrates several ways for obtaining feature parameters from the bitstream, where it is assumed that LSPs are the bitstream-based spectral parameters transmitted to the speech decoder. Note that there are three main approaches to transforming *LSPs*. The first is to convert LSPs into LPCs followed by a further transformation to obtain MFCC-type parameters. The second approach is to obtain the spectral magnitude from LSPs or LPCs and to apply conventional Mel-filterbank analysis and DCT to obtain MFCC parameters. The last approach is to directly convert LSPs into approximate cepstral coefficients, which are called pseudo-cepstral coefficients (PCEPs). Of course, a frequency-warping technique can be applied to LSPs prior to the pseudo-cepstral conversion, which results in a Mel-scaled PCEP (MPCEP).

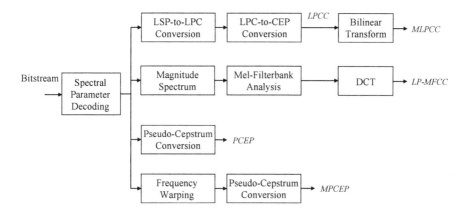

Fig. 3.5 Feature transforms from LSPs to ASR feature parameters

3.4.1 Mel-Scaled LPCC

Mel-scaled LPCCs (MLPCCs) can be derived using the following three steps: the conversion from LSP to LPC, the conversion from LPC to LPC cepstrum, and the frequency warping of LPC cepstrum using an all-pass filter. For a given set of LSP coefficients of order M, $\omega_1, \cdots, \omega_M$, where typically $M = 10$, the LPC coefficients, $a_1, \cdots, a_M$, can be obtained by using the following equations

$$A(z) = 1 + \sum_{i=1}^{M} a_i z^{-i} = \frac{(P(z)-1) + (Q(z)-1)}{2} \tag{3.1}$$

where

$$
\begin{aligned}
P(z)-1 &= \prod_{i=2,4,\cdots}^{M}(1+x_i z^{-1}+z^{-2}) - z^{-1}\prod_{i=2,4,\cdots}^{M}(1+x_i z^{-1}+z^{-2})-1 \\
&= z^{-1}\sum_{i=2,4,\cdots}^{M}\left[(x_i+z^{-1})\prod_{j=0,2,4,\cdots}^{i-2}(1+x_j z^{-1}+z^{-2})\right] - z^{-1}\prod_{i=2,4,\cdots}^{M}(1+x_i z^{-1}+z^{-2})
\end{aligned}
\tag{3.2}
$$

and

$$
\begin{aligned}
Q(z)-1 &= (1+z^{-1})\prod_{i=1,3,\cdots}^{M-1}(1+x_i z^{-1}+z^{-2})-1 = \prod_{i=1,3,\cdots}^{M-1}(1+x_i z^{-1}+z^{-2})+z^{-1}\prod_{i=1,3,\cdots}^{M-1}(1+x_i z^{-1}+z^{-2})-1 \\
&= z^{-1}\sum_{i=1,3,\cdots}^{M-1}\left[(x_i+z^{-1})\prod_{j=-1,1,3,\cdots}^{i-2}(1+x_j z^{-1}+z^{-2})\right]+z^{-1}\prod_{i=1,3,\cdots}^{M-1}(1+x_i z^{-1}+z^{-2})
\end{aligned}
\tag{3.3}
$$

with $x_i = -2\cos\omega_i$ for $i = 1, \cdots, M$, and $x_{-1} = -z^{-1}$ such that $1 + x_{-1}z^{-1} + z^{-2} = 1$.

Next, the real valued cepstrum for the spectral envelope can be defined by the inverse z-transform of the log spectral envelope represented by $\ln[1/A(z)]$. In other words,

$$\ln[1/A(z)] = \sum_{n=1}^{\infty} c_n z^{-n} \qquad (3.4)$$

where c_n is the n-th LPCC, and obtained from the recursion described in (Schroeder 1981). That is,

$$c_n = \begin{cases} -a_n - \dfrac{1}{n}\sum_{k=1}^{n-1} kc_k a_{n-k}, & 1 \le n \le M \\ -\dfrac{1}{n}\sum_{k=1}^{M} kc_k a_{n-k}, & n > M \end{cases} . \qquad (3.5)$$

It is common to truncate the order of LPCCs to 12–16 for ASR by applying a cepstral lifter (Juang et al. 1987; Junqua et al. 1993). This obtains a reasonable balance between spectral resolution and spectral smoothing and also largely removes the affects of the vocal tract excitation from the cepstrum.

In order to obtain MLPCCs, a *bilinear transform* is applied to the frequency axis of LPCCs (Oppenheim and Johnson 1972). Here, the n-th MLPCC, c_n^{MLPCC}, is obtained from the LPCCs $\{c_n\}$ by filtering the LPCCS with a sequence of all-pass filters such that

$$c_n^{MLPCC} = \sum_{k=-\infty}^{\infty} \frac{1}{n} k\psi_{n,k} c_k, \; n > 0 \qquad (3.6)$$

where $\dfrac{1}{n} k\psi_{n,k}$ is the unit sample response of the filter

$$H_n(z) = \frac{(1-\alpha^2)z^{-1}}{(1-\alpha z^{-1})^2}\left[\frac{z^{-1}-\alpha}{1-\alpha z^{-1}}\right]^{n-1}, \; n > 0 . \qquad (3.7)$$

In Eq. 3.7, the degree of frequency warping is controlled by changing α; a typical value of α for speech sampled at 8 kHz is 0.3624 (Wölfel and McDonough 2005).

3.4.2 LPC-Based MFCC (LP-MFCC)

A procedure for obtaining MFCC-type parameters from LSPs begins with the computation of the magnitude spectrum from LSPs. The squared magnitude spectrum of $A(z)$ evaluated at frequency ω is given by

$$\left|A(e^{j\omega})\right|^2 = 2^M\left[\sin^2(\omega/2)\prod_{i=2,4,\cdots,M}(\cos\omega - \cos\omega_i)^2 + \cos^2(\omega/2)\prod_{i=1,3,\cdots,M-1}(\cos\omega - \cos\omega_i)^2\right]. \qquad (3.8)$$

Thus, the n-th MFCC-type parameter, $c_n^{LP-MFCC}$, can be obtained by applying a conventional Mel-filterbank analysis (Davis and Mermelstein 1980) to the inverse of Eq. 3.8 and transforming the filterbank output using the DCT.

3.4.3 Pseudo-Cepstrum (PCEP) and Its Mel-Scaled Variant (MPCEP)

Pseudo-cepstral analysis has been proposed by (Kim et al. 2000) as a computationally efficient approach for obtaining ASR parameters from LSPs. Using this analysis, the n-th pseudo-cepstrum (PCEP), c_n^{PCEP}, is defined by

$$c_n^{PCEP} = \frac{1}{2n}(1+(-1)^n) + \frac{1}{n}\sum_{k=1}^{M}\cos(n\omega_k), \; n \geq 1. \tag{3.9}$$

It was shown that the spectral envelope represented by PCEPs is very similar to that represented by LPCCs, but can be computed with lower computational complexity. As such, PCEP can be further transformed to accommodate the characteristics of frequency warping. First, each LSP is transformed into its Mel-scaled version by using an all-pass filter (Gurgen et al. 1990), and then the i-th *Mel-scaled LSP* (MLSP), ω_i^{Mel}, can be obtained by

$$\omega_i^{Mel} = \omega_i + 2\tan^{-1}\left(\frac{\alpha\sin\omega_i}{1-\alpha\cos\omega_i}\right), \; 1 \leq i \leq M \tag{3.10}$$

where α controls the degree of frequency warping and is set as $\alpha = 0.45$ (Choi et al. 2000). Finally, a Mel-scaled version of PCEP or Mel-scaled (*MPCEP*) can be obtained by combining Eqs. 3.9 and 3.10 such that

$$c_n^{MPCEP} = \frac{1}{2n}(1+(-1)^n) + \frac{1}{n}\sum_{k=1}^{M}\cos(n\omega_k^{Mel}), \; n \geq 1. \tag{3.11}$$

MPCEP required lower computational resources than MLPCC but the ASR performance using MLPCC was better than that obtained using MPCEP. However, the two Mel-scaled ASR features, MPCEP and MLPCC, provided comparable ASR performance when the transmission errors of the network were under certain levels (Fabregas et al. 2005). In addition, when used in combination with techniques that will be described in the next section, the transformed ASR features obtained from LSPs, PCEP, and MPCEP have the potential to further improve ASR performance.

3.5 Enhancement of ASR Performance Over Mobile Networks

3.5.1 Compensation for the Effect of Mobile Systems

In performing ASR over cellular networks, ASR WER can be improved by applying techniques that have been developed for noise-robust ASR. For example, HMMs can be trained using a large amount of speech data collected from a range of different communications environments (Sukkar et al. 2002). This is similar to a kind of HMM multi-condition training. An alternative set of approaches is to train separate environment-specific HMMs and combine the models during ASR decoding (Karray et al. 1998). This approach can also incorporate dedicated models of specific non-stationary noise types, like impulsive noise or frame erasures, which can be trained from labeled examples of occurrences of these noise events in training data. Karray et al. (1998) proposed several examples of this class of approach, each differing in the manner in which the environment specific HMM models were integrated during search. Finally, the most widely discussed class of approaches for robust ASR in the cellular domain is the application of feature compensation techniques to ASR features. One of many examples is the work of Dufour et al. (1996) involving compensation for GSM channel distortion by applying non-linear spectral subtraction and cepstral mean normalization to root MFCCs.

Finally, HMM models can be adapted or combined with models of environmental or network noise to improve the performance of ASR over mobile networks. Linear transform-based adaptation methods such as maximum likelihood linear regression (MLLR), Bayesian adaptation, and model combination have been shown to be useful for compensating for the non-linear characteristics of mobile networks (Kim 2004; Zhang and Xu 2006). In particular, Kim (2004) exploited the relationship between the signal-to-quantization noise ratio (SQNR) measured from low-bit-rate speech coders in mobile environments and the signal-to-noise ratio (SNR) in wireline acoustic noise environments. This was motivated from the insight that the quantization noise introduced by the speech coder can be characterized as a white noise process. In order to obtain HMM acoustic models for use on decoded speech, a model combination technique was applied to compensate the mean and variance matrices of HMMs that were trained using uncoded speech. As a result, the ASR system using this compensation approach achieved a relative reduction in average WER of 7.5–16.0% with respect to a system that did not use any compensation techniques. Moreover, explicit knowledge of the characteristics associated with the mobile system can be used to improve the performance of model compensation procedures (Zhang and Xu 2006).

3.5.2 Compensation for Speech Coding Distortion in LSP Domain

In order to further improve the performance of ASR over mobile networks, we can also apply feature enhancement or model compensation in the speech coding parameter domain, i.e., LSP domain. Kim et al. (2002) proposed the enhancement of spectral parameters in the LSP domain at the decoder by estimating the level of background noise. Figure 3.6 shows the block diagram for *feature enhancement* in the LSP domain. Note that the objective of a speech enhancement algorithm is to obtain a smaller spectral distortion between clean speech and enhanced speech than that obtained between clean speech and noisy speech. Likewise, the purpose of the proposed feature enhancement algorithm is to obtain enhanced features that are close to the features obtained from clean speech. In the figure, the estimate of clean speech LSP, $\hat{\omega}_s$, is updated from the LSP decoded from the bitstream, ω_n, using two LSPs. The first is the LSP, ω_{q+e}, obtained from the enhanced version of the decoded speech signal, s_{q+e}, and the other is the LSP, ω_{n+q}, obtained from the LPC analysis of the decoded speech, s_{n+q}. The decoded speech is assumed to include background noise and speech coding distortion. The update equation for the estimated clean speech LSPs is

$$\hat{\omega}_{s,i} = \omega_{n,i} + \mu\left(\omega_{q+e,i} - \omega_{n+q,i}\right),\ 1 \le i \le M \tag{3.12}$$

where μ is the step size for the adaptive algorithm of Eq. 3.12 and is set to $\mu = 1/M - \varepsilon$. In practice, μ was set to 0.2.

To prevent the enhanced LSPs from being distorted by the feature enhancement algorithm, the update to the estimated LSPs in Eq. 3.12 is only applied at moderate SNR levels. The SNR of the decoded speech signal is estimated from the ratio be-

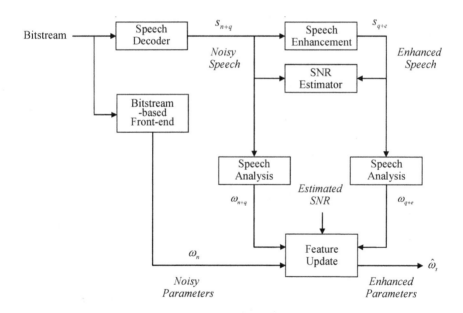

Fig. 3.6 Block diagram of the feature enhancement algorithm (After Kim et al. 2002)

tween the decoded speech, s_{n+q}, and its enhanced version, s_{q+e}. Eq. 3.12 is modified as so that it is only applied when the estimated SNR does not exceed a threshold

$$\hat{\omega}_{s,i} = \begin{cases} \omega_{n,i} + \mu(\omega_{q+e,i} - \omega_{n+q,i}), & if \quad Estimated \; SNR < SNR_{th} \\ \omega_{n,i}, & otherwise \end{cases} \tag{3.13}$$

where SNR_{th} is empirically determined according to the amount of SNR improvement by the speech enhancement and is set to 40 dB in Kim et al. (2002).

The performance of the proposed feature enhancement algorithm was evaluated on a large vocabulary word recognition task recorded by both a close-talking microphone and a far-field microphone and processed by the IS-641 speech coder. The twelve LP-MFCCs described in Sect. 3.4.2 were obtained for the bitstream-based front-end, and the log-energy obtained from the excitation information described in Sect. 3.3 was appended to the feature vector. The difference and second difference of this feature vector were concatenated with the static features to construct a 39-dimensional feature vector. It was subsequently determined that the bitstream-based front-end provided better performance than the front-end approach that extracted MFCCs from the decoded speech for the close-talking microphone speech recognition but not for the far-field microphone speech recognition. However, incorporating the feature enhancement algorithm into the bitstream-based front-end significantly improved ASR performance for far-field microphone speech recognition.

3.5.3 Compensation for Channel Errors

In speech coding, channel impairments can be characterized by bit errors and frame erasures, where the number of bit errors and frame erasures primarily depends on the noise, co-channel and adjacent channel interface, and frequency selective fading. Fortunately, most speech coders are combined with a channel coder so that the most sensitive bits are strongly protected by the channel coder. Protecting bits unequally has an advantage over protecting all the transmission bits when only a small number of bits are available for channel coding. In this case, a frame erasure is declared if any of the bits that are most sensitive to channel error are in error (Sollenberger et al. 1999). The bits for LSPs and gains are usually classified as the most sensitive bits (Servetti and de Martin 2002; Kataoka and Hayashi 2007), and they are strongly protected by the channel encoder. However, a method needs to be designed to deal with channel bit errors because ASR is generally more sensitive to channel errors than it is to channel erasures (Bernard and Alwan 2001b). The ASR problem regarding bit errors is usually overcome by designing a frame erasure concealment algorithm, whereas the ASR problem of frame errors is overcome by using *soft-decoding* in a Viterbi search for HMM-based ASR.

Frame erasure concealment algorithms can be classified into sender-based algorithms and receiver-based algorithms, based on where the concealment algorithm is implemented. Typically, sender-based algorithms, e.g., forward error correction (FEC), are more effective than receiver-based algorithms but require additional bits used for detecting or correcting errors in the decoder (Wah et al. 2000). Conversely, receiver-based algorithms such as repetition-based frame erasure concealment and interpolative frame erasure concealment (de Martin et al. 2000) have advantages over the sender-based algorithms since they do not need any additional bits, and thus we can use existing standard speech encoders without any modification. Likewise, such a receiver-based algorithm can be used for ASR, enabling the reconstruction of speech signals corresponding to the erased frames prior to the extraction of ASR features.

Gómez et al. (2006) proposed a linear interpolation method between feature vectors obtained from the first and last correctly received frames to reconstruct feature vectors for the erased frames. However, instead of using linear interpolation, Milner and Semnani (2000) introduced a polynomial interpolation method. Since delay was not critical for speech recognition, this interpolation method could provide better feature vectors, even under burst frame erasure conditions, than an extrapolation method. Moreover, Bernard and Alwan (2002) proposed a frame dropping technique that removes all feature vectors from the erased frames or any suspicious frames due to channel errors. The frame dropping technique worked reasonably well for random erasure channels, but provided poor performance when the channel erasure was bursty. It was also shown in Kim and Cox (2001) that the performance of bitstream-based front-end approaches employing frame dropping was better than that of the decoded speech-based front-end that included a frame erasure concealment algorithm.

Channel errors or frame erasures can be addressed by modifying the ASR decoder. The Viterbi algorithm can be modified to incorporate a time-varying weighting factor that characterizes the degree of reliability of the feature vectors (Bernard and Alwan 2001a; Siu and Chan 2006). The probability of a path terminating in HMM state j at time t, $\delta_t(j)$, in the Viterbi algorithm can be written to include a reliability measure

$$\delta_t(j) = \max_i [\delta_{t-1}(i)a_{ij}] [b_j(o_t)]^{\gamma_t} \qquad (3.14)$$

where $\gamma_t = P(o_t \mid y_t)$ is a time-varying weighting factor, and y_t is a received bitstream. Note that $\gamma_t = 1$ if the decoded ASR feature is completely reliable, and $\gamma_t = 0$ if it is completely unreliable. On the other hand, Siu and Chan (2006) proposed a robust Viterbi algorithm, where corrupted frames by impulsive noise would be skipped for the Viterbi path selection. This was implemented by expanding the search space of the Viterbi algorithm and by introducing a likelihood ratio threshold for the section.

3.6 Conclusion

This chapter has presented the major issues that must be addressed to facilitate robust automatic speech recognition over mobile networks. It has summarized new approaches for minimizing the impact of distortions introduced by speech coders, acoustic environments, and channel impairments. Obtaining ASR features directly from the bitstream of standardized speech coders was originally developed as a new paradigm for feature extraction over mobile communications networks. It was found that optimal ASR feature representations could be obtained by transforming spectral parameters transmitted with the coded bitstream. It was also found that feature parameter enhancement techniques that exploited the bitstream-based spectral parameters could result in more noise robust ASR. More recently, bitstream-based techniques have been applied to network-based ASR applications using many standard speech coding algorithms.

As mobile networks and the mobile devices that are connected to these networks evolve, it is likely that automatic speech recognition robustness over these networks will continue to be a challenge. With enhanced mobility and increased connectivity, the characteristics of future mobile networks are likely to be different from those existing today. They are also likely to lend themselves to new paradigms for novel distributed implementation of robust techniques that better configure ASR algorithms for these mobile domains. The work presented in this chapter contains several examples of new methods for implementing robust ASR processing techniques that exploit knowledge of the communications environment. This class of techniques will only become more important with time.

References

Atal, B. S. (1980). Predictive coding of speech at low bit rates. *IEEE Transactions on Communication*, vol. 30, no. 4, pp. 600–614.

Barcaroli, L., Linares, G., Costa, J.-P. and Bonastre, J.-F. (2005). Nonlinear GSM echo cancellation: Application to speech recognition. In *Proceedings of ISCA Tutorial and Research Workshop on Non-linear Speech Processing*, paper 021.

Bernard, A. and Alwan, A. (2001a). Joint channel decoding—Viterbi recognition for wireless applications. In *Proceedings Eurospeech*, pp. 2213–2216.

Bernard, A. and Alwan, A. (2001b). Source and channel coding for remote speech recognition over error-prone channels. In *Proceedings of ICASSP*, pp. 2613–2616.

Bernard, A. and Alwan, A. (2002). Channel noise robustness for low-bitrate remote speech recognition. In *Proceedings of ICASSP*, pp. 2213–2216.

Chang, H. M. (2000). Is ASR ready for wireless primetime: Measuring the core technology for selected applications. *Speech Communication*, vol. 31, no. 4, pp. 293–307.

Choi, S. H., Kim, H. K., Kim, S. R., Cho, Y. D. and Lee, H. S. (1999). Performance evaluation of speech coders for speech recognition in adverse communication environments. In *Proceedings of ICCE*, pp. 318–319.

Choi, S. H., Kim, H. K. and Lee, H. S. (2000). Speech recognition using quantized LSP parameters and their transformations in digital communication. *Speech Communication*, vol. 30, no. 4, pp. 223–233.

Cox, R. V., Kamm, C. A., Rabiner, L. R., Schroeter, J. and Wilpon, J. G. (2000). Speech and language processing for next-millennium communications services. *Proceedings of the IEEE*, vol. 88, no. 8, pp. 1314–1337.

Davis, S. B. and Mermelstein, P. (1980). Comparison of parametric representations for monosyllabic word recognition in continuously spoken sentences. *IEEE Transactions on Acoustics Speech and Signal Processing*, vol. 28, no. 4, pp. 357–366.

de Martin, J. C., Unno, T. and Viswanathan, V. (2000). Improved frame erasure concealment for CELP-based coders. In *Proceedings of ICASSP*, pp. 1483–1486.

Dufour, S., Glorion, C. and Lockwood, P. (1996). Evaluation of the root-normalized front-end (RN_LFCC) for speech recognition in wireless GSM network environments. In *Proceedings of ICASSP*, pp. 77–80.

Euler, S. and Zinke, J. (1994). The influence of speech coding algorithms on automatic speech recognition. In *Proceedings of ICASSP*, pp. 621–624.

Fabregas, V., de Alencar, S. and Alcaim, A. (2005). Transformations of LPC and LSF parameters to speech recognition features. *Lecture Notes in Computer Sciences*, vol. 3686, pp. 522–528.

Fingscheidt, T., Aalbury, S., Stan, S. and Beaugeant, C. (2002). Network-based versus distributed speech recognition in adaptive multi-rate wireless systems. In *Proceedings of ICSLP*, pp. 2209–2212.

Gallardo-Antolín, A., Díaz-de-María, F. and Valverde-Albacete, F. (1998). Recognition from GSM digital speech. In *Proceedings of ICSLP*, pp. 1443–1446.

Gallardo-Antolín, A., Peláez-Moreno, C. and Díaz-de-María, F. (2005). Recognizing GSM digital speech. *IEEE Transactions on Speech and Audio Processing*, vol. 13, no. 6, pp. 1186–1205.

Gómez, M., Peinado, A. M., Sánchez, V. and Rubo, A. J. (2006). Recognition of coded speech transmitted over wireless channels. *IEEE Transactions on Wireless Communications*, vol. 5, no. 9, pp. 2555–2562.

Gurgen, F. S., Sagayama, S. and Furui, S. (1990). Line spectrum frequency-based distance measure for speech recognition. In *Proceedings of ICSLP*, pp. 521–524.

Hermansky, H. (1990). Perceptual linear predictive (PLP) analysis of speech. *The Journal of the Acoustical Society of America*, vol. 87, no. 4, pp. 1738–1752.

Honkanen, T., Vainio, J., Järvinen, K., Haavisto, P., Salami, R., Laflamme, C. and Adoul, J.-P. (1997). Enhanced full rate speech codec for IS-136 digital cellular system. In *Proceedings of ICASSP*, pp. 731–734.

Huerta, J. M. and Stern, R. M. (1998). Speech recognition from GSM codec parameters. In *Proceedings of ICSLP*, pp. 1463–1466.

ITU-T Recommendation G.729 (1996). Coding of speech at 8 kbit/s using conjugate-structure algebraic-code-excited linear-prediction (CS-ACELP). March.

Juang, B.-H., Rabiner, L. R. and Wilpon, J. G. (1987). On the use of bandpass liftering in speech recognition. *IEEE Transactions on Acoustics Speech and Signal Processing*, vol. 35, no. 7, pp. 947–954.

Junqua, J.-C., Wakita, H. and Hermansky, H. (1993). Evaluation and optimization of perceptually-based ASR front-end. *IEEE Transactions on Speech and Audio Processing*, vol. 1, no. 1, pp. 39–48.

Karray, L., Jelloun, A. B. and Mokbel, C. (1998). Solutions for robust recognition over the GSM cellular network. In *Proceedings of ICASSP*, pp. 261–264.

Kataoka, A. and Hayashi, S. (2007). A cryptic encoding method for G.729 using variation in bit-reversal sensitivity. *Electronics and Communications in Japan (Part III: Fundamental Electronic Science)*, vol. 90, no. 2, pp. 63–71.

Kim, H. K., Choi, S. H. and Lee, H. S. (2000). On approximating line spectral frequencies to LPC cepstral coefficients. *IEEE Transactions on Speech and Audio Processing*, vol. 8, no. 2, pp. 195–199.

Kim, H. K. and Cox, R. V. (2001). A bitstream-based front-end for wireless speech recognition on IS-136 communications system. *IEEE Transactions on Speech and Audio Processing*, vol. 9, no. 5, pp. 558–568.

Kim, H. K., Cox, R. V. and Rose, R. C. (2002). Performance improvement of a bitstream-based front-end for wireless speech recognition in adverse environments. *IEEE Transactions on Speech and Audio Processing*, vol. 10, no. 8, pp. 591–604.

Kim, H. K. (2004). Compensation of speech coding distortion for wireless speech recognition. *IEICE Transactions on Information and Systems*, vol. E87-D, no. 6, pp. 1596–1600.

Lee, L.-S. and Lee, Y. (2001). Voice access of global information for broad-band wireless: Technologies of today and challenges of tomorrow. *Proceedings of the IEEE*, vol. 89, no. 1, pp. 41–57.

Lilly, B. T. and Paliwal, K. K. (1996). Effect of speech coders on speech recognition performance. In *Proceedings of ICSLP*, pp. 2344–2347.

Milner, B. and Semnani, S. (2000). Robust speech recognition over IP networks. In *Proceedings of ICASSP*, pp. 1791–1794.

Mohan, A. (2001). A strategy for voice browsing in 3G wireless networks. In *Proceedings of EUROCON*, pp. 120–123.

Nakano, H. (2001). Speech interfaces for mobile communications. In *Proceedings of ASRU*, pp. 93–95.

Nour-Eldin, A. H., Tolba, H. and O'Shaughnessy, D. (2004). Automatic recognition of Bluetooth speech in 802.11 interference and the effectiveness of insertion-based compensation techniques. In *Proceedings of ICASSP*, pp. 1033–1036.

Oppenheim, A. V. and Johnson, D. H. (1972). Discrete representation of signals. *Proceedings of the IEEE*, vol. 60, no. 6, pp. 681–691.

Paliwal, K. K. (1988). A perception-based LSP distance measure for speech re-cognition. *The Journal of the Acoustical Society of America*, vol. 84, no. S1, pp. S14–S15.

Peláez-Moreno, C., Gallardo-Antolín, A. and Díaz-de-María, F. (2001). Recognizing voice over IP: A robust front-end for speech recognition on the World Wide Web. *IEEE Transactions on Multimedia*, vol. 3, no. 2, pp. 209–218.

Peláez-Moreno, C., Gallardo-Antolín, A., Gómez-Cajas, D. F. and Díaz-de-María, F. (2006). A comparison of front-ends for bitstream-based ASR over IP. *Signal Processing*, vol. 86, no. 7, pp. 1502–1508.

Rabiner, L. R. (1997). Applications of speech recognition in the area of telecommunications. In *Proceedings of ASRU*, pp. 501–510.

Rose, R. C., Parthasarathy, S., Gajic, B., Rosenberg, A. E. and Narayanan, S. (2001). On the implementation of ASR algorithms for hand-held wireless mobile devices. In *Proceedings of ICASSP*, pp. 17–20.

Schroeder, M. R. (1981). Direct (nonrecursive) relations between cepstrum and predictor coefficient. *IEEE Transactions on Acoustics Speech and Signal Processing*, vol. 29, no. 2, pp. 297–301.

Servetti, A. and de Martin, J. C. (2002). Perception-based partial encryption of compressed speech. *IEEE Transactions on Speech and Audio Processing*, vol. 10, no. 8, pp. 637–643.

Siu, M. and Chan, A. (2006). A robust Viterbi algorithm against impulsive noise with application to speech recognition. *IEEE Transactions on Audio Speech and Language Processing*, vol. 14, no. 6, pp. 2122–2133.

Sollenberger, N. R., Seshadri, N. and Cox, R. (1999). The evolution of IS-136 TDMA for third-generation wireless services. *IEEE Personal Communications*, vol. 6, no. 3, pp. 8–18.

Sukkar, R. A., Chengalvarayan, R. and Jacob, J. J. (2002). Unified speech recognition for the landline and wireless environments. In *Proceedings of ICASSP*, pp. 293–296.

Tan, Z.-H., Dalsgaard, P. and Lindberg, B. (2005). Automatic speech recognition over error-prone wireless networks. *Speech Communication*, vol. 47, nos. 1-2, pp. 220–242.

Tan, Z.-H., Dalsgaard, P. and Lindberg, B. (2007). Exploiting temporal correlation of speech for error robust and bandwidth flexible distributed speech recognition. *IEEE Transactions on Audio Speech and Language Processing*, vol. 15, no. 4, pp. 1391–1403.

Tohkura, Y., Itakura, F. and Hashimoto, S. (1978). Spectral smoothing technique in PARCOR speech analysis-synthesis. *IEEE Transactions on Acoustics Speech and Signal Processing*, vol. 26, no. 6, pp. 587–596.

Vicente-Pena, J., Gallardo-Antolín, A., Peláez-Moreno, C. and Díaz-de-María, F. (2006). Band-pass filtering of the time sequences of spectral parameters for robust wireless speech recognition. *Speech Communication*, vol. 48, no. 10, pp. 1379–1398.

Wah, B. W., Su, X. and Lin, D. (2000). A survey of error-concealment schemes for real-time audio and video transmissions over the Internet. In *Proceedings of IEEE International Symposium on Multimedia Software Engineering*, pp. 17–24.

Wang, L., Kitaoka, N. and Nakagawa, S. (2005). Robust distance speech recognition based on position dependent CMN using a novel multiple microphone processing technique. In *Proceedings Eurospeech*, pp. 2661–2664.

Wölfel, M. and McDonough, J. (2005). Minimum variance distortionless response spectral estimation. *IEEE Signal Processing Magazine*, vol. 22, no. 5, pp. 117–126.

Yu, A. T. and Wang, H. C. (1998). A study on the recognition of low bit-rate encoded speech. In *Proceedings of ICSLP*, pp. 1523–1526.

Yu, A. T. and Wang, H. C. (2003). Channel effect compensation in LSF domain. *EURASIP Journal on Applied Signal Processing*, vol. 2003, no. 9, pp. 922–929.

Zhang, H. and Xu, J. (2006). Pattern-based dynamic compensation towards robust speech recognition in mobile environments. In *Proceedings of ICASSP*, pp. 1129–1132.

Zheng, F., Song, Z., Li, L., Yu, W., Zheng, F. and Wu, W. (1988). The distance measure for line spectrum pairs applied to speech recognition. In *Proceedings of ICSLP*, pp. 1123–1126.

4

Speech Recognition Over IP Networks

Hong Kook Kim

Abstract. This chapter introduces the basic features of speech recognition over an IP-based network. First of all, we review typical lossy packet channel models and several speech coders used for voice over IP, where the performance of a network speech recognition (NSR) system can significantly degrade. Second, several techniques for maintaining the performance of NSR against packet loss are addressed. The techniques are classified into client-based techniques and server-based techniques; the former ones include rate control approaches, forward error correction, and interleaving, and the latter ones include packet loss concealment and ASR-decoder based concealment. The last part of this chapter is devoted to explaining a new framework of NSR over IP networks. In particular, a speech coder that is optimized for automatic speech recognition (ASR) is presented, where it provides speech quality comparable to the conventional standard speech coders used in the IP networks. In addition, we compare the performance of NSR using the ASR-optimized speech coder to that using a conventional speech coder.

4.1 Introduction

The Internet is a worldwide publicly-accessible network of interconnected computer networks that transmits data by packet switching using standard Internet protocols (IP) (http://en.wikipedia.org/wiki/Internet). Currently, voice data is seen as one of the more important types of data, and the transfer of voice conversations over IP networks, referred to as voice over IP (VoIP), has significantly grown in recent years. In addition to the convergence of voice and traditional data, there has also been considerable convergence of IP networks with cellular/wireless networks such as GSM, WiMAX, WiFi, Bluetooth, etc. (Chandra and Lide 2007). This trend towards convergence has created quite a number of challenges associated with the architecture and implementation of automatic speech recognition (ASR) in convergent network environments.

In this chapter, we present issues related to ASR over IP networks in a framework of *network speech recognition* (NSR). In this framework, it is basically assumed that speech must be encoded for transmission at a client. However, an ASR server can make use of decoded speech or the bitstream prior to ASR decoding, as shown in Figs. 4.1a and b, respectively (Milner and James 2006; Kim and Cox 2001). Furthermore, when compared to ASR over mobile networks, speech transmitted over

IP networks is subject to degradation from sources based on the characteristics and limitations inherent to IP networks and end-to-end environments. In other words, in addition to speech coder distortion and acoustic environmental noises, IP networks primarily distort speech quality by network-oriented impairment factors such as jitter and packet loss.

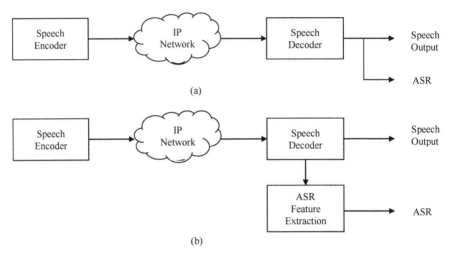

Fig. 4.1 Architecture for network speech recognition (NSR): (a) decoded speech based NSR and (b) bitstream-based NSR (After Kim and Cox 2001)

One can improve the quality of speech in an NSR framework by detecting jittered or lost packets and then recovering or concealing them. For this end, techniques for maintaining NSR performance against jitter or packet loss can be classified into either client-based techniques or server-based techniques. The former includes layered coding, forward error correction, and interleaving, among others. The latter includes techniques based on packet loss concealment and *ASR decoder-based concealment*. Since a number of these techniques are commonly used for ASR over mobile networks and/or distributed speech recognition (DSR), we only focus on the techniques dedicated to ASR over IP networks.

Due to limitations in bandwidth, a low-bit-rate speech coder is also applied in IP networks to compress speech. However, when compared to mobile networks, several speech coders can be selectively used in IP networks though a speech coder is exclusively standardized for mobile networks. Moreover, there is flexibility in delivering voice through the development of a new speech coder in IP networks. This implies that NSR performance would be significantly improved if a speech coder could be designed to optimize ASR performance rather than speech quality for speech compression. Of course, the speech coder should provide speech quality comparable to conventional speech coders currently used in IP networks.

Following this Introduction, Sect. 4.2 will briefly discuss the relationship between ASR performance and speech quality affected by IP network-oriented impairment factors including jitter and packet loss. Section 4.3 will classify robust techniques against such network-oriented impairment factors into client-based techniques and server-based techniques, and discuss the techniques dedicated to ASR over IP networks. Section 4.4 will explain such a speech coder and then show the effect of the new speech coder on ASR performance and speech quality in an NSR framework. Finally, this chapter is concluded in Sect. 4.5.

4.2 Speech Recognition and IP Networks

The deployment of ASR services over IP networks has been realized in several forms of architectures. Of these architectures, NSR using either the decoded speech or the bitstream of the encoded speech does not require any constraints to a client that supports VoIP. It, however, is well known that the performance of NSR over IP networks degrades due to sources of IP distortion, which include low-bit-rate speech coding, packet loss, and jitter. Among them, jitter can be ignored if the jitter buffer size of IP networks is allowed to be sufficiently large such that no speech packets are lost due to delay, which is a condition that does not harm overall ASR performance.

This chapter further discusses the two key points: speech coding distortion and packet loss. There have been many research previous works that have investigated the effect of IP networks on ASR performance (Milner and Semnani 2000; Milner 2001; Peláez-Moreno et al. 2001; Van Sciver et al. 2002; Falavigna et al. 2003; Mayorga et al. 2003; Mayorga and Besacier 2006). This section briefly summarizes some of these works, especially in terms of motivation for the development of the new speech coder described in Sect. 4.4.

4.2.1 Relationship Between ASR Performance and Speech Quality

There are several processing blocks required prior to successfully transmitting speech over IP networks. Fig. 4.2 shows the processing steps in VoIP at the client and at the server (Chandra and Lide 2007), where each processing block can be seen to contribute to the quality of speech. In this way, the conversation quality of speech becomes a function of factors such as distortion, loudness, delay, and echo. Actually, distortion is mainly caused by speech coding distortion and packet loss, which defines the listening quality of speech commonly measured in the *mean opinion score* (MOS) (Takahashi et al. 2004).

In general, it is known that the listening quality of speech degrades depending on the bit-rate of the speech coder used in an IP network and the condition of packet loss. Sun et al. (2004) demonstrated that ASR performance of noisy speech could be predicted using an objective speech quality measure, i.e., the perceptual evaluation of speech quality (PESQ) defined as ITU-T Recommendation P.861 (ITU-T Recommendation P.862 2001). This result provides evidence that ASR performance is highly associated with speech quality since PESQ can be used as a measure of estimating the quality of decoded speech even under a packet loss condition. Moreover, Hooper and Russell (2000) described the relationship between ASR performance and speech quality in *VoIP*, where speech quality was represented in terms of speech coder type, packet size, and *packet loss rate* (PLR), and concluded that ASR might be a viable quantitative measure of speech quality. Therefore, in order to improve ASR performance, we need to improve the quality of the decoded speech or develop a method that compensates for factors that can potentially degrade speech quality.

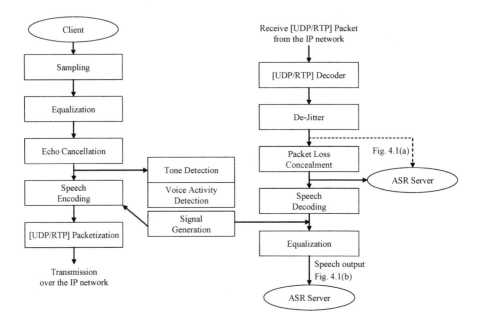

Fig. 4.2 Processing steps in VoIP and NSR scenarios; the first scenario of NSR, denoted as Fig. 4.1a, can be seen to include the use of the bitstream before or after the packet loss concealment block, and the second scenario, denoted as Fig. 4.1b, includes the application of a conventional ASR front-end to the speech output (From Chandra and Lide 2007)

4.2.2 Impact of Speech Coding Distortion

Speech is commonly transmitted over IP networks by one of the speech coders described in Table 4.1 (Walker and Hicks 2004). As shown in the table, the real bandwidth required by VoIP communications is higher than that of the speech coder. For example, speech is compressed by the G.729 coder with a bit-rate of 8 kbit/s, but the actual bitstream over IP is 32.2 kbit/s when the packet size is twice the analysis frame size of G.729. This size increase is due to the fact that headers are accumulated before the actual speech data, such as the *real time protocol* (RTP) header of 12 bytes, the *user datagram protocol* (UDP) of 8 bytes, the IP header of 20 bytes, and the Ethernet header of 18 bytes. Of course, the total overhead of 58 bytes per packet can be reduced by enlarging the packet size. However, it increases subsequent delays, and as a result the VoIP system is apt to be fragile to packet loss because single packet loss corresponds to a large number of consecutive frame losses (Hooper and Russell 2000).

Table 4.1 Five common speech coders used in VoIP (From Walker and Hicks 2004)[1]

Coder	Data Rate (kbit/s)	Typical Packet Size(ms)	Packetization Delay (ms)	Bandwidth (kbit/s)	Maximum MOS
G.711	64.0	20	1.0	87.2	4.41
G.726	32.0	20	1.0	55.2	4.22
G.723.1	5.3	30	67.5	20.8	3.69
G.723.1	6.3	30	67.5	21.9	3.87
G.729	8.0	20	25.0	31.2	4.07

There have also been many works investigating the effect of speech coding on ASR performance in terms of two different scenarios, as shown in Fig. 4.1 (Gallardo-Antolín et al. 1998; Milner and Semnani 2000; Kim and Cox 2001). In Van Sciver et al. (2002), the ASR performance of four different coders was compared based on the first scenario of Fig. 4.1a, where the coders were 6.3 kbit/s *G.723.1*, 6.4 kbit/s G.729D, 8 kbit/s *G.729*, and 11.2 kbit/s G.729E. Through this comparison, it was found that the performance of ASR over IP networks was always worse than that using uncoded speech at the client. Furthermore, the performance was more degraded when a lower bit-rate coder was used for speech coding, for whatever packet loss rate incurred in the recognition experiments. This coincides with the result suggesting that ASR performance is closely related to the decoded speech quality. From an ASR point of view, the reason why speech quality degrades with a low-bit-rate coder is that decoded speech can be distorted by the quantization distortion of spectral parameters in combination with excitation distortion (Peláez-Moreno et al. 2001). However, this problem can be overcome by using a bitstream-based approach, as shown in Fig. 4.1b.

On the other hand, the primary cause of performance degradation is the different frame rate of speech coding from that of ASR. In Falavigna et al. (2003), by obtaining speech recognition features with a frame rate of 7.5 ms from the bitstream of the G.723.1 speech coder, ASR performance was significantly improved compared to when they were obtained with a frame rate of 30 ms. For this end, Tan et al. (2007) investigated the relationship between ASR frame rate and the number of HMM states and showed that ASR performance was improved by matching the two factors such as by duplicating the frames under a low frame rate condition or by reducing the number of HMM states.

4.2.3 Impact of Network Channel Distortion

The effect of packet loss on ASR performance has been investigated in two NSR scenarios. In order to simulate the behavior of transmission models with memory,

---

[1] Walker/Hicks, TAKING CHARGE OF YOUR VOIP PROJECT, p. 86 Table 3-3 Default Attributes for Six Common Codes, © Cisco Systems, Inc. Reproduced by permission of Pearson Education, Inc. All rights reserved.

the Gilbert-Elliott channel model (Peláez-Moreno et al. 2001; Falavigna et al. 2003; ITU-T Recommendation G.191 2000) can be used for random and burst packet losses, as shown in Fig. 4.3a. A three-state packet loss model, shown in Fig. 4.3b, was further introduced to simulate the burst-like nature of packet loss with an intermediate state between packet loss free and *burst packet loss* (Milner 2001).

These models can be used to generate a packet loss pattern, and the received packet is declared either lost or not according to a binary number in the pattern. The probability of packet loss, i.e., the PLR, in the *Gilbert-Elliot model* is determined by

$$PLR = \frac{P_s}{1-\gamma}P_1 + \frac{P_t}{1-\gamma}P_2 \qquad (4.1)$$

where P_1 and P_2 are the probabilities of staying at the good and bad state, respectively, P_t and P_s are the transition probabilities from the good state to the bad state and vice versa, respectively, and $\gamma = 1-(P_s + P_t)$ controls the burstiness of packet loss. In Fig. 4.3b, PLR and burstiness are determined by P_{ef} and P_{eb}, respectively.

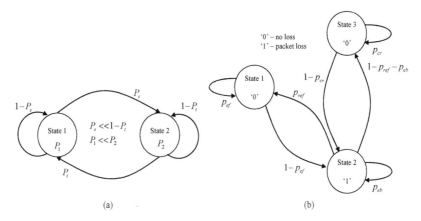

Fig. 4.3 Packet loss models: **(a)** Gilbert-Elliot model and **(b)** three-state burst-like packet loss model (From Milner 2001)

This packet loss pattern should reflect the characteristics of real voice traffic in IP networks. It was shown from the results reported in Borella (2000) that PLR was about 0.5–3.5% with a mean number of packets lost in a single burst of about 6.9, where around 90% of the bursts consisted of three packets or less for the G.723.1 coder, corresponding to a speech interval of 90 ms. Under this condition, the performance of NSR using the bitstream was always better than that using speech decoded by the coder for all PLRs and burstiness (Peláez-Moreno et al. 2001; Van Sciver et al. 2002; Mayorga and Besacier 2006), though it was significantly lower than that without any packet loss. This result is the basis of the motivation for encouraging the further development of techniques robust to packet loss.

4.3 Robustness Against Packet Loss

In this section, we address techniques associated with preventing, recovering, or concealing packet loss to improve ASR performance. Note that some techniques (Tan et al. 2005, 2007) are dedicated for application in other speech recognition scenarios, such as the ASR over mobile networks described in Chap. 3 and distributed speech recognition described in the following chapters. This section solely focuses on the techniques used for ASR over IP networks based on the NSR framework; Fig. 4.4 shows the basic taxonomy of error robust techniques applicable to NSR (Perkins et al. 1998; Tan et al. 2005).

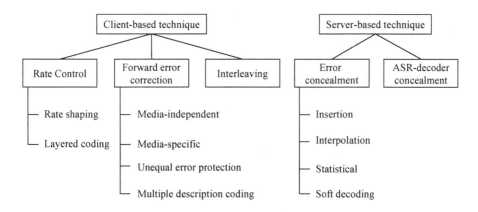

Fig. 4.4 Robustness techniques against packet loss for NSR

4.3.1 Rate Control

Maintaining a high quality of speech at the client is important for reliable ASR performance in an NSR framework. To this extent, speech quality can be improved by controlling QoS in the Internet via adaptive packet size or jitter buffer length and by optimizing network resources in active or passive ways. Rate shaping techniques are an active method of optimizing network resources and attempt to adjust the rate of speech encoding according to current network conditions. Seo et al. (2001) reported that the network condition was monitored based on the arriving time difference between a pair of packets by using the timestamp in an RTP header. As the time difference decreased, a higher rate of the adaptive multi-rate (AMR) coder (3GPP TS 26.090 1999) was preferred. As a result, the AMR coder could improve overall speech quality by trading off PLR and the bit-rate of the speech coding PLR, as compared to a fixed rate coder. A similar approach was proposed in Ruggeri et al. (2001) by modifying the G.729 coder into a multi-mode and multi-rate coder for rate shaping. In addition, Fingscheidt et al. (2002) used the AMR coder for rate shaping over GSM and showed that the performance of NSR using rate shaping of the AMR coder was comparable to that of DSR.

Layered encoding encodes speech into several layers, where a reasonable quality of speech can be obtained with the base layer. Then, when the network transmits layered speech, it can drop the higher layers in the event of network congestion; recently, the G.729 coder was extended with such a layered coding scheme (Ragot et al. 2007).

4.3.2 Forward Error Correction

Forward error correction (FEC) is a method by which the encoder sends extra information to help the decoder recover from packet loss. For example, media-independent channel coding is realized by using parity codes, cyclic redundancy codes, and Reed-Solomon codes, which enables the decoder to accurately repair lost packets without knowing the type of content. However, it requires additional delays and bandwidth (Shacham and McKenney 1990).

On the other hand, a media-specific FEC sends the same or similar contents in multiple packets. If a packet is lost, the packet may be recovered using a duplicate packet. For example, in Hardman et al. (1995) a current speech frame was basically encoded by the 13.2 kbit/s GSM coder and also encoded by a 4.8 kbit/s low-bit-rate LPC vocoder for a media-specific FEC. The actual information transmitted was composed of the 13.2 kbit/s GSM bitstream of the current frame and the 4.8 kbit/s LPC vocoder bitstream of the previous frame, thus speech could be decoded by using the LPC vocoder bitstream for the lost previous frame.

Another kind of media-specific FEC that attempts to make the decoder robust to bit error is unequal error protection (UEP), which protects only a part of the bits in each packet (Swaminathan et al. 1996). The bits are judged based on a bit sensitivity analysis (Servetti and De Martin 2002; Kataoka and Hayashi 2007).

4.3.3 Interleaving

The technique of interleaving aims at distributing the effects of the lost packets in such a way that the overall packet loss effects are reduced. For instance, burst packet loss affects the speech bitstream or speech quality as if it were a random packet loss. Moreover, compared to FEC techniques, interleaving does not increase the network load. For example, in Mayorga et al. (2003), each packet was divided into several units for PCM transmission. However, each packet of the bitstream of a speech coder, e.g., the G.729 or G.723.1 speech coder, consisted of 2–4 frames in an attempt to reduce the network overhead caused by the RTP/UDP/IP header (Mayorga and Besacier 2006). In this case, the unit corresponds to a single frame of the speech coder. Units are then combined in a different sequential order that is generated by a speech coder and rearranged into their original order at the decoder. Thus, packet loss results in the loss of several units distributed in the other packets. The *error concealment* (EC) techniques described in the next subsection will be applied to reconstruct the lost frames.

Multiple description coding (MDC) is an alternative to FEC for reducing the effects of packet loss by splitting the bitstream into multiple streams or paths, though this technique consumes a wider bandwidth (Anandakumar et al. 2000). To overcome the increased bandwidth demand, zero redundancy MDC can be designed by exploiting the characteristics of a CELP-type coder (Wah and Lin 2005). In other words, spectral parameters, LSPs in CELP-type coding, are temporally correlated such that LSPs can be interleaved in MDC while excitation is replicated along multiple descriptions in the interleaved units.

4.3.4 Error Concealment and ASR Decoder-Based Concealment

There have been a number of research works associated with EC based on insertion-based, interpolation-based, and statistical approaches reported in a DSR framework (Tan et al. 2005; Milner and James 2006). These approaches are further discussed in other chapters; here, we only discuss EC with respect to NSR.

In insertion-based EC techniques, lost frames are replaced with silence, noise, or estimated values. In general, the parameters of a lost frame are estimated by extrapolating those of a previous good frame. That is, the parameters of lost frames are estimated by repeating a down-scaled version of previous ones (*ITU-T Recommendation G.729* 1996). In particular, the specific steps taken for reconstructing a lost frame in G.729 are: (1) repeating the synthesis filter parameters, (2) attenuating the adaptive and fixed codebook gains, followed by attenuating the memory values of the gain predictor, and (3) randomly generating the excitation. This approach works well for speech communication, where delay is an essential issue as there is no time to wait for future good frames at the decoder.

Assuming that in a VoIP system a future good packet will be available in the playout buffer just after a series of lost packets, interpolation-based EC techniques can be applied (de Martin et al. 2000). This assumption can yield additional delays in speech decoding, though such delays do not affect ASR if the average time delay caused by burst packet loss is less than 100 ms as mentioned in Sect. 4.2.3. The interpolation-based EC algorithm has the potential to reconstruct a lost frame by applying a linear or polynomial interpolation technique between the parameters of the first and last correct speech frames before and after the burst packet loss. Such an interpolation-based EC algorithm has been successfully implemented for NSR (Mayorga et al. 2003; Gómez et al. 2006).

A novel approach was proposed in Mayorga et al. (2003), where different weights of the language model with respect to the acoustic model were assigned depending on PLR. From the NSR experiment using decoded speech from the G.723.1 coder, it was shown that the average word error rate could be relatively reduced by around 20% by changing the weight of the language model for a continuous French database when PLR was 10%.

4.4 Speech Coder for Speech Recognition Over IP Networks

In this section, a high-quality speech coder for NSR over IP networks is described, which has been proposed in Yoon et al. (2007). From the view of speech quality and speech recognition performance, the proposed speech coder is based on the use of *Mel-frequency cepstral coefficients* (MFCCs) for spectral envelope parameters instead of *linear prediction coefficients* (LPCs). In other words, MFCCs are directly transmitted to the decoder and used for ASR, where they are converted to LPCs for speech coding. Therefore, one of the major concerns in the proposed speech coder is how to efficiently compress or quantize MFCCs in terms of both speech coding and speech recognition.

4.4.1 MFCC-Based Speech Coder

We propose a CELP-type speech coder, where the spectral envelope is represented as MFCCs to maintain speech recognition performance at the server. In conventional CELP speech coders, the spectral envelope is represented as LPCs, and then the LPCs are quantized for transmission. However, since the proposed speech coder extracts and quantizes MFCCs, a conversion procedure from MFCCs to LPCs is required, as shown in Fig. 4.5. Since NSR can be performed with quantized MFCCs on the decoder side, the performance of MFCC quantization is closely related to NSR performance. Thus, we develop an efficient MFCC quantization method having a smaller number of bits, while maintaining speech recognition performance.

The proposed speech coder was developed by making use of the structure of the ITU-T Recommendation G.729 (1996). Here, the frame size is 10 ms, and each frame is divided into two subframes for long-term prediction and excitation modeling. However, it should be noted that MFCC extraction, MFCC-to-LPC conversion, and MFCC quantization are all different from G.729.

Figure 4.6 shows the procedure for obtaining MFCCs from the input speech. As can be seen from the figure, the speech signal is high-pass filtered with a cut-off frequency of 140 Hz, and then scaled down by a factor of 2 in the pre-processing block. Next, the pre-processed signal is windowed by an asymmetric window identical to the window used in G.729. Then, each frame is zero-padded to form an extended frame of 256 samples. A 256-point fast Fourier transform (FFT) is then applied to compute the magnitude spectrum of the windowed signal. The magnitude spectrum is subsequently passed through 23 triangular mel-filterbanks, and each mel-filtering output is transformed into a logarithmic scale. Finally, a discrete cosine transform (DCT) is applied to obtain the 13 MFCCs, ($c_0, c_1, \cdots, c_{12}$).

Figure 4.7 shows the procedure for obtaining LPCs from MFCCs. Note that the 13 MFCCs are first zero-padded to make 23 MFCCs. Then, an inverse DCT (IDCT) followed by the inverse logarithm is applied to these MFCCs, resulting in 23 frequency samples. Next, the 23 frequency samples are linearly interpolated to make 256 frequency samples. The power density spectrum is then computed by the square of the interpolated 256 frequency samples. A 256-point inverse FFT (IFFT) is applied to compute the autocorrelation coefficients, and the autocorrelation coefficients are subsequently smoothed by the application of a lag window. Finally, 10 LPCs can be obtained by using the Levinson-Durbin recursion.

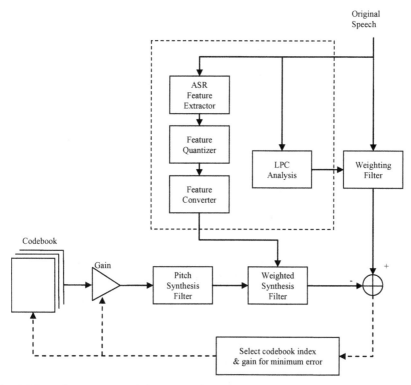

Fig. 4.5 Encoding structure of the proposed MFCC-based speech coder (From Yoon et al. 2007, ©2007 IEICE)

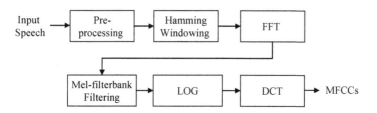

Fig. 4.6 Procedure for extracting MFCCs from speech signals

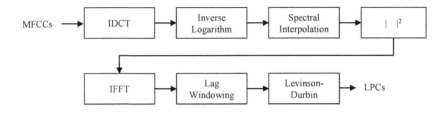

Fig. 4.7 Procedure for converting MFCCs to LPCs for CELP-type speech coding

4.4.2 Efficient Vector Quantization of MFCCs

In this section, we propose a *vector quantizer* (VQ) based on predictive VQ (*PVQ*) to reduce the bit-rate of the proposed speech coder by using the *interframe correlation* of MFCCs (Ramaswamy and Gopalakrishnan 1998). In addition, a *safety-net PVQ* is introduced to mitigate the effect of frame erasure on speech quality and speech recognition performance by minimizing error propagation (Eriksson et al. 1999).

Our proposed structure is based on the following investigation. First of all, we measure the interframe correlations to justify the use of PVQ, which are defined by

$$R(i,k) = \frac{\sum_{n=0}^{N-1-k} c_{i,n} c_{i,n+k}}{\sqrt{\sum_{n=0}^{N-1-k} c_{i,n}^2} \sqrt{\sum_{n=0}^{N-1-k} c_{i,n+k}^2}} \qquad (4.2)$$

where i is the quefrency index, k is the frame interval, N is the total number of frames, and $c_{i,n}$ is the i-th MFCC of the n-th frame. Figure 4.8 shows the interframe correlations of each MFCC according to a different number of intervals, where we used 3,200 frames collected from the utterances spoken by 2 males and 2 females. As shown in the figure, the MFCC of each frame was highly correlated with that of the previous frame. Moreover, it was found that c_0 had the highest correlation among all the MFCCs, with a correlation coefficient greater than 0.95. Accordingly, we divided MFCCs into two subvectors for quantization: a 1-dimensional vector $\mathbf{C}_1$, $[c_0]$, and a 12-dimensional vector $\mathbf{C}_2$, $[c_1, \cdots, c_{12}]^T$.

Second, a safety-net PVQ is introduced by combining a PVQ with a memoryless VQ, where the memoryless VQ plays a role in reducing the error propagation due to the prediction structure of the PVQ (Eriksson et al. 1999). For a given MFCC vector, selecting either PVQ or the memoryless VQ in the safety-net PVQ is required. To this end, we use the Euclidean distance measure to select one of the VQs; PVQ is selected if the distance from PVQ is smaller than that from the memoryless VQ, and vice versa.

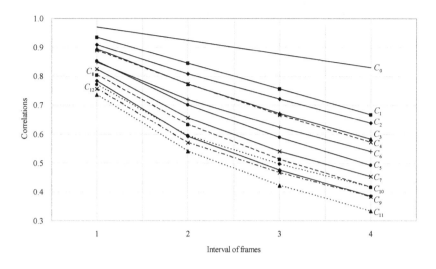

Fig. 4.8 Intraframe correlations of MFCCs (From Yoon et al. 2007, ©2007 IEICE)

Figure 4.9 shows the proposed VQ used in this paper. Here, an input MFCC vector of the n-th frame is split into two subvectors as

$$\mathbf{C}[n] = \begin{bmatrix} \mathbf{C}_1[n] \\ \mathbf{C}_2[n] \end{bmatrix} = \begin{bmatrix} c_0 \\ c_1 \\ \vdots \\ c_{12} \end{bmatrix} \tag{4.3}$$

where $\mathbf{C}_1[n]$ and $\mathbf{C}_2[n]$ are, respectively, a 1-dimensional subvector and a 12-dimensional subvector, as described above. Then, each subvector is quantized by its corresponding safety-net PVQ, where a selector chooses between either PVQ or the memoryless VQ depending on the Euclidean distance measure. In PVQ, the prediction is based on the quantized MFCC vector of the previous frame, such that

$$\mathbf{C}_{ip}[n] = \alpha_i \hat{\mathbf{C}}_i[n-1] \tag{4.4}$$

where α_i is the prediction coefficient of the previous frame of the i-th subvector in Eq. 4.3. Specifically, we construct the memoryless VQ and PVQ for $\mathbf{C}_2$ with a multi-stage VQ, as it is generally known to be efficient in the search and training of VQ for high dimensional vectors (Juang and Gray 1982). Finally, the number of bits assigned to each quantization index is as described in Table 4.2.

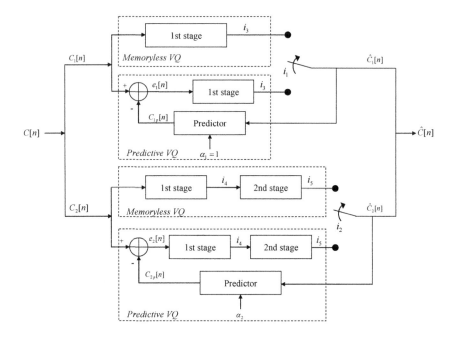

Fig. 4.9 Structure of a safety-net MFCC VQ combining PVQ and a memoryless VQ (From Yoon et al. 2007, ©2007 IEICE)

In order to select the optimal numbers of bits for i_3, i_4 and i_5, we divide the speech database into two parts. The first one, consisting of 172,800 American English and Korean frames, is used for training the proposed VQ; the other, consisting of 48,400 frames, is used for the evaluation of VQ. Typically, the number of bits for PVQ is closely related to the value of α_i. In fact, we first select an optimal α_i, and then assign the proper number of bits to each index when PVQ works with the selected optimal α_i.

Table 4.2 Bit allocation for the MFCC VQ

Index	No. of bits	Description
i_1	1	Prediction selector for $\mathbf{C}_1$
i_2	1	Prediction selector for $\mathbf{C}_2$
i_3	5	VQ index for $\mathbf{C}_1$
i_4	11	First stage VQ index for $\mathbf{C}_2$
i_5	7	Second stage VQ index for $\mathbf{C}_2$
Total	25	

As a criterion for selecting α_i and assigning the number of bits, we use the following Euclidean distance measure

$$D(\mathbf{C},\hat{\mathbf{C}}) = \frac{1}{N}\sum_{n=0}^{N-1}\sqrt{\sum_{i=0}^{K-1}\left(c_{i,n} - \hat{c}_{i,n}\right)^2} \qquad (4.5)$$

where K is the dimension of a subvector and is set as 1 for $\mathbf{C}_1$ and 12 for $\mathbf{C}_2$, N is the total number of frames, and $c_{i,n}$ and $\hat{c}_{i,n}$ are the i-th elements of unquantized and quantized subvectors of the n-th frame, respectively. Table 4.3 and Fig. 4.10 present the performance comparison measured from Eq. 4.5 by varying the prediction coefficient and the number of bits for $\mathbf{C}_1$ and $\mathbf{C}_2$, respectively. Note that as compared to the distance of SVQ, the proper number of bits for $\mathbf{C}_1$ should be set to 5 or more

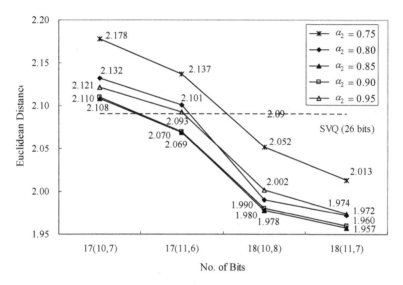

Fig. 4.10 Performance comparison by varying α_2 and the number of bits for $\mathbf{C}_2$, where (a,b) in the x-axis represents the number of bits for the first and second stage, respectively (From Yoon et al. 2007, ©2007 IEICE)

Table 4.3 Performance comparison of the Euclidean distance according to different values of the prediction coefficient and the different number of bits assigned for $\mathbf{C}_1$

α_1	Safety-net PVQ			SVQ
	4 bits	5 bits	6 bits	8 bits
1.0	0.72	0.36	0.18	
0.95	1.18	0.71	0.42	0.41
0.90	1.54	0.88	0.51	

Table 4.4 Bit allocation for the MFCC-based speech coder

Parameter	Subframe		Frame
	1st	2nd	
MFCC	–		25
Adaptive codebook index	8	5	13
Pitch parity	1	–	1
Fixed codebook index	13	13	26
Fixed codebook sign	4	4	8
Conjugate codebook gain	7	7	14
Total			87

when α_1 = 1. However, when α_1 is less than 1, we need to assign more bits to PVQ for C_1. For this reason, we set α_1 =1, and assign 5 bits to i_3, as shown in the third row of Table 4.2. Similarly, the proper number of bits for C_2 is determined to be 18 when α_2 is between 0.75 and 0.95, with α_2 = 0.85 giving the best result. Exhaustive experiments confirm that the best performance can be achieved using 18 bits, where 18 bits are split into 11 bits for i_4 and 7 bits for i_5. As a result, the 13 MFCCs are quantized with 25 bits, which is a reduction of 19 bits when compared with the SVQ quantizer. Finally, we summarize the bit allocation of the proposed speech coder with a bit-rate of 8.7 kbit/s, as shown in Table 4.4.

4.4.3 Speech Quality Comparison

We evaluated the performance of the proposed speech coder using the *perceptual evaluation of speech quality* (PESQ) measure (ITU-T Recommendation P. 862 2001). The experimental data consisted of 64 sentences spoken by four male and four female speakers. Each sentence was sampled with a rate of 8 kHz, and then filtered by the modified IRS filter (ITU-T Recommendation G.191 2000) to simulate the condition as if the recording were done through mobile devices.

Table 4.5 shows the mean opinion score (ITU-T Recommendation P. 862 2001) when the performances of the 8 kbit/s G.729 and the 8.7 kbit/s MFCC-based speech coders were evaluated under packet loss free conditions. It also shows the MOS score of the proposed speech coder according to different values of the prediction coefficient α_2 for C_2. Note that the prediction coefficient for C_1 was fixed as 1, as described in Sect. 4.4.2. The MOS score of the MFCC-based speech coder was about 0.02 higher than that of G.729 when α_2 was between 0.85 and 0.95. That is, by selecting an appropriate setting for α_2, the MFCC-based speech coder had a better performance than G.729. Fig. 4.11 and Table 4.5 further imply that 0.85 was the best choice for α_2 under a packet loss free condition.

In practice, it is essential for a coding scheme to cope with packet loss. For this reason, in order to evaluate the performance of the proposed speech coder under packet loss conditions, we used the error insertion algorithm defined by the ITU-T Recommendation G.191 (2000) to generate error patterns. Then, when a frame was erased, the proposed speech coder reconstructed the speech by using an extrapolation technique from a previous good frame, which is similar to the interpolation-based EC in G.729. Fig. 4.11 shows the MOS scores according to α_2 and *frame erasure rate* (FER), where FER is identical to PLR if each packet is composed of a single frame. Note that for the FER from 0% to 10%, it was found that the more α_2 decreased, the more robust the coder was to packet loss. Then, by considering the results shown in Table 4.5 and Fig. 4.11, it was concluded that $\alpha_2 = 0.85$ was again the best selection for the MFCC-based speech coder.

Table 4.5 Comparison of PESQ scores of G.729 and the MFCC-based speech coder with different α_2 with $\alpha_1 = 1$

G.729 (8 kbit/s)	MFCC-based speech coder (α_2) (8.7 kbit/s)				
	0.75	0.80	0.85	0.90	0.95
3.828	3.836	3.837	3.848	3.843	3.848

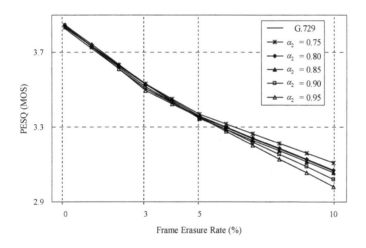

Fig. 4.11 PESQ scores under packet loss conditions obtained by varying the frame erasure rates against α_2 for C_2 when $\alpha_1 = 1$ (From Yoon et al. 2007, ©2007 IEICE)

4.4.4 ASR Performance Comparison

A. ASR Baseline and Task

We evaluated the performance of NSR using the MFCC-based speech coder. As a comparative experiment, a conventional client-based ASR, DSR, and another NSR using G.729 were also evaluated in this subsection. For the client-based ASR system, we used the ETSI DSR front-end (ETSI Standard ES 201 108, 2003) to extract MFCCs from the input speech signals but did not apply SVQ quantization to them; conversely, we used the ETSI DSR compression algorithm to quantize MFCCs for DSR. For the other NSR system using G.729, speech signals decoded by G.729 were used for the ETSI DSR front-end.

The Aurora 4 database (Hirsch 2002) was derived from the Wall Street Journal 5000-word closed-loop task (WSJ0) to evaluate the performance of large vocabulary continuous speech recognition (LVCSR). The database was divided into training and test sets, where all utterances were sampled at a rate of 8 kHz. The training set was constructed by adding six different noises (cars, babble, street traffic, train station, restaurant, and airport) to the 7,138 utterances recorded by a Sennheiser close talking microphone and several far talking microphones. Here, we performed the multi-condition training for acoustic models. In the Aurora 4 database, fourteen test sets were defined in order to evaluate speech recognition performance under the different microphone and noise conditions. For this evaluation, we selected seven test sets, where each set was composed of 330 utterances recorded by the Sennheiser close-talking microphone under one clean and six different noise conditions. The average signal-to-nose ratio (SNR) for the test utterances under noise conditions was measured at around 10 dB.

B. Loss-Free Condition

Table 4.6 shows the word error rates (WERs) of the ASR systems classified by three configurations. The second and third columns show the WERs of the ASR system under the client-based configuration and under the DSR configuration, respectively. The WERs of the NSR systems using the MFCC-based speech coder and G.729 are shown in the last two columns. As shown in Table 4.6, the client-based ASR system provided the best ASR performance. The client-based ASR system, however, is impractical for LVCSR because of the low power inherent in the small client devices.

Table 4.6 Comparison of the average word error rate (%) of different ASR configurations for the Aurora 4 database under multi-condition training

ASR configuration			NSR	
Test set	Client-based	DSR	G.729 coder	MFCC-based coder
Clean (Set 1)	18.21	18.92	19.39	18.87
Car (Set 2)	20.34	20.81	22.98	22.70
Babble (Set 3)	29.63	30.79	30.97	36.52
Restaurant (Set 4)	31.70	33.22	33.03	36.82
Street (Set 5)	32.51	32.71	34.19	36.41
Airport (Set 6)	28.21	28.73	29.93	32.36
Train station (Set 7)	32.84	33.79	35.36	37.18
Average WER	27.63	28.42	29.41	31.55

A report regarding the complexity of ASR conducted by ETSI on the Aurora 4 database Parihar and Picone (2001) showed that the computational amount was 85 times longer than real-time on an 800 MHz dual processor Pentium III with 1 GB RAM. This implies that the ASR system required a CPU time of 850 s to recognize an utterance if the utterance was 10 s long. In addition, it was further reported in Parihar and Picone (2001) that a memory size of around 300 MB to 650 MB was required to process the Aurora 4 database. This indicates that a client-based ASR approach is not yet realizable in terms of real-time processing, whereas the DSR approach is more desirable for small computing devices due to the heavy computing requirement of ASR. Thus, it was determined that our target performance should be that of the DSR system, especially if we take into consideration the feasible implementation of the LVCSR system. When compared to the DSR system, it was found that the average WER of the NSR system using G.729 significantly increased by 11.0%. On the other hand, contrary to the NSR system using G.729, the average WER of the NSR using the MFCC-based speech coder only increased by about 3.5%. Moreover, the relative WER of the NSR using the MFCC-based speech coder decreased by 6.8% compared with that of the NSR system using G.729.

C. Packet Loss Condition

We further evaluated the performance of ASR front-ends under packet loss conditions. The experimental setup for simulating the packet loss condition was identical to that used for speech quality, as shown in Fig. 4.11. Figure 4.12 then shows the average WERs of the three front-ends according to different frame erasure rates. It can be seen in the figure that NSR using the MFCC-based speech coder provided a more robust ASR performance than NSR using G.729, and it had comparable performance to DSR for all frame erasure rates.

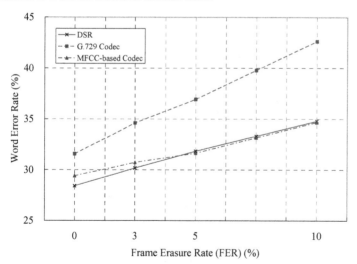

Fig. 4.12 Comparison of average word error rates (%) of each ASR configuration under different packet loss conditions (From Yoon et al. 2007, ©2007 IEICE)

4.5 Conclusion

In this chapter, we summarized the issues regarding ASR over IP networks in a framework of NSR. In this framework, it is basically assumed that speech must be encoded for transmission at a client and an ASR server can make use of decoded speech or the bitstream prior to speech decoding. Moreover, when compared to ASR over mobile networks, speech transmitted over IP networks is subject to degradation from sources inherent to the characteristics and limitations of IP networks and end-to-end environments. We then discussed methods for improving speech quality and the feature parameters for NSR according to speech coder distortion and packet loss. It was suggested that there was no single unique method to compensate for all the factors that could potentially degrade ASR performance, resulting in the combination of several techniques to solve such problems.

Next, we proposed a CELP-type speech coder using MFCC for NSR, where the spectral envelope was represented as MFCCs for speech recognition and speech reconstruction on the decoder side. To efficiently quantize MFCCs with a low bit-rate and make the proposed speech coder robust to packet loss, we then proposed a safety-net scheme that combined predictive VQ and memoryless VQ. Through the results of our experimental analysis, 25 bits per frame were assigned to MFCCs, and an 8.7 kbit/s speech coder was developed by using the proposed quantization. In addition, it was shown from the PESQ tests that the proposed MFCC-based speech coder provided slightly better speech quality under both packet loss free and packet loss conditions compared to the 8 kbit/s G.729 speech coder. Moreover, since the proposed speech coder directly transmitted MFCC, the word error rate of NSR using the proposed speech coder was relatively decreased by 6.8%, as compared to that of NSR using G.729.

References

3GPP TS 26.090 (1999). Mandatory speech codec speech processing functions; AMR speech codec; transcoding functions. v3.10, Dec.

Anandakumar, A. K., McCree, A. V. and Viswanathan, V. (2000). Efficient CELP-based diversity schemes for VoIP. In *Proceedings of ICASSP*, pp. 3682–3685.

Borella, M. S. (2000). Measurement and interpretation of Internet packet loss. *Journal of Communication and Networks*, vol. 2, no. 2, pp. 93–102.

Chandra, P. and Lide, D. (2007). Wi-Fi Telephony: Challenges and solutions for voice over WLANs. Elsevier Inc.

de Martin, J.C., Unno, T. and Viswanathan, V. (2000). Improved frame erasure concealment for CELP-based coders. In *Proceedings of ICASSP*, pp. 1483–1486.

Eriksson, T., Lindén, J. and Skoglund, J. (1999). Interframe LSF quantization for noisy channels. *IEEE Transactions on Speech and Audio Processing*, vol. 7, no. 5, pp. 495–509.

ETSI Standard ES 201 108 (2003). Speech processing, transmission and quality aspects; Distributed speech recognition; front-end feature extraction algorithm; compression algorithm. v.1.1.3, Sept.

Falavigna, D., Matassoni, M. and Turchetti, S. (2003). Analysis of different acoustic front-ends for automatic voice over IP recognition. In *Proceedings of ASRU*, pp. 363–368.

Fingscheidt, T., Aalbury, S., Stan, S. and Beaugeant, C. (2002). Network-based versus distributed speech recognition in adaptive multi-rate wireless systems. In *Proceedings of ICSLP*, pp. 2209–2212.

Gallardo-Antolín, A., Díaz-de-María, F. and Valverde-Albacete, F. (1998). Recognition from GSM digital speech. In *Proceedings of ICSLP*, pp. 1443–1446.

Gómez, M., Peinado, A. M., Sánchez, V. and Rubio, A. J. (2006). Recognition of coded speech transmitted over wireless channels. *IEEE Transactions on Wireless Communications*, vol. 5, no. 9, pp. 2555–2562.

Hardman, V., Sasse, A., Handley, M. and Watson, A. (1995). Reliable audio for use over the Internet. In *Proceedings of INET'95*, pp. 171–178.

Hirsch, G. (2002). Experimental framework for the performance evaluation of speech recognition front-ends on a large vocabulary task. *ETSI STQ Aurora DSR Working Group*.

Hooper, J. B. and Russell, M. J. (2000). Objective quality analysis of a voice over Internet protocol system. *Electronics Letters*, vol. 36, no. 22, pp. 1900–1901.

ITU-T Recommendation G.191 (2000). Software tools for speech and audio coding standardization. Nov.

ITU-T Recommendation G.729 (1996). Coding of speech at 8 kbit/s using conjugate-structure algebraic-code-excited linear-prediction (CS-ACELP). Mar.

ITU-T Recommendation P. 862 (2001). Perceptual evaluation of speech quality (PESQ), an objective method for end-to-end speech quality assessment of narrow-band telephone networks and speech codecs. Feb.

Juang, B. H. and Gray, A. H. (1982). Multiple stage vector quantization for speech coding. In *Proceedings of ICASSP*, pp. 597–600.

Kataoka, A. and Hayashi, S. (2007). A cryptic encoding method for G.729 using variation in bit-reversal sensitivity. *Electronics and Communications in Japan (Part III: Fundamental Electronic Science)*, vol. 90, no. 2, pp. 63–71.

Kim, H. K. and Cox, R. V. (2001). A bitstream-based front-end for wireless speech recognition on IS-136 communications system. *IEEE Transactions on Speech and Audio Processing*, vol. 9, no. 5, pp. 558–568.

Mayorga, P., Besacier, L., Lamy, R. and Serignat, J. Z. (2003). Audio packet loss over IP and speech recognition. In *Proceedings of ASRU*, pp. 607–612.

Mayorga, P. and Besacier, L. (2006). Voice over IP and vocal recognition. In *Proceedings of 3rd International Conference on Electrical and Electronics Engineering*, pp. 1–4.

Milner, B. and Semnani, S. (2000). Robust speech recognition over IP networks. In *Proceedings of ICASSP*, pp. 1791–1794.

Milner, B. (2001). Robust speech recognition in burst-like packet loss. In *Proceedings of ICASSP*, pp. 261–264.

Milner, B. and James, A. (2006). Robust speech recognition over mobile and IP networks in burst-like packet loss. *IEEE Transactions on Audio Speech and Language Processing*, vol. 14, no. 1, pp. 223–231.

Parihar, N. and Picone, J. (2001). DSR front end LVCSR evaluation – baseline recognition system description. *ETSI STQ Aurora DSR Working Group*.

Peláez-Moreno, C., Gallardo-Antolín, A. and Díaz-de-María, F. (2001). Recognizing voice over IP: A robust front-end for speech recognition on the World Wide Web. *IEEE Transactions on Multimedia*, vol. 3, no. 2, pp. 209–218.

Perkins, C., Hodson, O. and Hardman, V. (1998). A survey of packet loss recovery techniques for streaming audio. *IEEE Network*, vol. 12, no. 5, pp. 40–48.

Ragot, S., Kovesi, B., Trilling, R., Virette, D., Duc, N., Massaloux, D., Proust, E., Geiser, B., Garter, M., Schandl, S., Taddei, H., Gao, Y., Shlomot, E., Ehara, H., Yoshida, K., Vailancourt, T., Salami, R., Lee, M. S. and Kim, D. Y. (2007). ITU-T G.729.1: an 8-32 kbit/s scalable coder interoperable with G.729 for wideband telephony and voice over IP. In *Proceedings of ICASSP*, pp. 529–532.

Ramaswamy, G. N. and Gopalakrishnan, P. S. (1998). Compression of acoustic features for speech recognition in network environments. In *Proceedings of ICASSP*, pp. 977–980.

Ruggeri, G., Beritelli, E. and Casale, S. (2001). Hybrid multi-mode/multi-rate CS-ACELP speech coding for adaptive voice over IP. In *Proceedings of ICASSP*, pp. 733–736.

Seo, J. W., Woo, S. J. and Bae, K. S. (2001). Study on the application of an AMR speech codec to VoIP. In *Proceedings of ICASSP*, pp. 1373–1376.

Servetti, A. and De Martin, J. C. (2002). Perception-based partial encryption of compressed speech. *IEEE Transactions on Speech and Audio Processing*, vol. 10, no. 8, pp. 637–643.

Shacham, N. and McKenney, P. (1990). Packet recovery in high-speed networks using coding and buffer management. In *Proceedings of IEEE INFOCOM*, pp. 124–131.

Sun, H., Shue, L. and Chen, J. (2004). Investigations into the relationship between measurable speech quality and speech recognition rate for telephony speech. In *Proceedings of ICASSP*, pp. 865–868.

Swaminathan, K., Hammons Jr., A. T. and Austin, M. (1996). Selective error protection of ITU-T G.729 codec for digital cellular channels. In *Proceedings of ICASSP*, pp. 577–580.

Takahashi, A., Yoshino, H. and Kitawaki, N. (2004). Perceptual QoS assessment technologies for VoIP. *IEEE Communications Magazine*, vol. 42, no. 7, pp. 28–34.

Tan, Z.-H., Dalsgaard, P. and Lindberg, B. (2005). Automatic speech recognition over error-probe wireless networks. *Speech Communication*, vol. 47, nos. 1–2, pp. 220–242.

Tan, Z.-H., Dalsgaard, P. and Lindberg, B. (2007). Exploiting temporal correlation of speech for error robust and bandwidth flexible distributed speech recognition. *IEEETransactions on Audio Speech and Language Processing*, vol. 15, no. 4, pp. 1391–1403.

Van Sciver, J., Ma, J. Z., Vanpoucke, F. and Van Hamme, H. (2002). Investigation of speech recognition over IP channels. In *Proceedings of ICASSP*, pp. 3813–3815.

Wah, B. W. and Lin, D. (2005). LSP-based multiple-description coding for real-time low bit-rate voice over IP. *IEEE Transactions on Multimedia*, vol. 7, no. 1, pp. 167–178.

Walker, J. Q. and Hicks, J. T. (2004). Taking charge of your VoIP project: Strategies and solutions for successful VoIP deployments. Cisco Press, Indianapolis, IN.

Yoon, J. S., Lee, G. H. and Kim, H. K. (2007). A MFCC-based CELP speech coder for server-based speech recognition in network environments. *IEICE Transactions on Electronics, Communications and Computer Sciences*, vol. E90-A, no. 3, pp. 626–632.

Part II

Distributed Speech Recognition

5

Distributed Speech Recognition Standards

David Pearce

Abstract. This chapter provides an overview of the industry standards for Distributed Speech Recognition developed in ETSI, 3GPP and IETF. These standards were created to ensure interoperability between the feature extraction running on a client device and a compatible recogniser running on a remote server. They are intended for use in the implementation of commercial services for speech and multimodal services over mobile networks. In the process of developing and agreeing the standards substantial performance testing was conducted and these results are also summarised here. While other chapters provide more general information about feature extraction and channel error processing for DSR this chapter focuses on introducing the specifics of the standards.

5.1 Introduction

It is estimated that in 2007 there are over 2 billion mobile phone subscribers worldwide and the numbers continue to grow. The market was originally fuelled by person-to-person voice communications and this remains the dominant "application." Recently we have seen increasingly sophisticated devices packed with many new features including messaging, cameras, browsers, games and music. Alongside device developments the mobile networks have improved, giving increased coverage and widespread availability of the 2.5G packet data such as General Packet Radio Service (GPRS). There are also many new deployments of 3G networks, bringing much larger bandwidths to mobile users. The 2.5G and 3G data capabilities provide the opportunity to deliver a range of different audio and visual information to the user's device and enable access to "content" while on the move. The user interface for these devices has certainly improved but the small keypad remains a barrier to data entry. Reliable speech input holds the potential to help greatly. Alongside pure speech input and output, the benefits of a multimodal interface are well appreciated. The ability to combine alternative input modalities (e.g., speech and/or keypad) with visual (e.g., graphics, text, pictures) and/or audio output can greatly enhance the user experience and effectiveness of the interaction.

For some applications it is best to use a recogniser on the device itself (e.g., interfacing to the phone functions and voice dialling using personal address book) while for others it will be preferable to connect to a remote recognition server (e.g., directory assistance, voice search, information access). Although the computational power of these devices is increasing, the complexity of large vocabulary speech recognition systems is beyond the memory and computational resources of many devices. Also the associated delay to download speech data files (e.g., grammars, acoustic models, language models, vocabularies) may be prohibitive or be confidential (e.g., a corporate directory).

Server-side processing of the combined speech input and speech output can overcome many of these constraints by taking full advantage of memory and processing power as well as specialised speech engines and data files. New applications can also be more easily introduced, refined, extended and upgraded at the server.

So, with the speech input remote from the recognition engine in the server, we are faced with the challenge of how to obtain reliable recognition performance over the mobile network. In addition we would like to have an architecture that can provide a multimodal user interface. These have been two motivators that have led to the creation of the *standards* for *Distributed Speech Recognition* (DSR):

1. Improved recognition performance over wireless channels

The use of DSR avoids the degradations introduced by the speech codec and channel transmission errors over mobile voice channels:

    (a) By using a packet data channel (for example GPRS for GSM) to transport the DSR features, instead of the circuit switched voice channel that is normally used for voice calls, the effects of channel transmission errors are greatly reduced and consistent performance is obtained over the coverage area.

    (b) By performing the front-end processing in the device directly on the speech waveform, rather than after transcoding with a voice codec, the degradations introduced by the codec are avoided.

    (c) In addition the DSR Advanced Front-end is very *noise robust* and halves the error rate in background noise compared to the Mel-Cepstrum front-end, giving robust performance for mobile users who are often calling from environments where there is background noise.

2. Ease of integration of combined speech and data applications for multimodal interfaces.

In multimodal interfaces, different modes of input (including speech or keypad) may be used and different media for output (e.g., audio or visual on the device display) are used to convey the information back to the user. The use of DSR enables these to operate over a single wireless data transport rather than having separate speech and data channels. As such, DSR can be seen as a building block for distributed multimodal interfaces. See the chapter on "Software Architectures for Networked Mobile Speech Applications" for a detailed discussion on multimodal architectures.

5.2 Overview of the Set of DSR Standards

A comprehensive set of DSR standards has been developed and agreed within the international standards bodies (Pearce 2000). These cover the feature extraction algorithms with their floating point specification and software, their fixed-point specification and software, and the protocols and formats for feature transmission between client device and remote recognition server. A summary of the set of DSR related standards is given in Table 5.1.

Table 5.1 Summary list of the set of DSR standards

Standard no.	Description	Standards body
ES 201 108	Mel-Cepstrum Front-end	ETSI STQ-Aurora
ES 202 050	Advanced Front-end (AFE)	ETSI STQ-Aurora
ES 202 211	Extended Mel-Cepstrum Front-end (XFE)	ETSI STQ-Aurora
ES 202 212	Extended Advanced Front-end (XAFE)	ETSI STQ-Aurora
TS 26.243	Fixed point specifications for ES 202 050 and ES 202 212	3GPP
Rfc3557	RTP payload format for ES 201 108	IETF
Rfc4060	RTP payload formats for ES 201 050, ES 202 211 and ES 202 212	IETF

The feature extraction algorithms were developed and standardised within the *ETSI* STQ *Aurora* DSR working group (more commonly referred to as "Aurora"). The *Mel-Cepstrum* front-end was in widespread use for speech recognition systems but with many variations using different parameters in their implementations. So the first activity was to agree the specific parameters for a standard and to develop a feature compression algorithm to reduce the transmission bandwidth to 4.8 kbit/s. This resulted in the creation of the first ETSI Standard ES 201 108 2000 for the Mel-Cepstrum front-end, its compression and its circuit switched transmission format.

For mobile environments where there is often background noise it was desired to have a feature extraction standard that was more noise robust. So a new work item was created with the goal of halving the word error rate in background noise compared to the Mel-Cepstrum front-end standard. To compare the performance of different candidate algorithms a set of evaluation databases, and back-end HMM recogniser configurations together with an associated selection criteria were developed. A competitive selection process was organised that eventually resulted in agreement on the algorithms for the DSR *Advanced front-end* (AFE) (ETSI Standard ES 202 050 2002).

ETSI Aurora also saw the need in some applications to be able to reconstruct the speech signal and to have a fundamental frequency feature to assist with tonal language recognition. Rather than having a competition to develop this capability it was created by collaboration between the two companies that had candidate technologies. This produced the *extended front-end* standards. One standard (ETSI Standard ES 202 211 2003) provides the extension of the Mel-Cepstrum front-end while the other (ETSI Standard ES 202 212 2003) provides this functionality for the Advanced front-end.

For each of these ETSI standards, the algorithm is specified in floating-point form, and a reference implementation in C forms part of the specification.

At the time of the Aurora work, the wireless industry was also introducing packet data network capabilities, so in addition to the circuit switched payload formats specified in ETSI Aurora, appropriate protocols and payload definitions needed to be standardised for the DSR features. The IETF already had a framework to support many different payload types for applications like Voice over IP and streaming video within their *Real Time Protocols* (RTP), so this was a natural place to standardise the format and create a MIME type for DSR. An activity was therefore started in the IETF AVT working group to define the payload format for DSR in RTP and after following the appropriate processes in the IETF two specifications were published: the first being rfc3557 for ES 201 108 and the second, rfc4060, for the AFE and the two extended front-ends.

With the DSR front-end standards created in ETSI Aurora it was anticipated that these would be adopted and referenced within the specifications for the different wireless and wireline networks (for example 3GPP, 3GPP2 and ITU). By each network using the same DSR standard it would improve implementation efficiency and interoperability providing ubiquitous access to voice servers over different data transport networks.

The *3GPP* (3rd Generation Partnership Project) is the body responsible for the GSM and UMTS standards, and it was the first to consider the use of DSR for their requirement to support of "Speech Enabled Services." Before adopting DSR, the SA4 working group that looks after the specification of codecs wanted to be sure of the performance advantages and to compare with any other alternatives. So a lot more additional testing was performed within 3GPP working with commercial recognition vendors, IBM and Nuance (Speechworks at the time), using their commercial recognisers and testing with many larger speech databases (both public and private). The result of this process was the selection of the DSR Extended Advanced Front-end as the recommended codec for speech enabled services in 3GPP release 6. The fixed-point version of the AFE and XAFE were also specified as standards to ensure interoperability by having bit-exact implementation of the standard. This is published as standard 3GPP TS 26.243 (2004) which has the fixed-point C code specification software included. There are also a set of test vectors specified for testing bit-exactness to the standard (3GPP TS 26.177 2004).

Further details about each of these standards and their performance are given in the sections that follow.

5.3 Scope of the Standards

Figure 5.1 shows a block diagram of the processing stages of a DSR system. These are split into the terminal (or client) side processing and the server side processing.

Transmission between the client and server could be over either a wireless or a wireline communication network or a combination. While the standards are not restricted to this case it is anticipated that implementations will most likely use packet data protocols to support the end-to-end connectivity. The general principle that was applied when setting the standards was to specify the minimum to allow interoperability between client and server. Where there are blocks in the processing chain that vendors can further optimise in proprietary ways to obtain better overall performance then these are not mandated but left open for service providers to implement as they choose. For example, the standards only cover as far as the regeneration of static Mel-Cepstrum features and it is left to the ASR vendor to select which features to use in the recogniser and how to generate any derivative features as input to the decoder.

In the section below we progress through each of the blocks in the processing chain and the last digit of the section numbers used below correspond to the numbered blocks in Fig. 5.1. The grey filled boxes in the processing chain are those covered by the standard while the white filled ones are not.

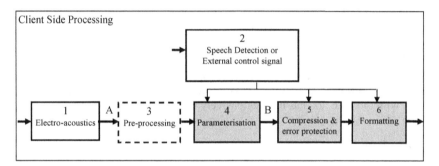

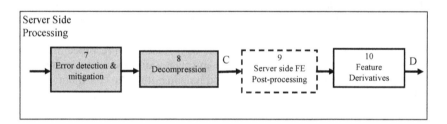

Fig. 5.1 Terminal/client side DSR processing chain

5.3.1 Electro-Acoustics

This block refers to everything that occurs during the conversion of the sound pressure waveform to a digitised signal. These include the microphone transducer, analogue filtering, automatic gain control, analogue to digital conversion.

The characteristics of the input audio parts of a DSR terminal will have an effect on the resulting recognition performance at the remote server. Developers of DSR speech recognition servers can assume that the DSR terminals will operate within the ranges of characteristics as specified in GSM 03.50 (GTS GSM 03.50). DSR terminal developers should be aware that reduced recognition performance might be obtained if they operate outside the recommended tolerances.

Sampling frequencies of 8, 11 and 16 kHz are supported in the original ETSI Aurora DSR standards but for 3GPP only sampling rates of 8 and 16 kHz were standardised.

5.3.2 Speech Detection or External Control Signal

In many applications a function performed at the terminal side will determine when the speech is to be processed and the DSR parameters transmitted over the network to the server. Three alternative ways in which this transmission control can be performed are:

1. speech detection—the input speech signal is used to determine when there is speech activity
2. push-to-talk—a user controlled button indicates when processing and transmission are to occur
3. a signal coming from another software module.

Speech detection is not part of the DSR front-end standard that is mandated. The AFE standard does include a Voice Activity Detector (VAD) that can be used in conjunction with the AFE and has been extensively tested but its use is not mandated.

5.3.3 Pre-Processing

This block is optional and in most implementations it will be absent. It is not part of the DSR standard. Implementers may apply proprietary pre-processing stages ahead of the DSR standard. When doing so it is a manufacturer's responsibility to ensure that any pre-processing does not degrade performance of a DSR service. The desired result of any pre-processing is to give a signal as if it had been recorded at a higher signal to noise ratio and it should not result in spectral distortion or clipping of the speech signal. The output of this stage should remain within the constraints of GSM 03.50.

5.3.4 Parameterisation

The frame based speech processing algorithm generates the feature vector representation (B). This is specified in the front-end processing part of the DSR standard. In the case of both the Mel-Cepstrum Front-end and the AFE it is the specification of the front-end feature vector extraction that produces the 14-element vector consisting of 13 Cepstral coefficients and log energy.

After further processing stages the corresponding feature vector is recreated at the server side (point C in Fig. 5.1).

5.3.5 Compression and Error Protection

The feature vector is compressed to reduce the data rate and error protection bits are added. This stage is specified as part of the DSR standards. In the DSR standards a split vector quantisation algorithm is used and error detection bits are added to each frame pair.

5.3.6 Formatting

The compressed speech frames are formatted into a bitstream for transmission. Both circuit data and packet data transmission are supported. The format is defined for a pair of 10 ms speech frames consisting of the quantised cepstral parameters. For circuit switched transmission a multiframe format with associated header and synchronisation bits is defined while for packet data a payload consisting of any number of frame pairs is specified in the IETF real-time protocol (RTP) payloads. The same frame pair format is used in both cases.

5.3.7 Error Detection and Mitigation

The formatted bitstream is received and unpacked at the remote server. Depending on the particular transmission channel the number and type of transmission errors will vary but for an unreliable channel (e.g., without retransmission) there will be errors in the received payload. For mobile channels these are often have burst characteristic. The standard therefore specifies a method for *error detection* and mitigation of these errors although there are situations where these may not be needed.

5.3.8 Decompression

Decompression is often performed in conjunction with the error mitigation using the quantisation tables to look up the corresponding cepstral features and recover the static feature vector.

5.3.9 Server Side Post Processing

This block is optional and often not present. It is to allow vendors the freedom to further process the received cepstral features and deliver any chosen representation (or subset of the features) to their back-end recogniser.

5.3.10 Feature Derivatives

It is common practice in speech recognition systems to extend the static cepstral feature parameters by adding derivative features (velocity and acceleration) before passing them to the back-end decoder. These have been found to give better recognition performance and it is usual to use 12 cepstral coefficients and either the log energy or C0 plus as the static features plus their first and second order derivatives to make a feature vector of dimension 39. Nevertheless, since this part of the processing is entirely at the server side and does not impact interoperability, it is left open to the implementer to choose whatever processing is most appropriate for their recognition system. Vendors may also choose whether or not to use the voice activity detection (VAD) bit in the AFE to drop non-speech frames between utterances and not pass these to the recogniser.

5.4 DSR Basic Front-End ES 201 108

The goal of the first standard was to agree the details of the processing for the widely used Mel-Cepstrum front-end features and produce a DSR standard relatively quickly while acknowledging it had weaknesses in background noise. The process of obtaining agreement on the details was based on starting with a software implementation proposed by one of the Aurora participants (Nokia) and then modified based on discussion and inputs from other organisations. Each change was justified by demonstrating performance gains on the Aurora-1 database (a predecessor to Aurora-2 database based on noisy connected digits recognition task—see below) and a publicly available and widely used recognizer called *HTK* (Hidden Markov Model Tool Kit) (http://htk.eng.cam.ac.uk) at that time produced by the company Entropic (and currently distributed by Cambridge University, UK).

5.4.1 Feature Extraction

Figure 5.2 shows a block diagram of the processing for the Mel-Cepstrum *feature extraction* algorithm. After pre-emphasis and windowing the short term spectrum is obtained by an FFT. This linear spectrum is then warped into a non-linear spectral distribution of 24 bins using triangular weighting filters on a Mel-scale. The 12 cepstral coefficients are obtained by retaining the 12 lowest quefrency coefficients after taking the cosine transform of the logarithm of the 24 Mel-spectrum bins. The chosen frame rate is 10 ms. The total energy of each frame is also computed before the pre-emphasis filter. The final output feature vector consists of 12 cepstral coefficients (C1-C12), log Energy and C0.

5.4.2 Compression

The requirement set for the target bit-rate was 4.8 kbit/s. The *feature compression* method selected uses split vector quantisation (SVQ). The 14 coefficients are split into 7 subvectors each consisting of a pair of cepstral coefficients. Ci and Ci+1, i=1,3…11 are quantised using a codebook size of 64 (6 bits) while the C0 and logE pair uses a larger codebook size of 256 (8 bits). The larger codebook was needed for C0 and logE to cover wider dynamic range without recognition performance degradation due to quantisation. The 7 subvectors at 6 bits each plus the one codebook with 8 bits gives a total of 44 bits per 10 ms frame. The chosen SVQ scheme provides a reasonable compromise between coding efficiency, computational complexity and error resilience. While other published papers have shown that it is possible to achieve greater compression without performance loss, the design requirement of 4.8 kbit/s was met and the small subvectors allow flexibility in alternative error mitigation strategies.

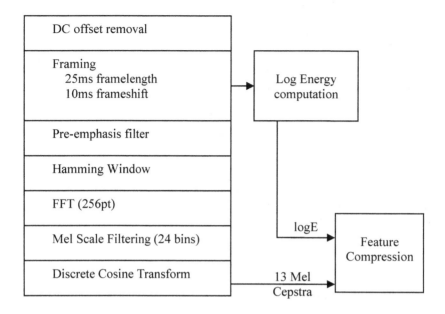

Fig. 5.2 Block diagram of the Mel-Cepstrum front-end algorithm

5.4.3 Error Detection and Mitigation

To assist with the detection of transmission errors 4 bits of error detection bits in the form of a Cyclic Redundancy Code (CRC) are added to each pair of speech frames (i.e., 44 bits for the first frame + 44 bits for the second + 4 bits of CRC).
The algorithm for error mitigation consists of two stages:

1. Detection of speech frames received with errors
2. Substitution of parameters when errors are detected.

To detect the speech frames received with errors the 4 error detection bits on each pair of frames are used first. Since errors may be missed due to overloading of the CRC a heuristic algorithm that looks at the consistency of the parameters in the decoded frames is also used. It measures the difference between cepstral coefficients for adjacent frames and flags them as errored if the difference is greater than expected for speech. The thresholds used are based on measurements of error free speech. If this algorithm was to run continuously then the number of misfirings could be too high, therefore it is only applied in the vicinity of detected CRC errors.

When a frame is flagged as having errors then the whole frame is replaced with a copy of the cepstral parameters for the nearest good frame received (occurring before or after the frame under consideration) (Pearce 2004b).

5.5 DSR Advanced Front-End ES 202 050

5.5.1 Feature Extraction

The AFE uses 10 ms frames and produces an output feature vector consisting of 12 cepstral features, C0 and log energy. Keeping the frame rate and parameters the same as for the Mel-Cepstrum front-end standard makes it relatively easy for server recognition engines to integrate the new robust DSR features without needing to change substantial aspects of the rest of the system. In most cases it is sufficient to retrain the recognition models from the source speech data and perhaps reoptimise a few control parameters.

The noise robustness of the AFE comes from the combination of a set of processing stages all of which contribute to the overall performance. At the heart of the algorithm are two stages of Wiener filtering that are performed first in the frequency domain before converting back to the time domain for a stage of waveform processing noise reduction. Finally, the cepstral features are computed and blind equalisation is applied to these. This stage helps to reduce the variability in the features and has a similar motivation to cepstral mean normalisation techniques.

The details of the algorithms are presented in the standard documents themselves (ES 202 050 2002) and readers may also find the explanations in the book by Peinado and Segura (Peinado and Segura 2006) helpful gaining a better understanding of the techniques used in the standard.

5.5.2 VAD

Compared to the DSR Mel-Cepstrum standard, one further enhancement coming from the Advanced Front-end is the inclusion of a bit in the bitstream to allow the communication of VAD. The VAD algorithm marks each 10 ms frame in an utterance as speech/non-speech so that this information can optionally be used for frame dropping at the server recogniser. During recognition, frame dropping reduces insertion errors in any pauses between the spoken words particularly in noisy utterances and can be used for end-pointing for training. It has been found that performance is particularly helped by model training with end-pointed data. The VAD information can also be used to reduce response time latencies experienced by users in deployed applications by giving early information on utterance completion.

5.5.3 Compression

The compression algorithm for the cepstral features uses the same split vector quantisation scheme as the earlier standard but with the quantiser tables retrained for the Advanced Front-end. To allow the VAD bit to be transmitted for each frame within the same payload size of 44 bits per 10 ms frame, the codebook size for the pair of highest order cepstral coefficients (C11 and C12) was reduced from 64 to 32. The frame pair transmission format is therefore very similar to that of the Mel-Cepstrum DSR standard with the only difference being that for each frame the 6 bit codebook for C11 and C12 in ES 201 108 is replaced by the 5 bit codebook for C11 and C12 plus the one bit for the VAD flag for the frame.

5.6 Recognition Performance of the DSR Front-Ends

5.6.1 Aurora Speech Databases and ETSI Performance Testing

Between 1999 and 2002 ETSI Aurora conducted a competitive selection process to create an Advanced DSR front-end standard that would provide improved robustness compared to the Mel-Cepstrum front-end. To support this, a new performance evaluation process and associated *speech databases* were created to allow the comparison between candidates. Three sets of noisy database were used for these performance evaluations:

1. Aurora-2 connected digits with simulated addition of noises (Hirsch, Pearce 2000)
2. Aurora-3 connected digits from real-world data collected in vehicle (5 languages)
3. Aurora-4 large vocabulary 5000 word Wall Street Journal dictation with simulated noise addition.

These databases have been made available for public distribution through the European Language Resource Association (ELRA) (www.elra.info) and are widely used in the speech research community to assess and compare new algorithm performance.

For the ETSI Aurora evaluations a reference back-end recogniser was defined for each database so that comparisons between different candidate front-ends could be made with the same fixed recogniser. For Aurora-2 and Aurora-3 the publicly available HTK was used with an agreed specific configuration for the model training and testing (number of states and mixtures per model, training iterations etc). For Aurora 4 an HMM recogniser framework suitable for this large vocabulary task was commissioned and prepared by the University of Mississippi. In each case the Mel-Cepstrum front-end in ES 201 108 provided a reference recognition performance on each database by which to measure the performance improvements from the alternative candidates. The performance was measured using word error rate.

A scoring procedure was agreed that gave appropriate weight to the results from each of the databases. The winning candidate that became the AFE standard gave an average of 53% reduction in word error rate compared to the DSR Mel-cepstrum standard (ES 202 108). Details of the Aurora-3 performance results are given below, while results on the other databases can be found in Macho et al. (2002).

5.6.2 Aurora 3: Multilingual SpeechDat-Car Digits—Small Vocabulary Evaluation

The purpose of the Aurora-3 tests was to evaluate the performance of the front-end on a database that has been collected from speakers in a real-world noisy environment.

It tests the performance of the front-end both when the training and testing conditions are well-matched as well as in mismatched conditions as often encountered in deployed DSR systems. The database also served to test the front-end on a variety of languages: Finnish, Italian, Spanish, German, and Danish. It is a small vocabulary task consisting of the digits selected from a larger database collection called SpeechDat-Car obtained from users in the real-world noise environment of the car. The databases each have 3 experiments consisting of training and test sets to measure performance with:

1. ***Well matched training and testing***—Train and test with the hands-free microphone over the range of vehicle speeds with the training and test sets covering a similar range of noise conditions.

2. ***Moderate mismatch training and testing***—Model training is performed on only of a subset of the range of noises present in the test set. The hands-free microphone for lower speed driving conditions is used for training and hands-free microphone at higher vehicle speeds for testing.

3. ***High mismatch training and testing***—Model training is performed with speech from the close-talking microphone and tested with the data from the hands-free microphone at range of vehicle speeds.

The results are presented below for the five languages making up the Aurora 3 database and using the HTK recogniser in its "simple" configuration i.e., 3 mixtures per state. The overall performance was computed as a weighted average of the different conditions i.e., 40% weight given to the well matched (W), 35% weight given to the medium mismatch (M) and 25% given to the high mismatch (H) results. These are shown in the row in the tables of results labelled "0.4W+0.35M+0.25H." Table 5.2 shows the absolute performance for the DSR Mel-Cepstrum Front-End as word accuracy, which then serves as a baseline for the performance comparisons with the Advanced Front-end.

Table 5.2 Baseline word accuracy performance of the Mel-Cepstum front-end ES 201 108 on the Aurora 3 database

Absolute performance						
Training mode	Italian	Finnish	Spanish	German	Danish	Average
Well matched	92.39%	92.00%	92.51%	91.00%	86.24%	90.83%
Medium mismatch	74.11%	78.59%	83.60%	79.50%	64.45%	76.05%
High mismatch	50.16%	35.62%	52.30%	72.85%	35.01%	49.19%
0.4W+0.35M+0.25H	75.43%	73.21%	79.34%	82.44%	65.81%	75.25%

The top half of Table 5.3 shows the absolute performance that is obtained when the speech is processed by the DSR Advanced Front End. The bottom half of the table shows the relative performance when compared to the Mel-Cepstrum baseline shown above in Table 5.2. The relative improvement is computed as the percentage reduction in the word error rate. On the Aurora 3 database the Advanced front-end provides an average improvement of 56%.

Table 5.3 Word accuracy performance of the Advanced front-end (ES 202 050) on the Aurora 3 Database

Absolute performance						
Training mode	Italian	Finnish	Spanish	German	Danish	Average
Well matched	96.90%	95.99%	96.66%	95.15%	93.65%	95.67%
Medium mismatch	93.41%	80.10%	93.73%	89.60%	81.10%	87.59%
High mismatch	88.64%	84.77%	90.50%	91.30%	78.35%	86.71%
0.4W+0.35M+0.25H	93.61%	87.62%	94.09%	92.25%	85.43%	90.60%

Performance relative to Mel-Cepstrum Front-End						
Training mode	Italian	Finnish	Spanish	German	Danish	Average
Well matched	59.26%	49.87%	55.41%	46.11%	53.85%	52.90%
Medium mismatch	74.55%	7.05%	61.77%	49.27%	46.84%	47.89%
High mismatch	77.21%	76.34%	80.08%	67.96%	66.69%	73.66%
0.4W+0.35M+0.25H	69.10%	41.50%	63.80%	52.68%	54.60%	56.34%

5.7 3GPP Evaluations and Comparisons to AMR Coded Speech

3GPP is the body that sets the standards for GSM and UMTS mobile communications. In 2002 3GPP conducted a study and produced a technical report on the feasibility of speech enabled services. The technical report (3GPP TR 22.977 2002) provides an overview of the speech and multimodal services envisaged and a new work item called Speech Enabled Services (SES) was started. The SA4 codecs group within 3GPP was the working group with responsibility for the selection and recommendation of the codec for SES. Following the usual process SA4 first agreed a selection procedure consisting of "design constraints" to set requirements on the SES front-end, "test and processing plan" to specify how to test and evaluate the performance of the candidates and "recommendation criteria" to define in advance what criterion would be used to select and recommend a "codec" standard for SES. Two candidates for the SES codec were considered: *AMR* and AMR-WB (being the existing voice codecs for 3GPP) and the DSR Extended Advanced Front-end. DSR would need to demonstrate substantial performance gains compared to the existing voice codec to justify the introduction of a new codec for SES services. Rather than using HTK for the performance evaluations it was decided that it would be best to use the talents of major server recognition vendors for the evaluations. By using commercial recognisers results would be indicative of what could be obtained from deployed commercial services. IBM and SpeechWorks (now Nuance) were the two ASR vendors who volunteered to undertake the extensive testing. The performance evaluations were conducted over a wide range of different databases some of which were brought in from 3GPP but also large proprietary databases owned by the ASR vendors. Testing covered many different languages (German, Italian, Spanish, Japanese, US English, Mandarin), environments (handheld, vehicle) and tasks (digits, name dialling, and place names). In addition, the codecs were tested under block transmission errors.

The results were reported at the SA4 meeting in February 2004 in Malaga and are summarised in Tables 5.4, 5.5 and 5.6. The average absolute performances are given as the percentage word error rates and the relative improvement as the reduction in word error rate provided by DSR compared to the AMR speech codec. Note that the results from both the ASR vendors have been averaged to preserve anonymity the source.

The comparisons between the AMR and DSR performances were made at two different categories of transmission bit rates. Speech enabled services need to operate over a variety of different packet data channels and consideration of these determined that it was appropriate to compare at a low data rate and at a high data rate. For example, the lowest bit rate was determined by considering the conversational class of service on a GPRS single slot uplink channel (coding scheme CS-1) the maximum source data rate is 5.6 kbit/s. The AMR narrow band speech codec can operate at a range of bit rates from 4.8 kbit/s to 12.2 kbit/s but to limit the number of experiments to a practical number it was decided to test at these two rates. Thus for the low data rate comparison, AMR 4.75 was compared to DSR (5.6 kbit/s). For the high data rate comparison, AMR 12.2 was compared to DSR (5.6 kbit/s). Evaluations were also made at higher sampling rate of 16 kHz; for this comparison the AMR wideband codec (AMR-WB) at 12.65 kbit/s was compared to 16 kHz DSR (5.6 kbit/s).

A detailed summary of the selection process followed, the testing procedures and the results can be found in the 3GPP Technical Report reference (3GPP TR 26.943 2004). The results are reproduced in the tables below. These results show a substantial performance advantage for DSR compared to AMR both at 8 kHz and at 16 kHz. DSR also shows particularly good robustness to channel errors with no degradation at 3% block error rate (BLER) and a further result obtained at 10% BLER also shows consistent performance for DSR whereas AMR performance falls substantially.

Based on these results DSR was selected as the recommended codec for Speech Enabled Services by SA4 and subsequently approved by 3GPP SA in June 2004 (Pearce 2004a).

Table 5.4 3GPP WER performance comparisons between DSR and AMR-NB 4.75

8 kHz	No. of databases tested	AMR4.75 average absolute WER	DSR average absolute WER	Average improvement (%)
Digits	11	13.2	7.7	39.9
Sub-word	5	9.1	6.5	30.0
Tone confusability	1	3.6	3.1	14.8
Channel errors	4	6.1	2.4	52.8
Weighted average				36

Table 5.5 3GPP WER performance comparisons between DSR and AMR-NB 12.2

8 kHz	No. of databases tested	AMR4.75 average absolute WER	DSR average absolute WER	Average improvement (%)
Digits	11	10.9	7.7	27.6
Sub-word	5	7.1	6.5	14.5
Tone confusability	1	3.8	3.1	19.7
Channel errors	4	5.5	2.4	40.9
Weighted average				25

Table 5.6 3GPP performance comparisons at 16 kHz between DSR and AMR-WB 12.65

16 kHz	No. of databases tested	AMR4.75 average absolute WER	DSR average absolute WER	Average improvement (%)
Digits	8	9	5.6	35
Sub-word	5	8.2	5.9	23.5
Channel errors	4	6.1	3.4	42.2
Weighted average				31

5.8 ETSI DSR Extended Front-End Standards ES 202 211 and ES 202 212

ES 202 211 is an extension of the Mel-Cepstrum DSR Front-end standard ES 201 108. In a similar way, ES 202 212 provides the extension of the DSR Advanced Front-end ES 202 050 to allow reconstruction for the AFE. The front-ends provide the features for speech recognition but these are not available for human listening. The purpose of the extension is to allow the reconstruction of the speech waveform from these features so that they can be replayed for human audition. The front-end feature extraction part of the processing is exactly the same as for ES 201 108. For *speech reconstruction* additional fundamental frequency (perceived as pitch) and voicing class (e.g., non-speech, voiced, unvoiced and mixed) information is needed. This is the extra information that is provided by the extended front-end processing algorithms at the device side that is compressed and transmitted along with the front-end features to the server. This extra information may also be useful for improved speech recognition performance with tonal languages such as Mandarin, Cantonese and Thai. The compressed extension bits need an extra 800 bps on top of the 4800 bps for the Mel-Cepstral features, as shown in Fig. 5.3.

One of the main use cases for the reconstruction is to assist dialogue design and refinement. During pre-deployment trials of services it is desirable to be able to listen to dialogues and check the overall flow of the application and refine the vocabulary used in the grammars. For this and other applications of the reconstruction the designer needs to be able to replay what was spoken to the system at the server (offline) and understand what was spoken. To test the *intelligibility* of the speech two evaluations were conducted. The first is a formal listening test for intelligibility called the *Diagnostic Rhyme Test* (DRT) that was conducted by Dynastat listening laboratories. The results of this are shown in Table 5.7. For comparison the MELP codec used for military communications was chosen as a suitable reference. The DSR reconstruction performs as well as MELP in the DRT tests giving confidence that the intelligibility is good. The *transcription* task is closer to the situation that would occur in an actual application. To measure and compare the transcription accuracy, a professional transcription house was used to transcribe sentences sourced from the Wall Street Journal that had been passed through the DSR reconstruction and the LPC-10 and MELP reference codecs. As well as clean speech, car, street and babble noises were added to the source sentences. Afterwards the number of errors was measured by counting the number of missed, wrongly transcribed or partially transcribed words. Table 5.8 shows the results of this assessment and the average percentage transcription error for each coder. The DSR reconstruction gave less than 1% transcription errors and fewer errors than for either LPC-10 or MELP reference codecs.

In ETSI Aurora, the pitch feature was also tested for tonal language recognition of Mandarin and Cantonese and shown to give better performance than proprietary

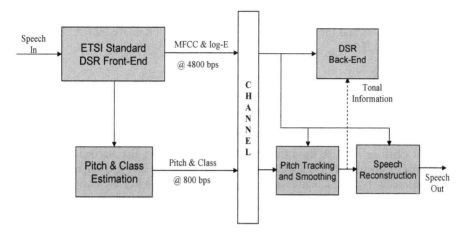

Fig. 5.3 Extended DSR front-ends

Table 5.7 Intelligibility listening tests using Diagnostic Rhyme tests (conducted by Dynastat listening laboratory)

	Clean	Car 10 dB	Street 15 dB	Babble 15 dB
Unprocessed	95.7	95.5	92.4	93.8
XFE reconstruction	93.0	88.8	85.0	87.1
XAFE reconstruction	92.8	88.9	87.5	87.9
LPC-10	86.9	81.3	81.2	81.2
MELP	91.6	86.8	85.0	85.3

Table 5.8 Listening test transcription task results: The list numbers in each cell of the table show the number of missed/wrongly transcribed/partially transcribed words

	Clean	Car	Street	Babble	Clean	Average error (%)
Uncoded (original)	1,1,2	1,0,1	0,2,4	3,9,3	0,4,1	0.6
XFE reconstruction	1,6,1	0,3,6	2,9,4	5,9,2	1,4,5	1.0
XAFE reconstruction	0,6,2	0,5,4	0,4,3	3,5,2	1,6,5	0.8
LPC-10 coder	8,18,6	62,26,7	67,22,7	47,12,3	18,10,9	5.5
MELP coder	0,3,1	1,6,3	4,6,2	16,10,3	1,9,5	1.2
No. of words in message	1166	1153	1155	1149	1204	Total: 5827

pitch extraction algorithms available at the time. Further information about the extension algorithms and their performance can be found in Ramabadran et al. (2004) and Sorin et al. (2004).

5.9 Transport Protocols: The IETF RTP Payload Formats for DSR

In addition to the standards for the front-end features themselves, protocols for the transport of these features from the device to the server are also needed. The IETF Real Time Protocol (RTP) is a well established mechanism for the transport of many different media types including video, VoIP, and music. Associated with RTP are also the SIP protocols for session initiation and codec negotiation. By defining a RTP format for the DSR features, services benefit from all of the added functionality of this set of protocols, as well as the support of other media types for multimodal applications. Formats for the RTP payloads for all the DSR standards have been published as at the *IETF* (IETF Xie 2003; IETF Xie and Pearce 2005).

Within these payloads any number of frame pairs may be sent within a packet. For the front-end features on their own this takes 12 bytes per frame pair and with the extension it takes 14 bytes per frame pair. The format allows an arbitrary number of frame pairs to send in each RTP payload. This allows the system designer flexibility with the choice depending on the latency and bandwidth of the channel available.

The total overhead for the protocol headers in the stack is quite high as shown in Table 5.9.

Table 5.9 RTP Protocol header sizes for packet data transport

Data	Size (bytes)
RTP	12
UDP	8
IP	20
Total	40

For low data rate channels such as GPRS it is appropriate to use multiple frames (DSR uplink or coded speech on downlink) per RTP payload to reduce the total bandwidth and therefore latency (e.g., four to ten). For higher data rate channels perhaps with residual packet loss such as UMTS a smaller number of frames per packet can be used (e.g., one or two). In testing a prototype implementation it has been found that even on GPRS the latencies are quite acceptable (less than two seconds) and for higher speed channels much less.

5.10 Conclusion

This chapter has presented the DSR standards that were created within ETSI, 3GPP and the IETF to enable the implementation of speech and multimodal services with the best possible performance. In particular they target services using remote speech recognition over narrow bandwidth mobile channels. For mobile device users accessing such services, the speech recognition performance in background noise, the robustness to channel errors and the response time are all important factors impacting the usability and quality of the user experience. As the capabilities of mobile speech services such as voice driven search progress, the enhanced performance from DSR can only help grow the take-up and popularity of these services and the satisfaction of users.

Acknowledgments

Many people participating in the Aurora and 3GPP working group have contributed to this work both directly and indirectly. There are too many people to list them all individually but I honour them collectively for their creativity, time and energy given to the creation of these DSR standards.

References

3GPP TR 22.977 (2002) Feasibility Study for Speech Enabled Services, Aug.

3GPP TS 26.243 (2004) ANSI C Code for the Fixed-Point Distributed Speech Recognition Extended Advanced Front-end.

3GPP TS 26.177 (2004) DSR Extended Advanced Front-end Test Sequences.

3GPP TR 26.943 (2004) Recognition Performance Evaluations of Codecs for Speech Enabled Services (SES).

ETSI Standard ES 201 108 (2000) Distributed Speech Recognition; Front-end Feature Extraction Algorithm; Compression Algorithm, April.

ETSI Standard ES 202 050 (2002) Distributed Speech Recognition; Advanced Front-end Feature Extraction Algorithm; Compression Algorithm.

ETSI Standard ES 202 211 (2003) Distributed Speech Recognition; Extended Front-end Feature Extraction Algorithm; Compression Algorithm, Back-end Speech Reconstruction Algorithm, Nov.

ETSI Standard ES 202 212 (2003) Distributed Speech Recognition; Extended Advanced Front-end Feature Extraction Algorithm; Compression Algorithm, Back-end Speech Reconstruction Algorithm, Nov.

GTS GSM 03.50: Digital cellular telecommunications system (Phase 2+); Transmission planning aspects of the speech service in the GSM Public Land Mobile Network (PLMN) system (GSM 03.50).

Hirsch, H. G. and Pearce, D. (2000) The Aurora Experimental Framework for the Performance Evaluation of Speech Recognition Systems Under Noisy Conditions. In *Proceedings of ISCA workshop on Automatic Speech Recognition*, Paris, France.

IETF Xie, Q. (2003) *RTP Payload Format for ETSI European Standard ES 201 108 Distributed Speech Recognition Encoding*, RFC 3557, July. http://www.ietf.org/rfc/rfc3557.txt

IETF Xie, Q and Pearce, D. (2005) *RTP Payload Formats for ETSI European Standard ES 202 050, ES 202 211, and ES 202 212 Distributed Speech Recognition Encoding*, RFC 4060, May. http://www.ietf.org/rfc/rfc4060.txt

Macho, D., Mauuary, L., Noe, B., Cheng, Y., Ealey, D., Jouvet, D., Kelleher, H., Pearce, D. and Saadoun, F. (2002) Evaluation of a Noise-Robust DSR Front-End on Aurora Databases. In *Proceedings of ICSLP 2002*, Denver, USA.

Pearce, D. (2000) Enabling New Speech Driven Services for Mobile Devices: An Overview of the ETSI Standards Activities for Distributed Speech Recognition Front-ends. *Applied Voice Input/Output Society Conference (AVIOS2000)*, San Jose, CA, May 2000.

Pearce, D. (2004a) Enabling Speech and Multimodal Services on Mobile Devices: The ETSI Aurora DSR standards and 3GPP Speech Enabled Services. *VoiceXML Review*, Nov/Dec 2004. http://www.voicexmlreview.org/Nov2004/features

Pearce, D. (2004b), Robustness to Transmission Channel—the DSR Approach. *COST278 & ICSA Research Workshop on Robustness Issues in Conversational Interaction*, Aug 2004.

Peinado, A. and Segura, J. (2006) *Speech Recognition over Digital Channels*. Book Publisher Wiley, Chichester.

Ramabadran, T., Sorin, A., McLaughlin, M., Chazan, D., Pearce, D. and Hoory, R. (2004) The ETSI Extended Distributed Speech Recognition (DSR) Standards: Server-Side Speech Re-construction. In *Proceedings of ICASSP*, Montreal, Canada.

Sorin, A., Ramabadran, T., Chazan, D., Hoory R., McLaughlin, M., Pearce, D., Wang, F. and Zhang, Y. (2004) The ETSI Extended Distributed Speech Recognition (DSR) Standards: Client Side Processing and Tonal Language Recognition Evaluation. In *Proceedings of ICASSP*, Montreal, Canada.

6

Speech Feature Extraction and Reconstruction

Ben Milner

Abstract. This chapter is concerned with feature extraction and back-end speech reconstruction and is particularly aimed at distributed speech recognition (DSR) and the work carried out by the ETSI Aurora group. Feature extraction is examined first and begins with a basic implementation of mel-frequency cepstral coefficients (MFCCs). Additional processing, in the form of noise and channel compensation, is explained and has the aim of increasing speech recognition accuracy in real-world environments. Source and channel coding issues relevant to DSR are also briefly discussed. Back-end speech reconstruction using a sinusoidal model is explained and it is shown how this is possible by transmitting additional source information (voicing and fundamental frequency) from the terminal device. An alternative method of back-end speech reconstruction is then explained, where the voicing and fundamental frequency are predicted from the received MFCC vectors. This enables speech to be reconstructed solely from the MFCC vector stream and requires no explicit voicing and fundamental frequency transmission.

6.1 Introduction

To perform automatic speech recognition from a terminal device, three architectures can be considered. The first is an embedded architecture where all speech processing is performed on the terminal device itself. Processing power limitations make this suitable only for small-scale speech recognition applications such as voice dialling. Part III of this book examines embedded speech recognition. The second architecture is *network speech recognition* (NSR) where the speech signal is encoded by a codec and transmitted to a remote speech recogniser for decoding. This is currently the most frequently used method of performing speech recognition from mobile devices. Part I of this book examines network speech recognition. Finally, the third method is *distributed speech recognition* (DSR), where feature extraction (or *front-end* processing) is performed on the terminal device and decoding (or back-end processing) is performed remotely. Figure 6.1 illustrates example architectures for NSR and DSR to highlight their differences.

The main difference between NSR and DSR is the location of feature extraction and the format of speech data that is transmitted from a terminal device to a remote recogniser. In NSR a speech codec is used to encode and decode speech for transmission across the network to the remote recogniser, whereupon feature extraction is applied. In DSR, feature extraction takes place on the terminal device and a stream of feature vectors is transmitted to the remote recogniser, as opposed to an encoded audio signal as in NSR. Chapters 2 and 5 provide detailed discussions into NSR and DSR architectures and examine their advantages and disadvantages.

In both NSR and DSR, the feature extraction components are often identical, as both have the task of transforming a time-domain speech signal into a series of feature vectors. With NSR, the input to feature extraction will have been compressed by a low bit-rate speech codec that will have distorted the speech in some way. In DSR, the original time-domain signal forms the input to the feature extraction, although both source coding and channel coding of the speech feature vectors must be applied in preparation for transmission to the remote recogniser. This leads to one of the problems of DSR which is that only parameterised speech feature vectors are received by the remote recogniser back-end. As no time-domain signal is received playback of the speech signal is not straightforward. This has been identified as a particular problem with DSR architectures, although several methods have been proposed to enable back-end speech reconstruction.

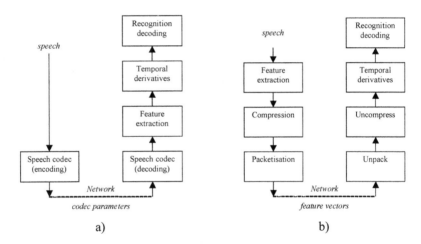

Fig. 6.1 Comparison of (**a**) network speech recognition (NSR) and (**b**) distributed speech recognition (DSR) architectures

The aim of this chapter is to first explain the operation of feature extraction, whether it be implemented for NSR or DSR. Some consideration is given to DSR applications where compression and error protection of the speech features is important. Practical issues such as robustness to acoustic noise and channel distortion are also examined. The discussion on feature extraction is strongly biased towards mel-frequency cepstral coefficients (MFCCs) as they are probably the most widely used speech feature in current speech recognition technology (Davis and Mermelstein 1980). They have also been adopted as the standardised speech feature for DSR by the ETSI Aurora group (ETSI 2003a,b, 2007). The second part of the chapter describes how a speech signal can be reconstructed at the back-end of a DSR system. Two methods to achieve this are explained with the first discussing the implementation proposed by the ETSI Aurora group which utilises the received MFCC vectors and additional source information (ETSI 2003b). The second method is an alternative that requires no additional information other than the MFCC vector stream itself (Milner and Shao 2007).

6.2 Feature Extraction

Feature extraction is the first stage of automatic speech recognition and transforms the input audio signal into a form suitable for classification. This typically involves several processing stages that output a stream of feature vectors which encode the spectral and temporal evolution of the speech signal. Some of the more successful feature extraction methods incorporate perceptual properties of human hearing and human speech production. For example, feature extraction methods such as *Mel-frequency cepstral coefficients* (MFCCs) and *perceptual linear prediction* (PLP) incorporate properties of human hearing at several stages of the feature extraction process (Davis and Mermelstein 1980; Hermansky 1990). Most feature extraction algorithms also consider properties of speech generation to maximise the discrimination between different speech sounds. For speech recognition, vocal tract information is considered more useful than source information. As a result, many feature extraction methods apply cepstral processing to separate vocal tract information from source information (Oppenheim and Schafer 1989).

Of all features proposed for speech recognition, MFCCs have been proved to be the most effective and widely used. Their use has been further re-enforced by their adoption by the ETSI Aurora group as the standard for DSR (ETSI 2003a). Due to their widespread deployment in DSR, MFCC features form the basis of the discussions into feature extraction in this section. A basic implementation of MFCC-based feature extraction is described first, which is suitable for clean, undistorted speech. Consideration is then given to the practical deployment of feature extraction which needs to take into account the undesirable affects of acoustic noise and channel distortion. The last stages of feature extraction, which take place on the server side, such as computation of temporal derivatives, are finally discussed.

6.2.1 Basic Terminal-Side Feature Extraction

Terminal-side feature extraction transforms the input audio signal into a stream of static feature vectors that are subsequently compressed and packetised for transmission to the recogniser back-end located on a remote server. This section describes basic MFCC feature extraction as specified in the first version of the ETSI Aurora DSR standard (ETSI 2003a). Figure 6.2 shows the processing stages for transforming a speech signal, $s(n)$, into MFCC vectors, $\mathbf{c}^x$.

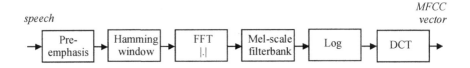

Fig. 6.2 ETSI Aurora standard for computing MFCC vectors

Pre-Emphasis

Feature extraction begins by pre-emphasising the speech signal using a high-pass filter. Speech signals tend to be low-pass in character and the application of high-pass filtering serves to spectrally balance the signal. Given an input speech signal, $s(n)$, the pre-emphasised output speech signal, $x(n)$, is computed,

$$x(n) = s(n) - \alpha s(n-1) \tag{6.1}$$

α is the filter coefficient and a suitable value, as used in the ETSI Aurora standard, is $\alpha = 0.9$. In practice the precise choice of α does not have a significant effect on recognition accuracy.

Hamming Window

A Hamming window, $h(n)$, is applied to the pre-emphasised speech signal to extract short-duration frames, $x_i(n)$, which will subsequently be transformed into feature vectors, where i indicates the frame number,

$$x_i(n) = x(n + Si \lfloor h(n \lfloor \quad 0 \le n \le W - 1 \tag{6.2}$$

where the Hamming window, $h(n)$, is defined,

$$h(n) = 0.54 - 0.46 \cos\left(\frac{2\pi(n+0.5)}{W}\right) \tag{6.3}$$

The time duration of the Hamming window varies for different feature extraction algorithms, but is typically in the range 10 ms to 50 ms which gives $W = 80$–$W = 400$ samples for 8 kHz sampled speech. This represents a time duration over which the speech can be assumed stationary, although for some sounds the speech remains stationary for much longer. A stream of short-duration frames of speech is extracted by sliding the Hamming window along the speech signal by S samples and output-ting a new window of samples. For the ETSI Aurora standard, the window width is 25 ms and the window slide is 10 ms to give a frame rate of 100 frames per second. Another commonly used frame slide is half the duration of the window. Many other windowing functions exist, such as Hanning, Bartlet and Kaiser, but for speech recognition applications the Hamming window is generally preferred.

Power Spectrum

The time-domain frames of speech are now converted to a power spectral representation, $|X_i(f)|^2$, using a discrete Fourier transform,

$$|X_i(f)|^2 = \left| \sum_{n=0}^{W-1} x_i(n) e^{\frac{-j2\pi fn}{W}} \right|^2 \tag{6.4}$$

Transforming the speech to a spectral representation reveals more structure in the speech signal, which is important for classification. In some implementations of MFCC extraction a magnitude spectrum is used rather than a power spectrum although this makes very little difference to classification accuracy. In practice the DFT is replaced by a *fast Fourier transform* (FFT) which gives considerable reductions in computation time, particularly for longer duration windows (Cooley and Tukey 1965).

Mel-Filterbank

The spectrally detailed power spectral representation is now non-linearly quantised in frequency through the application of a mel-scaled filterbank. The non-linear frequency spacing of the filterbank channels reflects the non-linear frequency sensitivity of human hearing and places a greater density of filterbank channels at low frequencies than at higher frequencies. Implementation of the mel-filterbank can take several forms although in this chapter a matrix transformation is adopted. A K-dimensional vector of mel-filterbank channel energies, $\mathbf{m}_i^x$, is computed as,

$$\mathbf{m}_i^x = \mathbf{M} \mathbf{p}_i^x \tag{6.5}$$

where $\mathbf{p}_i^x$ is a column vector containing the $W/2$ dimensional power spectrum of Eq. 6.4. The rows of matrix $\mathbf{M}$ are the frequency responses of the K channels in the mel-filterbank. For illustration, the frequency response of a $K=23$ mel-filterbank is shown in Fig. 6.3.

The mel-spacing of filterbank channels defines MFCCs, but other non-linear frequency scales exist, such as the Bark scale which is used in PLP feature extraction (Hermansky 1990). No strict rules exist for the number of filterbank channels or their spectral shape. In the ETSI Aurora MFCC standard, the number of filterbank channels is 23 and their shape is triangular. For 4 kHz bandwidth speech the lowest frequency channel is centred at 125 Hz and spans 125 Hz, while the highest frequency channel is centred at 3657 Hz and spans 656 Hz.

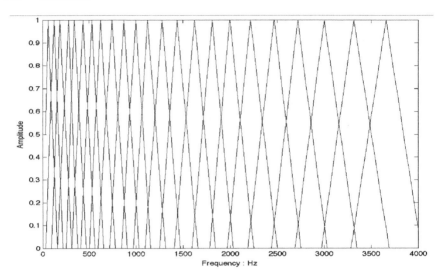

Fig. 6.3 Frequency responses of the 23 mel-filterbank channels

Log

Applying a log to the filterbank channel energies reduces their sensitivity to both very loud and very quiet sounds and models the non-linear amplitude sensitivity of human hearing. The effect on speech recognition accuracy is significant and without the log, recognition accuracy is severely reduced.

Discrete Cosine Transform

The final stage in extracting MFCC feature vectors, c_i^x, is to apply a discrete cosine transform (DCT) to the K log filterbank channel energies,

$$c_i^x(j) = \sum_{k=1}^{K} m_i^x(k) \cos\left(\frac{j\pi(k-0.5)}{K}\right) \qquad (6.6)$$

The DCT serves two purposes. First, the DCT performs the final part of a cepstral transformation which separates the slowly varying spectral envelope (or vocal tract) information from the faster varying speech excitation. Lower-order coefficients represent the slowly varying vocal tract while higher-order coefficients contain excitation information. For speech recognition, vocal tract information is more useful for classification than excitation information. Therefore, to create the final MFCC vector, the output vector from the DCT is truncated to retain only the lower-order coefficients. In the ETSI Aurora standard, the lower 13 coefficients are retained—$c_i^x(0)$ to $c_i^x(12)$.

The second purpose of the DCT is to decorrelate the elements of the feature vector. Elements of the log filterbank vector exhibit correlation due to both the spectral characteristics of speech and the overlapping nature of the filterbank. For statistical classifiers, such as HMMs, to accurately model correlated feature vectors requires full covariance matrices which are both computationally expensive and require large amounts of training data. To optimally decorrelate, or diagonalise, the log filterbank features requires a Karhunen-Loeve transform (KLT) which needs to be estimated from a set of training data. However, a good approximation to the KLT for log filterbank features is the DCT. This means that applying the DCT serves to decorrelate the elements of the feature vector, making it suitable for diagonal covariance matrix statistical classifiers.

Frame Energy

Including a measure of the energy of each frame of speech gives significant increases in speech recognition accuracy. The zeroth MFCC, $c_i^x(0)$, is the sum of the log energies from each filterbank channel and can be considered a geometric measure of frame energy. A common alternative is to compute the log energy, $\ln E_i$, of the time-domain frames of speech without pre-emphasis being applied. In practice these energy measures are similar and only one needs to be included in the feature vector. In the ETSI Aurora standard log energy is computed on the terminal device and transmitted to the back-end in addition to the 13 MFCCs. Rather than including both energy measures in classification, it is usual to select just one or to combine them through an appropriate weighing.

To illustrate the operations in MFCC feature extraction, Fig. 6.4 shows the transformation of both a voiced speech frame (left-hand column) and an unvoiced speech frame (right-hand column). Time-domain frames of speech are shown in the top panels and below are shown the resulting power spectra. For the voiced speech, harmonic structure is clearly visible and the fundamental frequency is seen to be ~250 Hz. The third row shows the resulting mel-filterbank feature. The effect of the non-linear spacing of filterbank channels is evident when examining the position of the spectral harmonics seen in the power spectrum of the voiced speech. In the mel-filterbank the first two harmonics (at frequencies ~250 Hz and 500 Hz) occur in channels 4 and 7 which shows the stretching of frequency in these lower channels made by the non-linear frequency spacing of the mel-scale. The bottom panels show the output of the DCT—for clarity the zeroth coefficient is not shown as its amplitude is very large. By considering the basis functions of the DCT, some spectral meaning can be given to the MFCCs. The basic function associated with MFCC 0 is a constant and as previously discussed represents the energy of the filterbank. The first basis function of the DCT is a half cosine wave which means that the first MFCC indicates the spectral slope. This is evident in the figure where the voiced filterbank shows significantly more energy at low frequencies than at higher frequencies and has a strongly positive value for MFCC 1. The unvoiced filterbank has an opposite spectral slope and has a strongly negative value for MFCC 1. This analysis can be continued—for example the second basis function is a full cosine wave and hence MFCC 2 indicates the proportion of mid-band spectral energy to outer-band spectral energy.

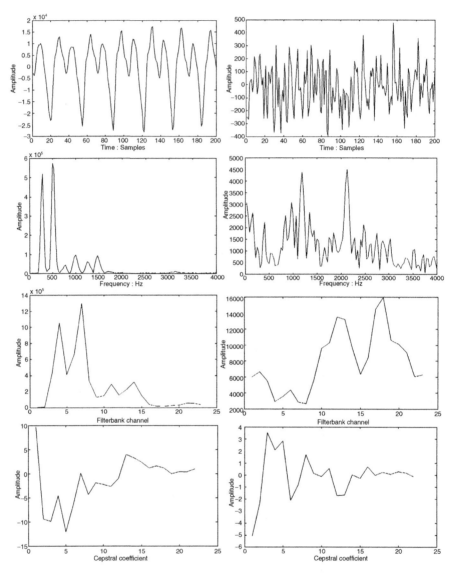

Fig. 6.4 Example of MFCC feature extraction for voiced speech (left-hand column) and un-voiced speech (right-hand column). Top row shows original frames of speech, then power spectrum followed by filterbank and finally DCT output

For the MFCC extraction described above, it is assumed that the speech is sampled at 8 kHz. However, speech sampled at other sampling frequencies can also be parameterised through appropriate modifications to the feature extraction algorithm. For example, at a sampling frequency of 16 kHz the ETSI Aurora standard specifies a Hamming window width of 400 samples and a frame slide of 160 samples, which retains the same duration frame width and slide as with 8 kHz sampled speech.

6.2.2 Advanced Terminal-Side Feature Extraction

The feature extraction algorithm described in the previous section is a basic method of extracting MFCC vectors and should give satisfactory performance in clean environments. However, to achieve good recognition accuracy in more realistic environments, where both acoustic noise and channel distortions are present, it is necessary to include extra processing. This includes noise reduction and channel equalisation stages. A later version of the ETSI Aurora standard, namely the Advanced Front-End (AFE), includes such processing (ETSI 2007).

Noise Reduction

Acoustic noise can severely corrupt the feature vectors produced by the front-end and cause large reductions in classification accuracy. Noise is usually considered additive in the time-domain and therefore has an additive effect in the power spectral domain and subsequent filterbank domain. Depending on the spectral character of the noise, the amplitudes of filterbank channels will be increased, leading to a distortion of the resulting MFCC vector. Even if the noise is narrowband, it will effect all elements of the MFCC vector as the DCT has the effect of smearing out the noise.

Many algorithms have been developed to reduce additive noise from the speech feature vectors. Most of these make an estimate of the contaminating noise during speech inactive periods and then remove this noise estimate during periods of speech activity. A reasonably successful noise reduction technique is spectral subtraction, which subtracts noise estimates in either the spectral domain or filterbank domain (Boll 1979). Performing noise reduction in the filterbank domain can take advantage of the spectral averaging made by the filterbank channels which reduces processing distortion resulting from excessive noise removal. Many extensions to spectral subtraction have been made since it was first proposed and these have reduced its sensitivity to noise type and power (Wu and Chen 2001). The spectral subtraction class of algorithms represent just one type of noise reduction method. Many other noise reduction methods for speech recognition have also been proposed and have varying levels of success. For example, in the ETSI AFE noise reduction is carried out during a pre-processing stage that is implemented before feature extraction. This is based on a two-stage Wiener filter which outputs a noise reduced speech signal that is input into feature extraction (ETSI 2007).

Blind Equalisation

Channel distortion, such as from a microphone in a handset, may cause a significant reduction in speech recognition accuracy. As such channel equalisation can play an important role in achieving robust speech recognition accuracy. For cepstral-based features, such as MFCCs, channel distortion is additive in the cepstral-domain. For example, consider a speech signal, $x(n)$, that is convolved with a channel distortion, $g(n)$, to give a channel distorted signal, $y(n)$. In the frequency domain this distortion becomes multiplicative, i.e. $Y(f) = X(f)G(f)$. After the log and DCT operations the channel distortion becomes an additive offset,

$$c_i^y(j) = c_i^x(j) + c_i^g(j) \tag{6.7}$$

where $c_i^y(j)$, $c_i^x(j)$ and $c_i^g(j)$ represent the jth MFCC of the distorted speech, clean speech and channel, respectively, at time frame i. If the channel distortion is stationary its time index can be ignored to give a constant offset distortion,

$$c_i^y(j) = c_i^x(j) + c^g(j) \tag{6.8}$$

Equalising the distortion becomes a process of removing the offset from the signal and several approaches have been developed to achieve this. A simple method is cepstral mean normalisation (CMN) (also known as cepstral mean subtraction (CMS)) which computes a mean cepstral vector from the stream of input vectors and subtracts it from the input vectors (Rosenberg 1994). This not only removes the channel but also removes the mean of the speech, although this has been found to be beneficial in terms of speech recognition accuracy. An alternative equalisation method is the RASTA filter (Hermansky and Morgan 1994). This uses a highpass filter to remove stationary and slowly time-varying components of the cepstral features which includes the channel distortion. The RASTA filter also includes a lowpass filter component which removes fast varying components of the cepstral vectors that improves robustness to noise. In the ETSI Aurora AFE, channel equalisation is achieved by least mean square (LMS) filtering, with a reference signal equal to the cepstrum of a flat spectrum (ETSI 2007).

6.2.3 Quantisation and Packetisation

Feature extraction generates a stream of static feature vectors that must be transmitted to the remote back-end for recognition. Before transmission they must first be compressed and formatted with the inclusion of appropriate error protection.

In the ETSI Aurora standard, 13 dimensional MFCC vectors and a log energy term are created at a rate of 100 vectors per second. Assuming the number representation used by HTK (4 byte floats for each element) this represents a bit rate of 44,800 bits per second, which is too high in terms of channel capacity for most applications (HTK 2007). Instead, source coding must be applied to reduce the bit-rate of the feature vector stream to an acceptable level. In the ESTI Aurora standard split vector quantisation is applied to pairs of coefficients to reduce the storage for each feature vector to 43 bits. A 1 bit voice activity detection (VAD) flag is also allocated to each frame which gives a source coded bit rate of 4400 bps for the MFCC feature vector stream. Chapter 7 describes the source coding of speech feature vectors in more detail.

The compressed feature vectors are next placed in an agreed framing structure and suitable error protection applied. In the ETSI Aurora front-end, pairs of feature vectors are grouped together and a 4-bit cyclic redundancy check (CRC) computed and included for error protection. Multiframes, which represent 240 ms of speech, are then formed by grouping together 12 pairs of feature vectors. The multiframe includes 48 bits of header information with the result that the final bit rate is 4800 bps. Chapter 8 discusses these channel coding methods and framing in more detail.

6.2.4 Server-Side Processing

At the recogniser back-end on the remote server the received feature vectors are unpacked and uncompressed. In the event that some feature vectors have either become lost or corrupt due to adverse network conditions, error concealment techniques can be applied. These may estimate the value of missing vectors or modify the decoding process of the speech recogniser to take into account the unreliability of parts of the feature vector stream. This is discussed in detail in Chap. 9.

Following any error correction, the unpacked stream of static feature vectors are augmented by their temporal derivatives (Furui 1986; Hanson and Applebaum 1990). Including temporal derivatives in the feature vector stream partially overcomes the assumption in HMM-based speech recognisers that the feature vectors are independent and identically distributed and gives substantial increases in recognition accuracy. Velocity derivatives, Δc_i^x, are computed as,

$$\Delta c_i^x = \sum_{d=1}^{D} \frac{d}{D} \left(c_{i+d}^x - c_{i-d}^x \right) \tag{6.9}$$

Similarly, acceleration derivatives, $\Delta\Delta c_i^x$, are computed as,

$$\Delta\Delta c_i^x = \sum_{a=1}^{A} \frac{a}{A} \left(\Delta c_{i+a}^x - \Delta c_{i-a}^x \right) \tag{6.10}$$

D and A specify the number of vectors used in computing the velocity and acceleration derivatives. Typical values range from $D=2$ and $A=1$ to $D=4$ and $A=4$, with the latter used in the ETSI Aurora standard.

6.3 Speech Reconstruction

In network speech recognition the time-domain speech signal is transmitted to the speech recogniser where feature extraction and classification take place. As the time-domain signal itself is transmitted to the speech recogniser, playback of speech on the server is straightforward. However, in distributed speech recognition only the speech feature vectors are received at the remote server. As no time-domain signal is transmitted, no readily available time-domain signal can be used for playback at the server. While this is not a problem for speech recognition, it may be desirable to listen to the speech. This is particularly true for automated services that are used for financial services. For example, a speech recognition error could lead to unwanted transactions in which case there may be a need to listen to the speech input to confirm what was actually said. Providing a back-end playback facility is a legal requirement in the US.

This section begins by examining the speech information present at the back-end through a received stream of MFCC vectors. This reveals the spectral envelope to be present but not source information needed for speech reconstruction. The method used in the ETSI Aurora extended front-end (XFE) for speech reconstruction is then discussed whereby source information is supplied to the back-end through additional feature extraction and data transmission from the terminal device (ETSI 2003b).

6.3.1 Analysis of Received Speech Information

The MFCC feature extraction process discards too much information to allow the features to be simply inverted back into a time-domain signal. Examining the MFCC extraction process of Fig. 6.2, shows some of the processing stages to be invertible while others are not. The effect of the pre-emphasis filter is invertible and can be equalised by a suitably designed lowpass filter. The log operation can also be inverted by a simple exponential operation. However, several stages in feature extraction are not invertible. Applying a magnitude operation to the complex frequency spectrum of the Fourier transform discards phase information which makes inversion of the power spectrum to a time-domain signal not possible. The quantisation of the power spectrum by the mel-filterbank loses spectral detail which cannot be recovered during inversion. Further spectral detail is also lost by truncating the DCT coefficients when forming the MFCC vector. Of course, for speech recognition purposes, these losses of spectral detail and phase are beneficial, but for playback their loss is serious.

The received MFCC vectors can provide a smoothed estimate of the speech power spectrum which encodes vocal tract information. Starting with an MFCC vector, an estimate of the mel-filterbank can be computed by zero padding the MFCC vector to the dimensionality of the filterbank and applying an inverse DCT followed by an exponential operation. A $W/2$ dimensional power spectrum can be estimated from the K mel-spaced filterbank channels (where $W/2 >> K$) using interpolation techniques (Vaseghi 2006). However, at this stage it is important to note that the resulting power spectrum is subject to high frequency tilt which arises from both the effect of pre-emphasis and the increasing mel-filterbank channel bandwidths. As was the case for channel distortion, discussed in Sect. 6.2.2, these effects are multiplicative in the frequency domain and can be equalised in the cepstral domain by subtracting their cepstral equivalents from the MFCC vector.

The area, and hence energy, w_k, of each filterbank channel increases with channel number due to the widening of channel bandwidths—see Fig. 6.3. Given a vector, w, that comprises the areas, w_k, of the K mel-spaced triangular filterbank windows, the resulting cepstral representation, c^w, can be computed through log and DCT operations. Similarly the cepstrum, c^p, of the pre-emphasis filter can be computed by passing its impulse response through the MFCC extraction algorithm.

An equalised MFCC vector, $\hat{c}_i^x$, can be estimated by subtracting the filterbank and pre-emphasis cepstra from the unequalised MFCC vector, c_i^y, produced by the feature extraction process,

$$\hat{\mathbf{c}}_i^x = \mathbf{c}_i^y - \mathbf{c}^w - \mathbf{c}^p \qquad (6.11)$$

The MFCC vector can be inverted to provide an equalised power spectrum esti-
mate, $\left|\hat{X}(f)\right|^2$. Figure 6.5 illustrates the effectiveness of recovery by showing the log
power spectrum (dotted line) of a frame of 200 speech samples and the log power
spectra recovered from MFCC vectors extracted from the same 200 speech samples.
The spectrum recovered from a non-truncated 23-D MFCC vector (solid line) closely
follows the spectral envelope of the original speech. When the inversion is applied to
a truncated 13-D MFCC vector a similar spectral envelope (dashed line) is produced
but the truncation of higher order cepstral coefficients removes some of the spectral
detail that was retained in the 23-D MFCC. In particular, the 13-D MFCC-derived
spectrum is unable to resolve the high frequency spectral peak at 3 kHz into two
separate formants as the 23-D MFCC-derived spectrum can.

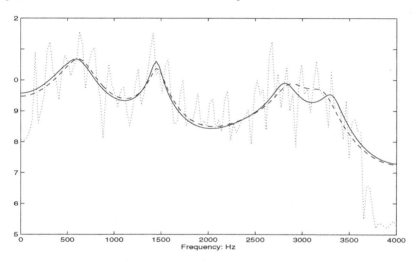

Fig. 6.5 Power spectrum reconstruction—the dotted line is the original log power spectrum
while the solid and dashed lines show the reconstructed log power spectrum from non-
truncated and truncated MFCC vectors, respectively

6.3.2 Speech Reconstruction

To reconstruct an audio speech signal the spectral envelope alone is insufficient as
important source information such as voicing and fundamental frequency (for voiced
speech) are missing. The ETSI Aurora extended front-end (XFE) addresses this
problem by estimating the voicing and fundamental frequency on the terminal device
and transmitting them to the back-end along with the MFCC vectors. This approach
delivers sufficient source information to the back-end to enable speech reconstruction but increases terminal-side processing and also increases bit-rate requirements
of the communication channel. This method of providing source information at the

back-end for speech reconstruction is discussed in the next section. An alternative to explicitly transmitting source information for back-end reconstruction has recently been proposed whereby the source information is predicted from the received MFCC vector stream (Milner and Shao 2007). This approach is discussed in Sect. 6.4.

Terminal-Side Voicing and Fundamental Frequency Estimation

Many algorithms for estimating the voicing and fundamental frequency of speech have been proposed in the last 40 years. These vary in many ways and operate in the time-domain, frequency-domain or cepstral-domain (de Cheveigne and Kawahara 2001). The fundamental frequency estimator used in the ETSI XFE operates in the frequency-domain and searches for spectral peaks that correspond to the fundamental frequency. The search begins in an upper band (200 Hz–420 Hz) and if no suitable fundamental frequency is found, the search moves to a middle band (100 Hz–210 Hz) and then to a low band (52 Hz–120 Hz). In implementation, the algorithm contains many processing stages that minimise estimation errors. A detailed discussion of these is can be found in (ETSI 2003b).

Once voicing and fundamental frequency have been estimated on the terminal device they must be transmitted to the back-end. In the ETSI Aurora XFE the fundamental frequency is converted into a fundamental period. This is measured in samples and is constrained to be in the range 19 samples to 140 samples which corresponds to fundamental frequencies from 57 Hz to 421 Hz. Even numbered frames are allocated 7 bits to represent the period while odd numbered frames are allocated 5 bits and represent the difference in period. The voicing class of each frame is also encoded and takes one of four different values—non-speech, unvoiced speech, mixed voiced speech and fully voiced speech. For non-speech and unvoiced speech, the 7 bit period value, or 5 bit differential period value, are set to zero and an additional single bit is used to identify non-speech or unvoiced speech. Mixed voiced and fully voiced speech are indicated by non-zero period values with the single bit indicating whether the frame is mixed or fully voiced.

For each pair of frames, 12 bits are used to represent the fundamental period and another 2 bits provide information to determine the voicing class. These 14 data bits are protected by a 2 bit CRC. Therefore, with 50 frame pairs per second, the transmission of voicing and fundamental frequency requires 800 bits per second of channel capacity. This is in addition to the 4800 bits per second used by MFCC vector transmission which give an overall bit rate for the ETSI XFE of 5600 bps.

Sinusoidal Modelling of Speech

The MFCC vectors, voicing and fundamental frequency provide sufficient information to enable back-end speech reconstruction. Several models of speech production have been developed that are suitable for reconstructing, or synthesising, a speech waveform. These include the linear predictive (LP) model, the sinusoidal model and the harmonic plus noise (HNM) model (Rabiner and Schaeffer 1978; McAulay and Quatiery 1986). The ETSI Aurora XFE speech reconstruction is based on the sinusoidal model. The principles of sinusoidal model speech reconstruction from MFCCs are presented next and specific implementation details can be found in (ETSI 2003b). The sinusoidal model synthesises a speech signal, $x(n)$, as a sum of L sinusoids with amplitudes, A_l, frequencies, f_l, and phases, θ_l,

$$x(n) = \sum_{l=1}^{L} A_l \cos(2\pi f_l n + \theta_l) \tag{6.12}$$

The sinusoid frequencies are selected to be equal to the fundamental frequency and its harmonics. Given only the fundamental frequency, the frequencies of the sinusoids, f_l, can be approximated as multiples of the fundamental frequency, f_0,

$$f_l = l f_0 \tag{6.13}$$

The amplitude, A_l, of each sinusoid can be estimated from the smoothed spectral envelope provided by inverting the MFCC vector, at frequency, f_l,

$$A_l = \left| \hat{X}(f_l) \right| \tag{6.14}$$

The phase offset, θ_l, is calculated as the sum of phase components from the speech excitation, φ_l, and the vocal tract, ϕ_l,

$$\theta_l = \varphi_l + \phi_l \tag{6.15}$$

The excitation phase component at the fundamental frequency is estimated using a linear phase model and maintains continuity of the phase at frame boundaries. The phases at harmonic frequencies are calculated by multiplying the harmonic number

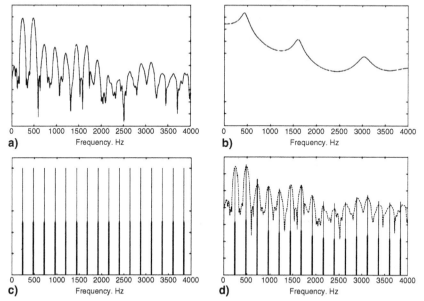

Fig. 6.6 Illustration of sinusoidal modelling of a frame of speech. Panel (**a**) shows the original log power spectrum of a frame of speech, (**b**) shows its spectral envelope and (**c**) a set of harmonically spaced sinusoids and (**d**) the sinusoidal model of log power spectrum

with the phase at the fundamental frequency. The phase from the vocal tract is calculated by assuming a minimum phase system. This allows the phase at each harmonic frequency to be computed from the spectral envelope using a Hilbert transform.

To illustrate speech reconstruction, Fig. 6.6a shows the log power spectrum of a 25 ms segment of phoneme /u/. Figure 6.6b shows the spectral envelope of the same frame of speech and Fig. 6.6c shows a series of sinusoids that are placed at harmonics of the fundamental frequency (in this example the fundamental frequency is 240 Hz). These provide the vocal tract and excitation information needed for speech reconstruction and multiplying the two results in the synthesised log power spectrum shown in Fig. 6.6d. For comparison, the original log power spectrum is shown as the dashed line which reveals the assumption of harmonicity in the excitation signal to be valid.

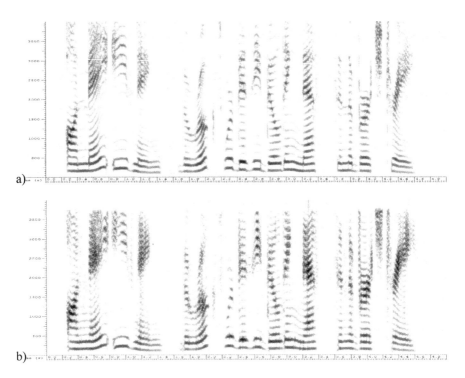

Fig. 6.7 Spectrograms showing (**a**) original and (**b**) reconstructed speech of the sentence "*On May evenings the rooks were busy building nests in the birch tree*"

From each MFCC vector and fundamental frequency estimate, a frame of reconstructed speech can be generated. For unvoiced speech the sinusoid frequencies are chosen randomly to provide a wideband excitation source. Continuous speech is reconstructed by extending the duration of each frame of speech by a half at both sides with a triangular windowing function. This allows the overlap-and-add algorithm to combine frames of speech and smooth discontinuities at frame boundaries

(George and Smith 1992). To demonstrate the effectiveness of sinusoidal model-based reconstruction, Fig. 6.7a shows a spectrogram of the utterance "*On May evenings the rooks were busy building nests in the birch tree.*" The spectrogram of the same utterance, but reconstructed from its MFCC vector representation and estimates of voicing and fundamental frequency is shown in Fig. 6.7b.

Comparing the two spectrograms shows the MFCC-based reconstruction to be highly effective in reproducing the original speech signal. The reconstructed harmonic tracks follow closely the original harmonics which is due to the accuracy of fundamental frequency estimation. Similarly, formant frequencies tend to be well preserved and these are provided by the MFCC vectors.

6.4 Prediction of Voicing and Fundamental Frequency

This section describes how speech can be reconstructed solely from the MFCC vector stream without explicitly transmitting voicing and fundamental frequency (Milner et al. 2007; Shao and Milner 2004). This is achieved by predicting the voicing and fundamental frequency of each frame of speech from the received MFCC vectors. Prediction of fundamental frequency is based on forming a model of the joint density of MFCCs and fundamental frequency. This model can then be used to predict the fundamental frequency associated with an MFCC vector. Similarly, the voicing associated with an MFCC vector is predicted from two models, one modelling voiced speech and the other modelling unvoiced speech and non-speech.

6.4.1 Fundamental Frequency Prediction from MFCC Vectors

Fundamental frequency is predicted from MFCC vectors using a model of the joint density of MFCC vectors and fundamental frequency. To begin, a joint feature vector, $\mathbf{y}_i$, is defined which comprises the MFCC vector, $\mathbf{x}_i$, and the fundamental frequency f_i, of frame i,

$$\mathbf{y}_i = [\mathbf{x}_i \ f_i]^T \tag{6.16}$$

For unvoiced frames the fundamental frequency is set to zero.

Phoneme-Independent Prediction of Fundamental Frequency

A simple method to model the joint density of MFCC vectors and fundamental frequency is to use a single model for all voiced speech sounds, making no distinction between different phonemes. Using a training data set, Z, joint vectors corresponding to voiced speech can be pooled into a voiced vector set, Ω^v (the superscript v indicates voiced speech),

$$\Omega^v = \{\mathbf{y}_i \in Z : f_i \neq 0\} \tag{6.17}$$

Expectation-maximisation (EM) clustering can be applied to this data to create a Gaussian mixture model (GMM), Φ^v, that models the joint density of MFCC vectors and fundamental frequency using a set of K^v clusters that localise the correlation between MFCCs and fundamental frequency in the joint feature vector space,

$$p(\mathbf{y}_i) = \Phi^v(\mathbf{y}_i) = \sum_{k=1}^{K^v} \alpha_k^v N(\mathbf{y}_i : \boldsymbol{\mu}_k^v, \boldsymbol{\Sigma}_k^v) \tag{6.18}$$

Each cluster is represented by a prior probability, α_k^v, and a Gaussian probability density function (PDF), N, with mean vector, $\boldsymbol{\mu}_k^v$, and covariance matrix, $\boldsymbol{\Sigma}_k^v$. The mean vector comprises two components, the mean vector of the voiced MFCC vectors in cluster k and the mean of the fundamental frequency in cluster k. Similarly, the covariance matrix comprises four components; the covariance matrix of the MFCC vectors, the variance of the fundamental frequency and the covariances of the MFCCs and fundamental frequency. This allows the mean and variance associated with the kth cluster to be decomposed as,

$$\boldsymbol{\mu}_k^v = \begin{bmatrix} \boldsymbol{\mu}_k^{v,x} \\ \boldsymbol{\mu}_k^{v,f} \end{bmatrix} \text{ and } \boldsymbol{\Sigma}_k^v = \begin{bmatrix} \boldsymbol{\Sigma}_k^{v,xx} & \boldsymbol{\Sigma}_k^{v,xf} \\ \boldsymbol{\Sigma}_k^{v,fx} & \boldsymbol{\Sigma}_k^{v,ff} \end{bmatrix} \tag{6.19}$$

Knowledge of the joint density of MFCCs and fundamental frequency in the GMM can be used to predict the fundamental frequency of a frame of speech from the MFCC vector representing that frame. From the kth cluster in the GMM, ϕ_k^v, a MAP prediction of fundamental frequency, $\hat{f}_i^k$, from MFCC vector $\mathbf{x}_i$ can be made,

$$\hat{f}_i^k = \arg\max_f \left(p(f|\mathbf{x}_i, \phi_k^v) \right) \tag{6.20}$$

This leads to the prediction of the fundamental frequency in terms of the statistics of the kth GMM cluster as,

$$\hat{f}_i^k = \boldsymbol{\mu}_k^{v,f} + \boldsymbol{\Sigma}_k^{v,fx} \left(\boldsymbol{\Sigma}_k^{v,xx} \right)^{-1} \left(\mathbf{x}_i - \boldsymbol{\mu}_k^{v,x} \right) \tag{6.21}$$

The predicted fundamental frequencies from all of the GMM clusters can be combined according to the posterior probability of the MFCC coming from that cluster, $h_k(\mathbf{x}_i)$,

$$\hat{f}_i = \sum_{k=1}^{K^V} h_k(\mathbf{x}_i) \left(\boldsymbol{\mu}_k^{v,f} + \boldsymbol{\Sigma}_k^{v,fx} \left(\boldsymbol{\Sigma}_k^{v,xx} \right)^{-1} \left(\mathbf{x}_i - \boldsymbol{\mu}_k^{v,x} \right) \right) \tag{6.22}$$

where the posterior probability, $h_k(\mathbf{x}_i)$, is given as,

$$h_k(\mathbf{x}_i) = \frac{\alpha_k^v p(\mathbf{x}_i|\phi_k^{v,x})}{\sum_{k=1}^{K^v} \alpha_k^v p(\mathbf{x}_i|\phi_k^{v,x})} \tag{6.23}$$

Phoneme-Dependent Prediction of Fundamental Frequency

An alternative to using a single GMM to model the joint density of MFCCs and fundamental frequency over all speech sounds is to allow a phoneme-dependent prediction. With this method phoneme-specific models of the joint density of MFCCs and fundamental frequency are created and subsequently used to provide phoneme-specific fundamental frequency predictions. This method is more complex than the phoneme-independent method, as a phoneme decoding for the MFCC vectors is required, but does provide more detailed modelling of the joint density.

The first stage in phoneme-dependent prediction of fundamental frequency is to train a set of phoneme HMMs that will be used to decode the input MFCC stream into a phoneme sequence. Assuming a set of W phonemes in the vocabulary (a typical value is $W = 44$ phonemes), then a set of HMMs, $\Lambda = \{ \lambda_1, \lambda_2,.., \lambda_W \}$, must be trained using the MFCC component, $\mathbf{x}$, of the augmented feature vector, $\mathbf{y}$.

The next stage in training is to use the phoneme HMMs to supply a model and state allocation to the MFCC vectors in the training data to allow phoneme-dependent GMMs to be trained. The resulting GMMs provide more localised modelling of the joint density of MFCCs and fundamental frequency. The state-dependent GMMs are created by force aligning the training data vectors to the correct sequence of HMMs using Viterbi decoding. The correct sequence of HMMs can be taken from phoneme-level annotations of the training database that may be created manually or automatically through forced word-level decoding with an appropriate pronunciation dictionary. This provides for each training data utterance $\mathbf{X} = [\mathbf{x}_1, \ldots, \mathbf{x}_i, \ldots, \mathbf{x}_M]$ a model allocation, $\mathbf{m} = [m_1, \ldots, m_i, \ldots, m_M]$, and a state allocation, $\mathbf{q} = [q_1, \ldots, q_i, \ldots, q_M]$, for each MFCC vector. This indicates the state, q_i, and model, m_i, that the ith MFCC vector, $\mathbf{x}_i$, is allocated, where $m_i \in \{1, .., W\}$ and $q_i \in \{1, .., S_{m_i} \}$ where S_{m_i} indicates the number of states in model m_i. This provides sufficient information to allow state-dependent clustering of the voiced vectors to take place. Voiced vectors allocated to each state, s, of each model, w, are pooled to form state and model dependent subsets of voiced feature vectors, $\Omega^v_{s,w}$,

$$\Omega^v_{s,w} = \left\{ \mathbf{y}_i \in Z : f_i \neq 0, q_i = s, m_i = w \right\} \quad 1 \leq s \leq S_w \quad 1 \leq w \leq W \qquad (6.24)$$

Unvoiced vectors allocated to each state of each model can also be pooled to form subsets of unvoiced vectors, $\Omega^u_{s,w}$,

$$\Omega^u_{s,w} = \left\{ \mathbf{y}_i \in Z : f_i = 0, q_i = s, m_i = w \right\} \quad 1 \leq s \leq S_w \quad 1 \leq w \leq W \qquad (6.25)$$

At this stage the state-dependent unvoiced vectors pools are not used but they will be used later for voicing prediction.

The state and model specific joint densities of MFCCs and fundamental frequency can now be modelled by applying EM clustering to the voiced vector pools. This creates a set of model and state-dependent voiced GMMs, $\Phi^v_{s,w}$, that are represented by mean vectors, $\mu^v_{k,s,w}$, covariance matrices, $\Sigma^v_{k,s,w}$, and prior probabilities, $\alpha^v_{k,s,w}$, corresponding to the kth cluster of the GMM associated with state s of model w.

To predict the fundamental frequencies associated with a stream of MFCC vectors their model and state sequence must first be determined. These are obtained by decoding the MFCC vectors into a model and state sequence, $\mathbf{m} = [m_1, \ldots, m_i, \ldots, m_M]$ and $\mathbf{q} = [q_1, \ldots, q_i, \ldots, q_M]$, using the set of HMMs trained previously together with an appropriate grammar. For unconstrained speech input the grammar should be an unconstrained phoneme grammar, while for specific tasks a more constrained grammar may be appropriate. For each MFCC vector the decoding provides a state and model specific GMM, $\Phi^v_{q_i,m_i}$, from which fundamental frequency can be predicted. Utilising the MAP prediction, as applied previously, yields a state and model specific fundamental frequency prediction, $\hat{f}_i$, from MFCC vector, $\mathbf{x}_i$, as,

$$\hat{f}_i = \sum_{k=1}^{K^V} h_{k,q_i,w_i}(\mathbf{x}_i)\left(\mu^{v,f}_{k,q_i,w_i} + \Sigma^{v,fx}_{k,q_i,w_i}\left(\Sigma^{v,xx}_{k,q_i,w_i}\right)^{-1}\left(\mathbf{x}_i - \mu^{v,x}_{k,q_i,w_i}\right)\right) \qquad (6.26)$$

where the posterior probability, $h_{k,q_i,w_i}(\mathbf{x}_i)$, of the MFCC vector in cluster k of state q_i and model w_i is given as,

$$h_{k,q_i,w_i}(\mathbf{x}_i) = \frac{\alpha^v_{k,q_i,w_i}\, p\left(\mathbf{x}_i\middle|\phi^{v,x}_{k,q_i,w_i}\right)}{\sum_{k=1}^{K^v}\alpha^v_{k,q_i,w_i}\, p\left(\mathbf{x}_i\middle|\phi^{v,x}_{k,q_i,w_i}\right)} \qquad (6.27)$$

where $p\left(\mathbf{x}_i\middle|\phi^{v,x}_{k,q_i,w_i}\right)$ is the marginal distribution of the MFCC vector in the kth cluster of GMM ϕ^v_{k,q_i,m_i}.

6.4.2 Voicing Prediction from MFCC Vectors

Fundamental frequency should be predicted only from MFCC vectors corresponding to voiced speech. To classify MFCC vectors as voiced or unvoiced a prior voicing probability is first computed from voicing information present in the states of the HMMs used for fundamental frequency prediction. The prior voicing probability can then be incorporated into a posterior voicing probability which classifies the MFCC vectors as being either voiced or unvoiced.

Prior Voicing Probabilities

The phoneme HMMs that provide localisation for fundamental frequency prediction contain useful prior voicing information. From the number of MFCC vectors allocated to the voiced and unvoiced vector pools in each state, s, and model, w, a prior voicing probability, $v_{s,w}$, can be computed,

$$v_{s,w} = \frac{n\left(\Omega^v_{s,w}\right)}{n\left(\Omega^v_{s,w}\right) + n\left(\Omega^u_{s,w}\right)} \qquad 1 \le s \le S_w, \quad 1 \le w \le W \qquad (6.28)$$

where the function $n(.)$ returns the number of vectors in the set. To examine the prior voicing probabilities for different phonemes, Table 6.1 shows the prior voicing probabilities for the 3 states of phonemes /ow/, /uw/, /s/ and /f/.

Table 6.1 Prior voicing probabilities of 3-state phoneme HMMs for models (a) /ow/, (b) /uw/, (c) /s/, (d) /f/

Phoneme	State 1	State 2	State 3
/ow/	0.84	0.98	0.96
/uw/	0.81	0.97	0.95
/s/	0.39	0.04	0.06
/f/	0.37	0.07	0.09

Voiced phonemes (/ow/ and /uw/) have very high prior voicing probabilities while the unvoiced phonemes (/s/ and /f/) have very low probabilities. The first state, and to a lesser extent the final state, are not as strongly voiced or unvoiced as the centre state. The first and last states are transitional states and minor errors in state alignment contribute to this effect.

Posterior Voicing Probabilities

A simple method of determining the voicing is to select the voicing class (voiced or unvoiced) in the state that the MFCC vector is allocated to that has the highest prior voicing probability. For states that are strongly voiced or strongly unvoiced this gives satisfactory results, but for states with less strong voicing or for MFCC vectors with inaccurate state alignment, this method is likely to introduce voicing classification errors. A better solution is to compute the posterior voicing probability for an MFCC vector allocated to a particular state. In Sect. 6.4.1, state-dependent voiced GMMs, $\Phi_{s,w}^{v}$, were trained from the sets of voiced augmented vectors, $\Omega_{s,w}^{v}$, within each state and model. For voicing classification a further set of GMMs, $\Phi_{s,w}^{u}$, (the superscript u indicates unvoiced) each comprising K^{u} clusters, can be trained from the sets of unvoiced vectors, $\Omega_{s,w}^{u}$, associated with each state of each model.

This produces a set of unvoiced means, $\mu_{k,s,w}^{u}$, covariances, $\Sigma_{k,s,w}^{u}$, and priors, $\alpha_{k,s,w}^{u}$, associated with each cluster, k, state, s, and model, w. The probability of an MFCC vector, $\mathbf{x}_i$, allocated to state, q_i, and model, m_i, belonging to the voiced GMM, Φ_{q_i,m_i}^{v}, can be computed as,

$$p(voiced|\mathbf{x}_i) = \frac{\sum_{k=1}^{K^v} \alpha_{k,q_i,m_i}^{v,x}\, p\left(\mathbf{x}_i \Big| \Phi_{q_i,m_i}^{v,x}\right)}{p(\mathbf{x}_i)} \tag{6.29}$$

Similarly the probability of the MFCC vector belonging to the unvoiced GMM, Φ_{q_i,m_i}^{u}, can be computed,

$$p(unvoiced|\mathbf{x}_i) = \frac{\sum_{k=1}^{K^u} \alpha_{k,q_i,m_i}^{u,x}\, p\left(\mathbf{x}_i \Big| \Phi_{q_i,m_i}^{u,x}\right)}{p(\mathbf{x}_i)} \tag{6.30}$$

Using these two probabilities the voicing of an MFCC vector allocated to state, q_i, of model, m_i, can be determined,

$$voicing_i = \begin{cases} voiced & p(voiced|\mathbf{x}_i) \geq p(unvoiced|\mathbf{x}_i) \\ unvoiced & p(voiced|\mathbf{x}_i) < p(unvoiced|\mathbf{x}_i) \end{cases}$$
(6.31)

For the purposes of the voicing prediction, the probability of the MFCC vector, $p(\mathbf{x}_i)$, can be ignored in Eqs. 6.29 and 6.30.

6.4.3 Speech Reconstruction from Predicted Fundamental Frequency and Voicing

The predicted voicing and fundamental frequency can be applied to the sinusoidal model based speech reconstruction described in Sect. 6.3.2. In this case the speech is reconstructed solely from the MFCC vector stream and uses no explicit fundamental frequency or voicing information. Figure 6.8 shows the spectrogram of the sentence *On May evenings the rooks were busy building nests in the birch tree*" reconstructed solely from 13-D MFCC vectors.

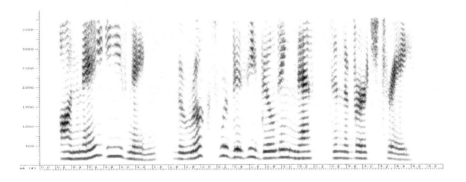

Fig. 6.8 Spectrogram of sentence "*On May evenings the rooks were busy building nests in the birch tree*" reconstructed solely from a stream of 13-D MFCC vectors

Comparing Fig. 6.8 with Fig. 6.7b (which shows speech reconstructed from estimated fundamental frequency and voicing) reveals very little difference between the two speech signals. This suggests that the MFCC vectors contain source information which has traditionally not been thought the case. The presence of fundamental frequency information in the MFCC features is also highlighted in Fig. 6.4. Examining the mel-filterbank shows that the first two harmonics shown in the power spectrum are preserved by the relatively close spacing of filterbank channel.

6.5 Conclusion

This chapter has examined the feature extraction and speech reconstruction components associated with distributed speech recognition and has placed emphasis on the standards specified by ETSI Aurora DSR group. The first ETSI front-end standard (FE) provided a basic MFCC feature which was superseded by the advanced front-end (AFE) that included noise reduction and channel equalisation. An extended front-end (XFE) was also standardised and provided voicing and fundamental frequency information to enable back-end speech reconstruction.

An examination of back-end reconstructed speech shows it to be a good approximation of the original speech in both its harmonic and formant structure. An alternative to the XFE is to predict the fundamental frequency and voicing from the MFCC vectors themselves. This approach has also led to a good approximation of the original speech although not quite as good as in the XAFE. However, the prediction method has the significant advantage that no source information needs to be transmitted to the back-end which saves 800 bps and allows speech reconstruction solely from the MFCC vector stream.

References

Boll, S.F. (1979) Suppression of acoustic noise in speech using spectral subtraction. *IEEE Trans. Acoust. Speech Signal Process.*, vol. 27, pp. 113–120.

Cooley, J.W. and Tukey, J.W. (1965) An algorithm for the machine calculation of complex Fourier series. *Math. Comput.*, vol. 19, pp. 297–301.

Davis, S.B. and Mermelstein P. (1980) Comparison of parametric representations for monosyllabic word recognition in continuously spoken sentences. *IEEE Trans. Acoust. Speech Signal Process.*, vol. 28, pp. 357–366.

de Cheveigne, A. and Kawahara, H. (2001) YIN, a fundamental frequency estimator for speech and music. *J. Acoust. Soc. Am.* Vol. 111, no. 4, pp. 1917–1930.

ETSI Standard ES 201 108 (2003a) Speech Processing, Transmission and Quality Aspects (STQ); Distributed speech recognition; Front-end feature extraction algorithm; Compression algorithms, version 1.1.3, September 23rd 2003.

ETSI Standard ES 202 211 (2003b) Speech Processing, Transmission and Quality Aspects (STQ); Distributed speech recognition; Extended front-end feature extraction algorithm; Compression algorithms; Back-end speech reconstruction algorithm, version 1.1.1, November 14th 2003.

ETSI Standard ES 202 050 (2007) Speech Processing, Transmission and Quality Aspects (STQ); Distributed speech recognition; Advanced front-end feature extraction algorithm; Compression algorithms, version 1.1.5, January 11th 2007.

Furui, S. (1986) Speaker-independent isolated word recognition using dynamic features of speech spectrum. IEEE Trans. ASSP, vol. 34, pp. 52–59.

George, E.B. and Smith, M.J.T. (1992) An analysis-by-synthesis approach to sinusoidal modelling applied to the analysis and synthesis of musical tones. *J. Audio Eng. Soc.*, vol. 40, pp. 467–516.

Hanson, B.A. and Applebaum, T.H. (1990) Robust speaker-independent word features using static, dynamic and acceleration features. In *Proc. ICASSP*, pp. 857–860.

Hermansky, H. (1990) Perceptual linear predictive (PLP) analysis of speech, *J. Acoust. Soc. Am.*, vol. 87, no. 4, pp. 1738–1752.

Hermansky, H. and Morgan, N. (1994) RASTA processing of speech. *IEEE Trans. Speech Audio Proc.,* vol. 2, no. 4, pp. 578–589.

HTK (2007) Hidden Markov model toolkit, http://htk.eng.cam.ac.uk/

McAulay, R.J. and Quatiery, T.F. (1986) Speech analysis/synthesis based on a sinusoidal representation. *IEEE Trans. ASSP* vol. 34, pp. 744–754.

Milner, B.P. and Shao, X. (2007) Prediction of Fundamental Frequency and Voicing from Mel-Frequency Cepstral Coefficients for Unconstrained Speech Reconstruction. *IEEE Trans. Audio Speech Lang. Process.,* vol. 15, pp. 24–33.

Oppenheim, A.V. and Schafer, R.W. (1989) *Discrete-time signal processing,* Prentice-Hall, New Jersey, USA.

Rabiner, L.R. and Schaeffer, L.W. (1978) Digital processing of speech signals, Prentice Hall, New Jersey, USA.

Rosenberg, A.E., Lee, C.-H. and Soong, F.K. (1994) Cepstral channel normalization techniques for HMM-based speaker verification. In *Proc. ICSLP,* pp. 1835–1838.

Shao, X. and Milner, B.P. (2004) Pitch prediction from MFCC vectors for speech reconstruction. In *Proc. ICASSP.*

Vaseghi, S.V. (2006) Advanced digital signal processing and noise reduction, John-Wiley.

Wu, K. and Chen, P. (2001) Efficient speech enhancement using spectral subtraction for car hands-free application. *Int. Conf. Consum. Electron.,* vol. 2, pp. 220–221.

7

Quantization of Speech Features: Source Coding

Stephen So and Kuldip K. Paliwal

Abstract. In this chapter, we describe various schemes for quantizing speech features to be used in distributed speech recognition (DSR) systems. We analyze the statistical properties of Mel frequency-warped cepstral coefficients (MFCCs) that are most relevant to quantization, namely the correlation and probability density function shape, in order to determine the type of quantization scheme that would be most suitable for quantizing them efficiently. We also determine empirically the relationship between mean squared error and recognition accuracy in order to verify that quantization schemes, which minimize mean squared error, are also guaranteed to improve the recognition performance. Furthermore, we highlight the importance of noise robustness in DSR and describe the use of a perceptually weighted distance measure to enhance spectral peaks in vector quantization. Finally, we present some experimental results on the quantization schemes in a DSR framework and compare their relative recognition performances.

7.1 Introduction

With the increase in popularity of wireless devices such as personal digital assistants (PDAs) and cellular phones, there has been a growing interest in incorporating automatic speech recognition (ASR) technology into mobile communication systems. Speech recognition can facilitate consumers in performing common tasks, which have traditionally been accomplished via buttons and/or pointing devices.

Distributed speech recognition (DSR) is a mode of client-server-based ASR, where speech features are extracted on the client device and then transmitted to the server, which performs the recognition task, as shown in Fig. 7.1. Let us calculate the bitrate that is required to transmit uncoded feature vectors. If feature vectors of 13 Mel frequency-warped cepstral coefficients (MFCCs) are extracted at a frame rate of 100 Hz and that each MFCC is represented as a 32 bit floating point value, then the required bitrate is 41.6 kbps. As we shall see later on, current state-of-the-art quantization schemes used in DSR can operate at *bitrates* as low as 300 bps.

In this chapter, we are interested in the lossy coding of feature vectors for DSR applications. The ultimate aim is to quantize feature vectors using the least amount of bits, while maintaining a recognition performance that is as close as possible to that of ASR. Note that when we use the term *ASR performance*, we are referring to the

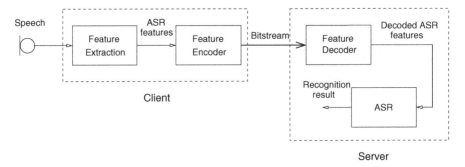

Fig. 7.1 Block diagram of a typical distributed speech recognition system (From So and Paliwal 2006)

recognition performance achieved when no lossy coding has been applied to the feature vectors, as opposed to *DSR performance*, where feature vectors have been coded in a lossy fashion. It is reasonable to assume that, using the same features, DSR performance will generally be less than and upper bounded by the ASR performance, hence the latter serves as a useful baseline for evaluating quantization schemes.

This chapter is divided into four sections. In the first section, we will review some basic concepts of source coding and quantization as well as outline some quantization schemes that will be evaluated later in the chapter. In the second section, we examine the statistical properties of the MFCC feature vectors as well as determine the relationship between mean squared error and recognition accuracy. In the third section, we present a brief review of the literature on the topic of quantizing feature vectors. Following this, we will present some results of recent quantization schemes that we have investigated in our laboratory (So and Paliwal 2005, 2006). We then conclude the chapter in the final section.

7.2 Quantization Schemes

7.2.1 Brief Introduction to Quantization Theory

Source coding schemes can be broadly classified into two categories: lossless and lossy coding. While lossless coding incurs no loss of information (that is, the decoded output data is exactly the same as the input data), the amount of compression is limited by the Shannon entropy of the data (Gersho and Gray 1992). Examples of lossless coding schemes (often referred to as *entropy coders*) include Huffman coding, arithmetic coding and runlength encoding.

It is common for an entropy coder to be cascaded on the output of a lossy coder to further reduce the bitrate (Gray and Neuhoff 1998). An example of this is in the JPEG image coder, where the output coefficients of the lossy scalar quantization stage are coded using a runlength encoder and a Huffman coder (Wallace 1991).

While it is possible to apply entropy coding on the output of the quantization schemes discussed in this chapter to reduce the bitrate further, various complications arise, such as the resulting bitrate being variable over time. Therefore, buffering is often required to handle the variable bitrates, which adds to the complexity of the overall DSR system.

On the other hand, lossy coding schemes have no constraints on the amount of compression that can be achieved, hence they are often more useful in scenarios where channel capacity is low and limited. The bitrate of lossy coding schemes can be made fixed, thus removing the requirement for buffering. The challenge with lossy coding schemes is minimizing the distortion given a fixed bitrate, or given an allowed and fixed distortion, minimizing the bitrate required—this is often referred to as the *rate-distortion trade-off*.

Quantization is a fundamental process for information reduction in lossy coding schemes and is generally the source of information loss. It is defined as the mapping of individual (scalar) or a vector of input samples to a codebook of a finite number of *codewords*. Each codeword has a unique binary word or index associated with it so each input sample is substituted with this binary word before transmission. The mapping is done in such a way that the distortion incurred by substituting the input sample by its corresponding codeword is minimized. The input samples may be quantized individually (referred to as *scalar quantization*), or as vectors (referred to as *vector quantization*). Figure 7.2 shows where the quantization scheme 'fits' in the DSR feature encoder.

The rate-distortion (RD) efficiency of any quantizer is influenced by the properties of the signal source, such as statistical dependencies (otherwise known as memory) and the probability density function (PDF) (Makhoul et al. 1985). Furthermore, it has been shown that vector quantizers always have a better RD efficiency than scalar quantizers, and therefore are optimal quantizers (Lookabaugh and Gray 1989). The properties of the speech features used in DSR will be discussed in the following subsections. However, before we move further, we will present popular distortion measures that have been used in speech processing as well as describe the quantization schemes that will be evaluated later in the chapter.

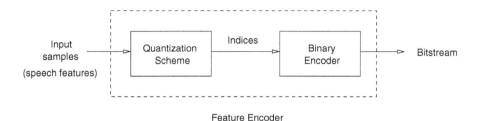

Feature Encoder

Fig. 7.2 Block diagram of the 'Feature encoder' in Fig. 7.1, showing the quantization scheme and binary encoder

7.2.2 Distortion Measures for Quantization in Speech Processing

It is important to define the distortion measure to be used in quantizers as different applications may require the minimization of an error calculation that incorporates some signal-based or perceptual properties in order to improve the overall fidelity. The simplest distortion measure that is commonly used in the coding literature is *mean squared error* (MSE), d_{MSE}, which is defined below:

$$d_{MSE}(\mathbf{x}, \hat{\mathbf{x}}) = E[(\mathbf{x} - \hat{\mathbf{x}})^T (\mathbf{x} - \hat{\mathbf{x}})] \tag{7.1}$$

In this equation, $E[\bullet]$ is the expectation operator, x and $\hat{x}$ are the input vector and quantized vector, respectively, and $\bullet^T$ is the transpose operator. The error contribution of each vector component is weighted the same.

Weighted distortion measures are often used to perform *quantization noise shaping*, which can improve the overall fidelity by exploiting signal-based properties. For example, in speech coding applications, line spectral frequency (LSF) vectors can be quantized using a weighted mean square error, where the error contributions of each LSF are non-uniformly weighted based on the relative spectral power at that particular frequency (Paliwal and Atal 1993). For components that have a higher weighting, the quantization error will be less. This weighted mean squared error (WMSE) can be expressed as:

$$d_{WMSE}(\mathbf{x}, \hat{\mathbf{x}}) = E[(\mathbf{x} - \hat{\mathbf{x}})^T \mathbf{W}(\mathbf{x} - \hat{\mathbf{x}})] \tag{7.2}$$

In this equation, $\mathbf{W}$ is a square diagonal weighting matrix whose diagonal elements consist of the relative weightings of each vector component.

Another common distortion measure that is used for evaluation in speech coding is the *logarithmic spectral distortion* (this is often simply referred to as *spectral distortion*). It is defined as the root mean squared error between the log power spectral density estimates of the original and quantized frame of speech:

$$d_{SD} = \sqrt{\frac{1}{F_s} \int_0^{F_s} [10\log_{10} P(f) - 10\log_{10} \hat{P}(f)]^2 \, df} \tag{7.3}$$

In Eq. 7.3, F_s is the sampling frequency, $P(f)$ and $\hat{P}(f)$ are the power spectral density estimates of the input and quantized speech frame, respectively. It can be shown that the MSE distortion measure in the cepstral domain is equivalent to the spectral distortion (Rabiner and Juang 1993).

Other distortion measures that have been used in speech processing include the Itakura-Saito distortion, Itakura distortion, COSH distance, etc. (Rabiner and Juang 1993). For distributed speech recognition, the quantization distortion measure should be somewhat correlated to the desired performance metric—recognition accuracy. We will discuss this further in Sect. 7.3.4. Because of their relatively low computational complexity, we will mostly focus on MSE-based distortion measures as these need to be computed multiple times in quantization schemes such as VQ.

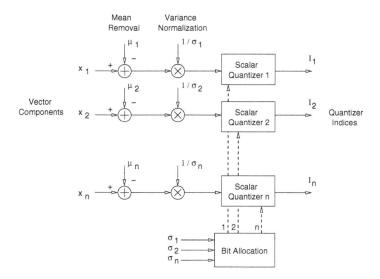

Fig. 7.3 Block diagram of scalar quantization of vectors, with mean removal, variance normalization and bit allocation

7.2.3 Scalar Quantization

The simplest quantizer is the *scalar quantizer* (SQ), where input samples are mapped individually to scalar codewords, which are also referred to as code-points or reproduction values (Gersho and Gray 1992). The number of reproduction values or quantization levels, n, is given by $n = 2^b$, where b is the number of bits.

For input samples that have a non-uniform probability density function, such as Gaussian or Laplacian, it has been shown that non-uniform scalar quantizers incur less distortion than uniform scalar quantizers, where quantization levels are uniformly spaced (Max 1960). The quantization levels for Gaussian and other arbitrary distributions (with zero-mean and unity variance) have been reported in the coding literature (Max 1960; Paez and Glisson 1972; Lloyd 1982). The input samples should have zero-mean and normalized variance before quantization, as shown in Fig. 7.3.

When quantizing a vector of input samples using scalar quantizers, we need to allocate the bit budget among the individual SQs. For example, if the vector dimensionality is n and the bitrate is fixed at b bits/sample, then a total of nb bits need to be allocated to the n SQs. The objective is to determine the best bit allocation such that the quantization distortion is minimized. We discuss two methods for bit allocation in scalar quantization: high resolution-based optimization (HRO) and the greedy-based heuristic algorithm.

In HRO bit allocation, which was first presented in relation to block quantization (Huang and Schultheiss 1963), the average distortion incurred by the overall scalar quantization scheme is expressed in terms of the high resolution approximation of the non-uniform scalar quantizer:

$$d_{avg} = \frac{1}{n}\sum_{i=1}^{n} K\sigma_i^2 2^{-2b_i}$$ (7.4)

In this equation, n is the vector dimensionality, K is a constant which varies for different PDFs (for Gaussian PDFs, $K = \frac{\pi\sqrt{3}}{2}$), σ_i^2 is the variance of the ith vector component, and b_i is the number of bits allocated to the ith scalar quantizer. This expression is to be minimized using the fixed bitrate constraint,

$$b_{tot} = \sum_{i=1}^{n} b_i$$ (7.5)

We are then left with the following bit allocation formula (for the full derivation, see Huang and Schultheiss 1963):

$$b_i = \frac{b_{tot}}{n} + \frac{1}{2}\log_2 \frac{\sigma_i^2}{\left[\prod_{i=1}^{n}\sigma_i^2\right]^{\frac{1}{n}}}$$ (7.6)

Let us consider an example of scalar quantizing vectors of dimension 4 using a total of 20 bits, given the following variances: $\sigma_i^2 = \{2, 30, 10, 52\}$. Using Eq. 7.6, we calculate a bit allocation of $b_i = \{3.634, 5.587, 4.7948, 5.984\}$ bits. We note that, firstly, more bits have been allocated to vector components with higher variances; and secondly, the formula gives fractional (and even negative in some cases) bit allocations. One may truncate these fractional bit allocations though this generally leads to a total bitrate that is less than the target. A method is presented in Paliwal and So (2005) for handling fractional bit allocations so that more of the bit budget is utilized. A further constraint that enforces the b_i to be always positive may also be applied to the optimization process (Segall 1976).

The greedy-based heuristic algorithm for allocating bits is simpler than the HRO algorithm and is more readily applicable to vectors with non-standard PDFs, where deriving closed-form expressions may be difficult or impossible. Allocation is performed one bit at a time for each vector component, with the one resulting in the largest drop in quantization distortion to be selected to receive the bit. The process continues until all bits have been allocated, where the resulting solution may only be locally optimal. Greedy-based heuristic bit allocation has been investigated in DSR in the literature (Digalakis et al. 1999).

7.2.4 Block Quantization

In *block quantization*, also known as *transform coding*, an orthogonal linear transformation P, whose columns consist of the basis vectors, is applied to a zero-mean input vector, x, before scalar quantization (Huang et al. 1963):

$$y = P^T x \qquad (7.7)$$

where y is the transformed vector containing the transform coefficients, $\{y_i\}_{i=1}^n$. The inverse linear transformation is expressed as:

$$x = Py \qquad (7.8)$$

The covariance matrix of the transformed vectors is given by:

$$\begin{aligned}
\Sigma_y &= E[yy^T] \\
&= E[P^T x (P^T x)^T] \\
&= P^T E[xx^T] P \\
&= P^T \Sigma_x P
\end{aligned}$$

When scalar quantizing input samples, the statistical dependencies between these samples are not exploited and this leads to wasted bits and thus inefficient quantization. In block quantization, the linear transformation serves to decorrelate the samples before scalar quantization, which will improve the coding efficiency. The correlation is 'added' back in the decoding stage via the inverse transformation of Eq. 7.8.

The decorrelating transformation also tends to pack the energy or variance into the first few coefficients. When using the HRO bit allocation formula of Eq. 7.6, the skewed variance distribution of the transformed coefficients will cause more bits to be allocated to the scalar quantizers of the first few coefficients. Typical transformations used in coding include the Karhunen-Loève transform (KLT) and the discrete cosine transform (DCT).

7.2.5 Vector Quantization

The basic definition of a vector quantizer Q of dimension n and size K is a mapping of a vector from n dimensional Euclidean space, $\Re^n$, to a finite set, C, containing K reproduction *codevectors*:

$$Q : \Re^n \to C \qquad (7.9)$$

where $C = \{y_i; i \in I\}$ and $y_i \in \mathfrak{R}^n$. Associated with each reproduction codevector is a partition of $\mathfrak{R}^n$, called a region or cell, $S = \{S_i; i \in I\}$.

The most popular form of vector quantizer is the *Voronoi* or *nearest neighbour* vector quantizer (Gersho et al. 1992), where for each input source vector x, a search is done throughout the entire codebook to find the nearest codevector y_i, which has the minimum distance:

$$y_i = Q[\mathbf{x}] \quad \text{if } d(\mathbf{x}, \mathbf{y}_i) < d(\mathbf{x}, \mathbf{y}_j) \quad \text{for all } i \neq j \tag{7.10}$$

where $d(x, y)$ is the distortion measure between the vectors, x and y. Generally, the most common distortion measure used in vector quantizers is the MSE.

The VQ codebook is designed using a large number of training vectors, which are representative of the set of vectors that will be quantized by the VQ. The iterative Linde-Buzo-Gray (LBG) algorithm (Linde et al. 1980) is applied to the training vectors and the resulting K centroids or codevectors constitute the VQ codebook. The bitrate of the vector quantizer is $\log_2 K$ bits/vector.

Though the *unconstrained VQ* (that is, the VQ codebook has no structural constraints) is theoretically the optimal quantizer that one can design, its computational complexity and memory requirements may become prohibitive at high bitrates. Furthermore, designing a high bitrate VQ codebook requires a large amount of training data. Therefore, the application of unconstrained VQ is often constrained to low bitrates, while structurally constrained forms, such as multistage, split, and tree-structured VQ are used when higher bitrates are required. Constrained VQs sacrifice rate-distortion performance for lower computational and memory requirements.

7.2.6 GMM-Based Block Quantization

The GMM-based block quantizer (Subramaniam and Rao 2003) is an improved version of the Gaussian block quantizer (Huang et al. 1963). Rather than assume the PDF of the input vectors to be Gaussian, Gaussian mixture models (GMMs) are used to approximate the PDF and each mixture component is quantized using a Gaussian block quantizer. These modifications result in better RD performance as the GMM-based block quantizer is designed to match the PDF more closely, assuming that there is minimal overlap between the mixture components.

Compared with vector quantizers, the GMM-based block quantizer has the advantages of: fixed computational and memory requirements that are independent of the bitrate; and *bitrate scalability*, where any bitrate can be used without the need to redesign the codebook (Subramaniam and Rao 2003). Bitrate scalability is a desirable feature in DSR applications, since one may need to adjust the bitrate adaptively, depending on the network conditions (So and Paliwal 2006).

This quantization scheme can be broken down into three stages: PDF estimation, bit allocation and minimum distortion block quantization. Each stage will be described in the following subsections.

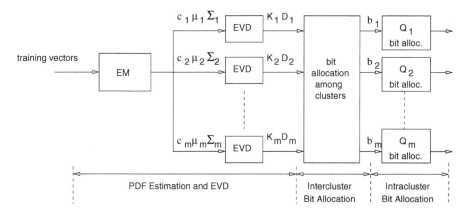

Fig 7.4 PDF estimation and bit allocation from training data (From So and Paliwal 2006)

PDF Estimation using Gaussian Mixture Models

The PDF model and Karhunen-Loève transform (KLT) orthogonal matrices are the only static and bitrate-independent parameters of the GMM-based block quantizer. These only need to be calculated once during the training stage and stored at the client encoder and server decoder. The bit allocations for different bitrates can be calculated 'on-the-fly' using the common PDF model stored on both client and server. The PDF estimation procedure is shown in Fig. 7.4.

The PDF model, G, as a mixture of multivariate Gaussians, $N(x; \mu, \Sigma)$, can be expressed as:

$$G(x \mid M) = \sum_{i=1}^{m} c_i N(x; \mu_i, \Sigma_i) \tag{7.11}$$

$$M = [m, c_1, ..., c_m, \mu_1, ..., \mu_m, \Sigma_1, ..., \Sigma_m] \tag{7.12}$$

$$N(x; \mu, \Sigma) = \frac{1}{(2\pi)^{\frac{n}{2}} |\Sigma|^{\frac{1}{2}}} e^{-\frac{1}{2}(x-\mu)^T \Sigma^{-1}(x-\mu)} \tag{7.13}$$

where x is a source vector, m is the number of mixture components, and n is the dimensionality of the vector space. c_i, μ_i Σ_i are the weight, mean, and covariance matrix of the ith mixture component, respectively.

The parametric model, M, is initialized by applying the LBG algorithm (Linde et al. 1980) on the training vectors where m mixture components are produced, each represented by a mean or centroid, μ, a covariance matrix, Σ, and a mixture component weight, c. These form the initial parameters for the GMM estimation procedure. Using the expectation-maximization (EM) algorithm (Dempster et al. 1977), the maximum-likelihood estimate of the parametric model is computed iteratively and a final set of means, covariance matrices, and weights are produced.

An eigenvalue decomposition (EVD) is calculated for each of the m covariance matrices. The eigenvectors form the rows of the orthogonal transformation matrix, K, of the KLT.

Bit Allocation

Assuming there are a total of b_{tot} bits available for quantizing each vector, these need to be allocated to each of the block quantizers of each mixture component in an optimal fashion. Using Lagrangian minimization (Subramaniam et al. 2003), the following formula is derived:

$$2^{b_i} = 2^{b_{tot}} \frac{(c_i \Lambda_i)^{\frac{n}{n+2}}}{\sum_{i=1}^{m} (c_i \Lambda_i)^{\frac{n}{n+2}}} \quad \text{for } i = 1, 2, ..., m \quad (7.14)$$

$$\Lambda_i = \left[\prod_{j=1}^{n} \lambda_{i,j} \right]^{\frac{1}{n}} \quad (7.15)$$

In Eqs. 7.14 and 7.15, $\lambda_{i,j}$ is the jth eigenvalue of mixture component i and b_i is the number of bits allocated to the block quantizer of mixture component i.

Once bits have been allocated to the block quantizer of each mixture component, these need to be further allocated to the scalar quantizers within the block quantizer. The bit allocation was presented in Sect. 7.2.3 and the formula for allocating bits is given by Eq. 7.6.

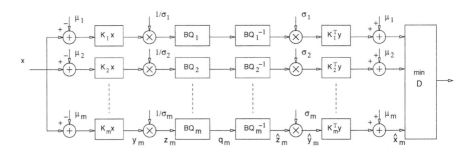

Fig. 7.5 Minimum distortion block quantization (*BQ*, block quantizer) (From So and Paliwal 2006)

Minimum Distortion Block Quantization

Figure 7.5 shows the minimum distortion block quantization stage, whose operation is described in more detail in Subramaniam et al. (2003). At first glance, it can be seen to consist of m independent block quantizers, BQ_i, each with their own orthogonal matrix, K_i, and bit allocations, $\{b_{i,j}\}_{j=1}^{n}$. A vector, x, is quantized m times and the kth block quantizer is chosen such that it incurs the least distortion.

$$k = \arg\min_{i} d(x, \hat{x}_i) \tag{7.16}$$

7.3 Quantization of ASR Feature Vectors

7.3.1 Introduction and Literature Review

So far, we have only discussed quantization and the various schemes in general with no reference made to quantizing ASR feature vectors. In this section, we discuss the task of quantizing ASR feature vectors as well as examine some statistical properties that may affect the quantization and recognition performance. We will also examine the performance of the DSR system in the presence of background noise. Unless otherwise specified, we will be mostly focusing on *Mel frequency-warped cepstral coefficients* (MFCCs) (Davis and Mermelstein 1980) as the ASR feature set.

Various schemes for quantizing the ASR features have been proposed in the literature. Digalakis et al. (1999) evaluated the use of uniform and non-uniform scalar quantizers as well as product code vector quantizers for coding MFCCs at rates of between 1.2 and 10.4 kbps. They used the greedy-based bit allocation algorithm for the scalar quantizers, where the component, which resulted in the largest improvement in recognition performance, was chosen to receive the allocated bit. They concluded that split vector quantizers achieved word error rates (WER) similar to that of scalar quantizers while requiring fewer bits. A bitrate of 2 kbps was the required bitrate for split vector quantization to achieve ASR recognition performance. Also scalar quantizers with non-uniform bit allocation performed better than those with uniform bit allocation.

In Ramaswamy and Gopalakrishnan (1998), the authors investigated the application of tree-searched multistage vector quantizers (MSVQ) with first-order linear prediction operating at a bitrate of 4 kbps. The current MFCC feature vector was subtracted from the previous quantized frame to give a residual vector. The first 12 coefficients of the residual vector were then quantized using a two-stage MSVQ, while the last coefficient, c_0, was scalar quantized. Their system achieved near identical recognition performance as the ASR recognition performance, with only minor degradation.

Transform coding, based on the DCT, was investigated in Kiss and Kapanen (1999) at a bitrate of 4.2 kbps. In this scheme, feature vectors of dimension 14 (13 MFCCs plus the energy coefficients, c_0 and *log E*) were processed. For each cepstral coefficient, eight temporally consecutive coefficients were grouped together and processed by the DCT, which exploited temporal correlation. The energy coefficient was encoded separately.

In Zhu and Alwan (2001), 12 successive MFCC frames were stacked together to form a block of 12×12 and a two-dimensional DCT was applied. Zonal sampling was performed, where a fraction of the lowest energy components was set to zero and the remaining coefficients were scalar quantized and entropy coded. The advantage of this scheme compared to that of Kiss and Kapanen (1999) is that both within-frame and across-frame correlation is exploited by the 2D-DCT. Noise-robust feature sets, such as peak isolated MFCCs (MFCCP) (Strope and Alwan 1997) and variable frame-rate peak isolated MFCCs (VFR_MFCCP) (Zhu and Alwan 2000) were also tested. Their results showed that, firstly, the DSR recognition performance always performed slightly worse than the ASR recognition performance at all signal-to-noise (SNR) levels. Secondly, the quantized noise-robust features at 624 bps resulted in recognition accuracies that even surpassed the ASR performance at low SNRs.

The ETSI DSR standard (2003) uses split vector quantizers to compress the MFCC vectors at 4.4 kbps. Feature vectors of dimension 14 (13 MFCCs and log E) are split into pairs of subvectors, with the energy parameters, c_0 and log E belonging to the same pair. A weighted MSE distortion measure is used for the energy parameter subvector.

In Srinivasamurthy et al. (2006), correlation across consecutive MFCC features was exploited by a differential pulse coded modulation (DPCM) scheme followed by entropy coding. Their scheme is a scalable one, where the bitstream is embedded. That is, a coarsely quantized base layer is transmitted. If higher recognition performance is required, the client can transmit further enhancement layers, which are combined with the base layer by the server to obtain higher quality features.

7.3.2 Statistical Properties of MFCCs

The statistical properties of the MFCC vectors have a direct influence on the rate-distortion performance of any quantization scheme. According to Makhoul et al. (1985), these properties are:

linear dependency (i.e. correlation);
non-linear dependency;
probability density function shape; and
dimensionality (i.e. quantizing vectors is more efficient than scalars).

We will investigate properties 1 and 3 of MFCC vectors in the following subsection. In particular, the correlation across successive vectors will be examined as this property is exploited by interframe schemes such as multiframe/matrix and prediction-based quantizers.

Correlation within MFCC Vectors (Intraframe Dependencies)

We examine the amount of correlation between cepstral coefficients within a feature vector by computing the covariance matrix of MFCCs from the training speech set of the Aurora-2 database (Hirsch and Pearce 2000). The MFCCs consist of 13 cepstral coefficients, $\{c_i\}_{i=0}^{12}$. The log energy coefficient $log\ E$, which is often concatenated with the MFCC feature set in ASR, has not been included. Rather than presenting a 13×13 matrix of coefficients, we have plotted the absolute value of the covariance coefficients in Fig. 7.6. Because of the large difference in magnitude of the variance of c_0 compared with those of the other cepstral coefficients, we have applied a square root operation to the covariance coefficients to compress the dynamic range. Therefore, the coefficients on the diagonal represent the standard deviation of each cepstral coefficient rather than the variance.

We can see that a large percentage of the energy is contained in the zeroth cepstral coefficient, c_0. Recall that the final stage of MFCC computation comprises a discrete cosine transform (DCT), which tends to compact most of the energy into the zeroth cepstral coefficient or DC component. In addition, most of the off-diagonal covariance coefficients have low magnitude, which indicates that the cepstral coefficients are weakly correlated with each other—apart from c_0, where the cross-variance with the other cepstral coefficients appears to be higher. This suggests that the other cepstral coefficients $\{c_i\}_{i=1}^{12}$ contain some information of the zeroth cepstral coefficient. Hence, in most speech recognition systems, c_0 is not included in the feature set.

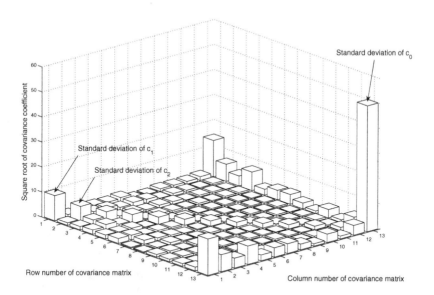

Fig. 7.6 Graphical representation showing the absolute value of the covariance coefficients of MFCCs within a single vector with compressed dynamic range (log energy is not included)

Because the efficiency of scalar quantization is generally optimal when the vector components are not correlated (which is the basis of block quantization), the covariance statistics of MFCCs (shown in Fig. 7.6) suggest that directly scalar quantizing the MFCCs may not be optimal. In which case, a further transform (such as the KLT) may be required to remove the remaining correlation and henceforth improve the rate-distortion performance.

This improvement will be become apparent when comparing the results between the scalar quantizer and the block quantizer.

Correlation across Successive MFCC Vectors (Interframe Dependencies)

In order to examine the correlation across successive MFCC vectors, we concatenate these vectors to form higher dimensional vectors and compute the covariance matrix of this new vector set. Any linear dependencies between MFCCs in successive vectors will be shown by large off-diagonal coefficients in the corresponding rows and columns of the covariance matrix. Figure 7.7 is similar to Fig. 7.6, where the covariance matrix is graphically represented in a three dimensional representation. We also present the graphical covariance matrix representation for two, three, four, and five

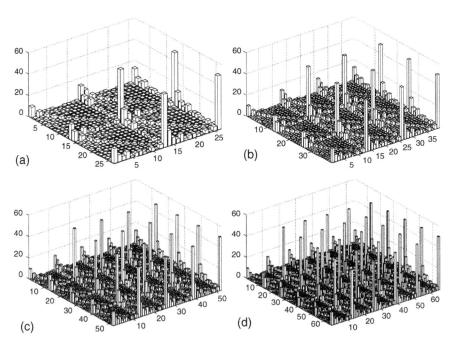

Fig. 7.7 Graphical representation showing the coefficients of the covariance matrix of MFCCs within a multiple successive vectors with compressed dynamic range: **a** two vectors, **b** three vectors, **c** four vectors, and **d** five vectors

concatenated MFCC vectors in order to show the amount of correlation between MFCCs across these successive vectors. As before, the log energy coefficient has not been included and an absolute value followed by a square root operation has been applied to all covariance coefficients in order to compress the dynamic range.

Looking at Fig. 7.7a, where two vectors have been concatenated together, we notice a large number of off-diagonal covariance coefficients that have a large magnitude, which indicates a high degree of correlation between the MFCCs across successive frames.

This is to be expected, as the speech frames used to compute the MFCCs are highly overlapped. When we look at the covariance coefficients for three, four, and five vectors, in Fig. 7.7b–d, we notice greater numbers of off-diagonal elements with large magnitude. Therefore, it is expected that quantization schemes, which exploit memory across multiple successive, will be more efficient in the rate-distortion sense, than memoryless schemes.

We should point out that this method of vector concatenation does not capture all of the dependencies. For example, if we represent four successive MFCC vectors as x_1, x_2, x_3, x_4, then concatenating them will produce: $[x_1, x_2], [x_3, x_4]$. The covariance matrix will capture the dependencies between MFCCs in both x_1 and x_2 and between MFCCs in both x_3 and x_4, but not the dependences between x_2 and x_3.

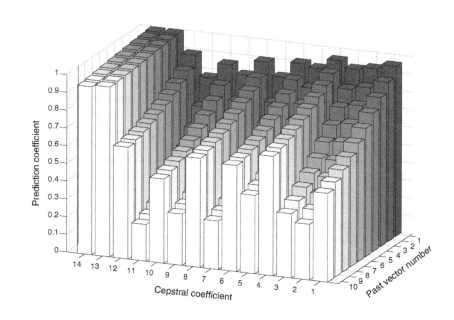

Fig. 7.8 Graphical representation showing the prediction coefficients from a single-step linear prediction of MFCC vectors (c_0 and $log\ E$ are represented as cepstral coefficient 13 and 14, respectively)

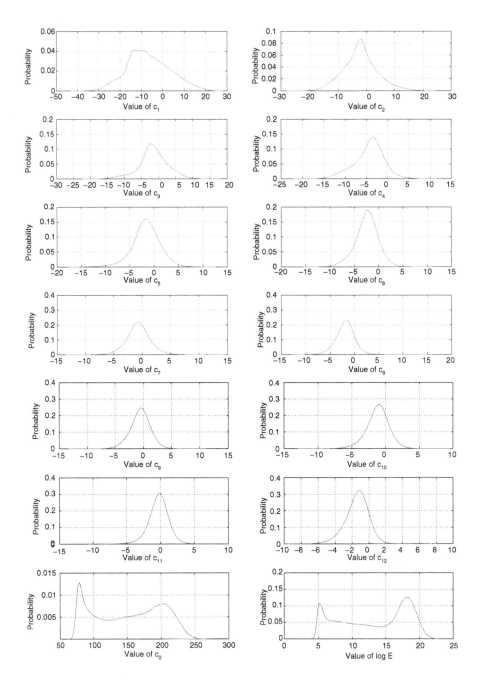

Fig. 7.9 Probability density function estimates of MFCCs

As a further method of capturing the correlation that exists across successive frames, we apply a single-step backward prediction analysis using the covariance method over the MFCC feature vector set to compute prediction coefficients. Both the energy coefficients, c_0 and $log\ E$ have been included. Up to 10 past vectors were used in the analysis. The closer the prediction coefficients are to unity, the higher the degree of correlation between any MFCC vector and a past vector. Figure 7.8 shows a graphical representation of the prediction coefficients for each cepstral coefficient. We can see that consecutive vectors (past vector number equal to one) are highly correlated as is shown by the prediction coefficients being closest to unity. The coefficients decrease in value as vectors further away in the past are used to predict the current vector, with some MFCCs decreasing faster than others. It is interesting to point out that the energy coefficients across 10 frames are highly correlated. This observation suggests that the energy coefficients could be efficiently quantized using prediction-based schemes.

Probability Density Functions of MFCCs

The probability density function (PDF) of MFCCs are particularly important when we consider scalar quantization-based schemes. Figure 7.9 shows the probability density function (PDF) estimates of the MFCCs in addition to the $log\ E$ coefficient.

The PDFs of the MFCCs, apart from c_0 and $log\ E$, resemble unimodal Gaussians, which suggests that they are amenable to non-uniform scalar quantization optimized for Gaussian sources as well as block quantization. This is to be expected as the MFCCs were formed from linear combinations of vector components during the DCT operation. According to the central limit theorem, as the dimension of the vectors increases, the distributions of the transform coefficients approach a Gaussian (Chen and Smith 1977). In contrast, the c_0 and $log\ E$ coefficients possess a bimodal distribution, which suggests that custom-designed scalar quantizers would be needed here.

We conclude this section on the statistical properties of MFCCs by noting the differences in the statistics of the energy coefficients (c_0 and $log\ E$) when compared with those of $c_1...c_{12}$, in terms of the correlation and PDF. It is for this reason that the energy coefficients are often quantized independently from the rest of the cepstral coefficients. Because of this, the issue of bit allocation arises. That is, how much of the bit budget should be allocated for quantizing energy coefficients in order to maximize the recognition performance? The majority of the quantization schemes reported in the literature have arbitrarily allocated bits to the energy coefficients, rather than utilising a formula obtained from constrained minimization. The problem is that it is not entirely clear how much impact quantization errors in the energy coefficients have on the recognition performance, compared with errors in the other cepstral coefficients. In order to isolate the uncertainty associated with energy coefficient quantization as well as to present a simple and consistent bit allocation framework, we have performed all DSR experiments where the energy coefficients are not included as part of the MFCC feature set. For the Aurora-2 recognition task, the ASR performance dropped from 99% to 98% as a result of not including the energy coefficients.

7.3.3 Use of Cepstral Liftering for MFCC Variance Normalization

The variances of each MFCC are shown in Fig. 7.10. The variances of c_0 and $log\ E$ (not shown in Fig. 7.10), are 2,530 and 260, respectively. The non-uniform variance distribution of the MFCCs is a result of the energy-packing characteristics of the discrete cosine transform. It is also well known that the lower order cepstral coefficients are particularly sensitive to undesirable variations caused by factors such as transmission, speaker characteristics, vocal efforts, etc. (Juang et al. 1987).

According to the HRO bit allocation formula for scalar quantization in Eq. 7.6, bits are allocated to vector components on the basis of variance, in order to minimize the mean squared error. This can be seen in the first row of Table 7.1, which shows the number of bits that are allocated to each MFCC, using HRO bit allocation. Because c_1 has the highest variance, it has been allocated the most number of bits.

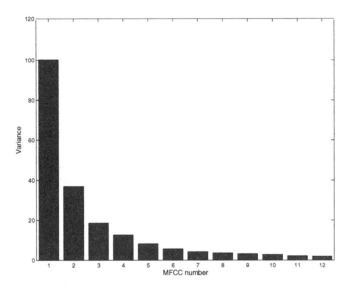

Fig. 7.10 Variances of MFCCs (c_0 and $log\ E$ are not included)

Table 7.1 Number of bits allocated to each MFCC with and without the application of cepstral liftering (computed using Eq. 7.6)

	Total bits	c_1	c_2	c_3	c_4	c_4	c_6	c_7	c_8	c_9	c_{10}	c_{11}	c_{12}
Without liftering	15	3.1	2.4	1.9	1.6	1.3	1.1	0.9	0.7	0.7	0.6	0.4	0.3
With liftering	15	2.4	2.3	2.2	2.2	2.0	1.8	1.5	1.3	1.0	0.5	-0.4	-2

From a quantization point of view, where the mean squared error between the original and reconstructed MFCC vectors is minimized, finely quantizing the first few MFCCs makes sense since they have higher variance. As will be shown in the next section, the relationship between MSE and recognition accuracy is monotonic and non-linear. However, if the operating bitrate is low, there may be a shortage of bits to allocate to the important middle-order MFCCs.

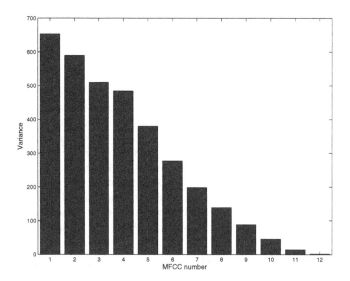

Fig. 7.11 Variances of MFCCs after cepstral liftering (c_0 and $log\ E$ are not included)

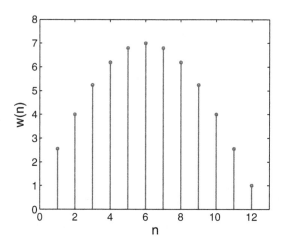

Fig. 7.12 Lifter window function of Eq. 7.16

If the shortage of bits that is due to a low operating bitrate, is found to cause a performance degradation, then one may normalize the variances of the MFCCs so that the bit allocation is not too highly skewed. This normalization can be done via the use of *liftering*, which performs 'filtering' in the cepstral domain. Cepstral liftering was a technique that was investigated in the literature to improve the recognition performance (Paliwal 1982), where cepstral coefficients were linearly weighted. Another method of cepstral liftering (Juang et al. 1987) uses the following sinusoidal lifter window function:

$$w(n) = 1 + \frac{L}{2}\sin\left[\frac{\pi n}{L}\right] \qquad (7.16)$$

where L is the dimensionality of the MFCCs. This window function is plotted in Fig. 7.12, where we can see an emphasis on the middle order cepstral coefficients. The effect of the liftering operation on the MFCC variances and the bit allocation are shown in Fig. 7.11 and Table 7.1, respectively, where bits are allocated more uniformly to the middle order MFCCs. In our experiments, we have used cepstral liftering for the purpose of variance normalization. Further work is needed to determine the benefits that it may provide to the recognition performance as well as noise-robustness in a DSR scenario. This is in light of the results presented in Paliwal (1999), where cepstral liftering on MFCCs was shown to improve the noise robustness for dynamic time warping-based speech recognizers, which use Euclidean distance measures.

7.3.4 Relationship Between the Distortion Measure and Recognition Performance

All quantization schemes attempt to minimize the error between the original and quantized samples. For instance, the HRO bit allocation formula of Eq. 7.6 for scalar quantizing vector components was obtained from a constrained minimization of the average MSE. In vector quantization, the codebook vector that minimizes the distortion is selected.

The direct application of these quantization schemes to distributed speech recognition readily assumes that decreasing the MSE between the original and quantized MFCC features will guarantee that the degradation in recognition performance due to the quantization decreases as well. We will validate this assumption by applying unconstrained vector quantization on MFCCs at varying bitrates, measuring the average MSE and recognition rates for each bitrate. Figure 7.13 shows the average recognition rate plotted against the average MSE incurred by the vector quantizer.

We can see from Fig. 7.13 that the recognition rate appears to decrease monotonically as the average MSE increases. Therefore, this shows that a quantization scheme that minimizes the MSE is also guaranteed to improve the recognition accuracy. Furthermore, we note that it is a non-linear relationship, where if the average MSE was large, a decrease in quantization distortion leads to a larger improvement in recognition rate than if the MSE were low.

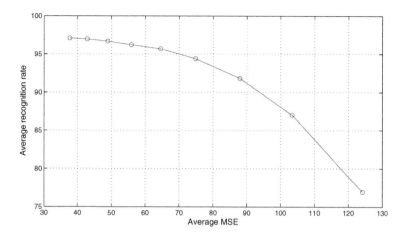

Fig. 7.13 Relationship between average recognition rate and average MSE

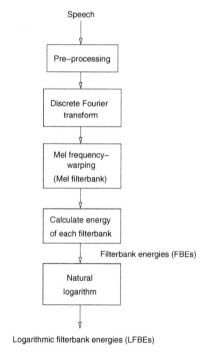

Fig. 7.14 Extraction of logarithmic filterbank energies from speech

7.3.5 Improving Noise Robustness: Perceptual Weighting of Filterbank Energies

Noise-robustness is an important consideration in DSR since the user at the client end will mostly be immersed in various environmental sounds. It is well known in the ASR literature that noise has a detrimental effect on the recognition performance when using conventional MFCC features. Much of the current work in ASR research involves finding speech features that are robust to the effects of noise. These speech features can be used in DSR as well.

Using a two-dimensional discrete cosine transform coder, Zhu and Alwan (2001) improved the robustness of DSR to noise by using peak-isolated MFCCs (MFCCPs). MFCCPs are derived by applying half-wave rectification to the spectrum reconstructed from a bandpass liftered cepstral vector (Strope and Alwan 1997). They are robust to noise because of the preservation and emphasis of power spectral peaks, whose frequency locations are known to be important for the discrimination of vowels. The idea is that accuracy in the location of spectral peaks is more important than the location of spectral valleys.

Another method of exploiting this idea is to quantize the logarithmic filterbank energies (LFBEs) (see Fig. 7.14) rather than the MFCCs themselves (So and Paliwal 2005). The advantage of working with LFBEs is their correspondence with the power spectrum. That is, a strong peak in the power spectrum would generally lead to a large LFBE coefficient in the same critical band. On the other hand, the frequency location information of this spectral peak is not readily available in the MFCC representation as each MFCC consists of a linear combination of all LFBEs. By quantizing the LFBEs, we can apply noise-shaping techniques to quantize LFBEs that correspond to spectral peaks more finely than those that correspond to spectral valleys. The disadvantage of using LFBE vectors is that they have a higher dimensionality than MFCC vectors.

In order to achieve quantization noise shaping, we apply a perceptually-weighted distance measure to vary the emphasis of the quantization, which can easily be incorporated into a vector quantizer (So et al. 2005). The weighted distance measure $d_w(E, \hat{E})$ between the original LFBE vector E and the LFBE $\hat{E}$ is defined as:

$$d_w(E, \hat{E}) = \sum_{i=1}^{n} [w_i(E_i - \hat{E}_i)]^2 \qquad (7.17)$$

where n is the vector dimensionality, w_i is the weight of the ith component, E_i and $\hat{E}_i$ are the ith component of the original and code-vector, respectively. In order to emphasize a vector component, E_i, such that it is quantized more finely, the weight w_i should be made larger. In the LFBE vector quantizer, it is desirable to emphasize the LFBEs that represent the spectral peaks. Therefore, w_i is set to be a scaled version of the FBE, e^{E_i}:

$$w_i = [e^{E_i}]^r \qquad (7.18)$$

Through experimentation, we have found 0.5 to be a good value for r.

7.4 Experimental Results

7.4.1 ETSI Aurora-2 Distributed Speech Recognition Task

The purpose of the ETSI Aurora-2 experiment is to provide a common framework for evaluating noise-robust speech recognition systems. It consists of a clean speech database, a noise database, a standard MFCC-based frontend, and scripts for performing the various training and test sets. The recognition engine that is used is the HMM Toolkit (HTK) software (Young et al. 2002).

The TIDigits database (Leonard 1984) forms the basis of the clean speech database, where the original 20 kHz speech was downsampled to 8 kHz and filtered using the frequency characteristic of ITU G.712 (300–3,400 Hz). Aurora-2 also provides a database of eight background noises, which were deemed to be commonly encountered in real-life operating conditions for DSR. These noises were recorded at the following places (Hirsch and Pearce 2000):

- Suburban train (subway)
- Crowd of people (babble)
- Car
- Exhibition hall (exhibition)
- Restaurant
- Street
- Airport
- Train station

This noise is added to the filtered clean speech at various SNRs to simulate noise corruption.

There are two training modes: training with clean speech only and training with clean and noisy (multicondition) speech. In multicondition training, the noises added are subway, babble, car, and exhibition. When training with clean speech only, the best recognition performance is achieved in matched conditions, i.e. when testing with clean speech as well. However, when the speech to be tested has background noise, then multicondition training is desirable, as it includes the distorted speech in the training data.

For the testing, there are three test sets, known as test set A, B, and C. In test set A and B, 4,004 test utterances from the TIDigits database are divided into four subsets of 1,001 utterances each and four different types of noises are added to each subset at varying levels of SNRs (∞, 20, 15, 10, 5, 0, −5 dB). Therefore, there are a total of $4 \times 7 = 28$ recognition accuracies reported in test set A and B. In test set C, only two subsets of 1001 utterances and two noises are used, giving a total of 14 recognition accuracies.

In test set A, the subway, babble, car, and exhibition noises are added to each subset and these are the same noises used in multicondition training, hence test set A evaluates the system in matched conditions. In test set B, the other four noises, namely restaurant, street, airport, and train station, are used instead. Because these noises were not present in the multicondition training, then test set B evaluates the system in mismatched conditions (mismatched noise). Test set C contains two utterance subsets only (of the four) with the noises, subway and street, added. Both the speech and noise are filtered using the MIRS frequency characteristic before they are added, hence test set C evaluates the system in mismatched conditions (mismatched frequency characteristic).

Whole word HMMs are used for modelling the digits with the following parameters:

- 16 states per word (with 2 additional dummy states at beginning and end);
- left-to-right topology without skips over states;
- 3 Gaussian mixtures per state; and
- diagonal covariance matrices.

7.4.2 Experimental Setup

We have evaluated the recognition performance of various quantization schemes version 3.2.1 of the HMM Toolkit (HTK) software. Training was done on clean data only (no multicondition training) and testing was performed using test set A. In order to see the recognition performance as a function of bitrate, we focus on the results of testing on clean speech, where the four word recognition accuracies for each type of noise are averaged to give the final score for the specific quantization scheme. In addition to this, the effect of different types of noise at varying levels of SNR on the recognition performance is also investigated at the bitrates of 1.2 kbps and 0.6 kbps for each quantization scheme.

The ETSI DSR standard Aurora frontend (2003) was used for the MFCC feature extraction. MFCCs are extracted at a frame rate of 100 Hz. As a slight departure from the ETSI DSR standard, we have used 12 MFCCs (excluding the zeroth cepstral coefficient, c_0, and logarithmic frame energy, $log\ E$) as the feature vectors to be quantized. We have applied the cepstral liftering technique (Juang et al. 1987) to the MFCC vectors. Cepstral mean subtraction (CMS) is applied to the decoded 12 MFCC features, which are concatenated with their corresponding delta and acceleration coefficients, giving the final feature vector dimension of 36 for the ASR system. The HTK parameter type is MFCC_D_A_Z. The baseline average recognition accuracy or ASR accuracy using unquantized MFCC features derived from clean speech is 98.0 %.

7.4.3 Non-Uniform Scalar Quantization Using HRO Bit Allocation

For the scalar quantization experiment, each MFCC was quantized using a non-uniform Gaussian Lloyd-Max scalar quantizer whose bit allocation was calculated using the HRO bit allocation formula of Eq. 7.6. We have chosen this method over the WER-based greedy algorithm (Digalakis et al. 1999) because of its computational simplicity and this allows us to scale any bitrate with ease. Table 7.2 shows the average recognition accuracy of the non-uniform scalar quantizer. It can be seen that the accuracy decreases linearly in the range of 4.4 to 1.2 kbps and drops rapidly below this range.

Table 7.2 Average DSR word recognition accuracy as a function of bitrate for non-uniform scalar quantizer (ASR accuracy = 98.0%)

Bitrate (kbps)	Average recognition accuracy (in %)
0.6	38.2
0.8	72.3
1.0	86.7
1.2	93.3
1.5	95.5
1.7	96.2
2.0	97.0
2.2	97.2
2.4	97.4
3.0	97.8
4.4	98.0

7.4.4 Unconstrained Vector Quantization

An unconstrained, full-search vector quantizer was used to quantize single MFCC frames. The distance measure used was MSE. In terms of minimizing quantization distortion, the vector quantizer is considered the optimum coding scheme, hence it will serve as an informal upper recognition bound for single frame quantization. Table 7.3 shows the average recognition accuracies at several bitrates.

Table 7.3 Average DSR word recognition accuracy as a function of bitrate for the unconstrained vector quantizer (ASR accuracy = 98.0%)

Bitrate (kbps)	Average recognition accuracy (in %)
0.4	76.9
0.6	91.8
0.8	95.7
1.0	96.9
1.2	97.0

When comparing with Table 7.2, we can see that the superior rate-distortion efficiency of the vector quantizer translates to better recognition rates as well. For example, at 600 bps, which corresponds to 6 bits in total for quantizing 12 coefficient MFCC vectors, the recognition rate for the vector quantizer is 53.6% higher than that for the scalar quantizer. With such a small bit budget, the scalar quantizer cannot allocate bits to some MFCCs, thus in the decoding, they would simply be replaced by the mean value. On the other hand, the vector quantizer codebook, which contains 64 code-vectors, exploits linear and non-linear dependencies between the MFCCs, matches the joint PDF, and uses optimal quantization cell shapes (Lookabaugh and Gray 1989).

7.4.5 GMM-Based Block Quantization

Table 7.4 shows the average recognition accuracies for the GMM-based block quantizer with 16 mixture components. We can see that for this quantization scheme, the recognition accuracy decreases gracefully to about 800 bps. Comparing it with Table 7.2, we notice higher recognition accuracies in the GMM-based block quantizer, which may be attributed to better PDF matching as well as the use of a decorrelating transformation. At 600 bps, the GMM-based block quantizer is 49.4% better than the scalar quantizer. However, it is not as high as the recognition performance achieved with the vector quantizer at 600 bps (Table 7.3). This is consistent in the rate-distortion sense since the vector quantizer should be the optimum single-frame quantizer. However, in practice, the vector quantizer suffers from high computational complexity, while the GMM-based block quantizer has fixed requirements as well as possessing the feature of bitrate scalability.

7.4.6 Multi-frame GMM-Based Block Quantization

The multi-frame GMM-based block quantizer is similar to the matrix quantizer (Tsao and Gray 1985). Five successive MFCC frames are concatenated to form a vector of dimension 60 and these larger vectors are then quantized. Table 7.5 shows the average word recognition accuracy of the 16 mixture component, five frame multi-frame GMM-based block quantizer for different bitrates.

It can be observed that this quantizer achieves an accuracy that is close to the unquantized, baseline system at 1 kbps or 10 bits/frame, which is half the bitrate of the single-frame GMM-based block quantizer. For bitrates lower than 600 bps, the performance gradually rolls off.

In terms of quantizer distortion, the multi-frame GMM-based block quantizer generally performs better as more frames are concatenated together because inter-frame memory can be exploited by the KLT. Furthermore, because the dimensionality of the vectors is high, the block quantizer operates at a higher rate.

Compared with the results of the single frame GMM-based block quantizer in Table 7.4, the multi-frame scheme does not suffer from a dramatic drop in recognition accuracy at low bitrates. Unlike the single frame scheme, where there was a shortage of bits to distribute among mixture components, the multi-frame GMM-based block quantizer is able to provide enough bits, thanks to the increased dimensionality of the vectors. For example, at 300 bps, a 16-mixture component, single frame GMM-based block quantizer has a total bit budget of 3 bits. On the other hand, a 16-mixture component, five-frame scheme has a total bit budget of 15 bits. Therefore, the multi-frame GMM-based block quantizer can operate at lower bitrates while maintaining good recognition performance.

The multi-frame GMM-based block quantizer also outperforms the vector quantizer since the latter is only a single frame scheme. As we have seen previously, successive MFCC frames are highly correlated with each other so it is expected that quantization schemes that exploit multiple frame dependencies will perform much better in the rate-distortion sense. The disadvantage of this scheme is the inherent delay that is introduced.

Table 7.4 Average DSR word recognition accuracy as a function of bitrate for the GMM-based block quantizer with 16 mixture components (ASR accuracy = 98.0%)

Bitrate (kbps)	Average recognition accuracy (in %)
0.3	8.1
0.4	23.3
0.6	87.6
0.8	93.7
1.0	95.5
1.2	96.4
1.5	97.2
1.7	97.3
2.0	97.6
2.2	97.7
2.4	97.9
3.0	97.8
4.4	98.0

Table 7.5 Average word recognition accuracy as a function of bitrate for the multi-frame GMM-based block quantizer with 16 mixtures and 5 frames (ASR accuracy = 98.0%)

Bitrate (kbps)	Average recognition accuracy (in %)
0.2	82.9
0.3	93.0
0.4	95.4
0.6	96.8
0.8	97.5
1.0	97.7
1.2	97.9
1.5	97.8
1.7	98.0
2.0	98.0

7.4.7 Perceptually-Weighted Vector Quantization of Logarithmic Filterbank Energies

We can see from Fig. 7.15 that the proposed perceptually weighted vector quantization scheme operating on logarithmic filterbank energies (PWVQ-LFBE) is more robust to noise than the unweighted vector quantization of MFCCs (VQ-MFCC). At SNRs of 10 and 15 dB, the PWVQ-LFBE scheme achieves up to 6 to 10% improvement over VQ-MFCC. This may be attributed to the use of the weighted distance measure to emphasize the spectral peaks. However, for low SNRs, the PWVQ-LFBE

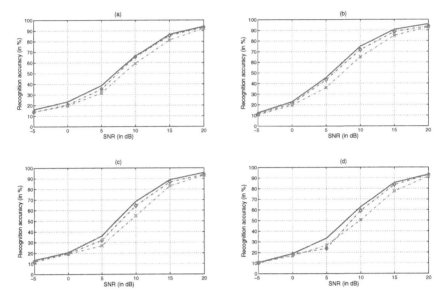

Fig. 7.15 Word recognition accuracy for speech corrupted with noise at varying SNRs (in dB) at 1.2 kbps using the perceptually weighted vector quantizer on LFBEs (PWVQ-LFBE) (solid line represents the ASR accuracy; squares represent PWVQ-LFBE and crosses represent VQ-MFCC): **a** corrupted with subway noise, **b** corrupted with babble noise, **c** corrupted with car noise, **d** corrupted with exhibition noise

scheme fails to improve the noise robustness, when compared with VQ-MFCC. Furthermore, this quantization scheme appears to be bounded by the ASR recognition accuracy (shown as the solid line in Fig. 7.15). We should point out that higher bitrates were not investigated due to computational constraints.

7.5 Conclusion

In this chapter, we have described a series of quantization schemes for coding MFCC feature vectors that are to be used for distributed speech recognition. These include the scalar quantizer, vector quantizer, perceptually weighted vector quantizer and GMM-based block quantizer. These quantization schemes have been described in detail in the coding literature but their application to quantizing MFCC feature vectors has been a relatively recent development. It is important to note that the objective measure in DSR that is to be optimized is the recognition accuracy, rather than the mean squared error. Therefore, quantization in the context of DSR deserves further investigation.

We have discussed the statistical properties of MFCCs that are relevant to quantization. In particular, we have shown that successive MFCC vectors are highly correlated with each other. Because of this property, multi-frame and predictive quantization schemes should perform more efficiently. In relation to the energy coefficients (c_0 and $log\ E$), which were shown to possess different statistical properties, we concluded that they should be quantized independently from the rest of the cepstral coefficients. We have also shown via empirical results that the recognition rate increases monotonically as MSE decreases. That is, optimizing quantizers to minimize the MSE, in general, should guarantee an improvement in recognition rate. However, the relationship is a non-linear one.

Next, we presented a brief review of the distributed speech recognition literature, where various schemes for quantizing MFCCs were investigated. The Aurora-2 database used for evaluating the performance of our MFCC quantization schemes as well as the parameters for the recognition task were described in detail. Following this, we presented our results on MFCC quantization in a DSR framework using non-uniform scalar quantization with HRO bit allocation, vector quantization, and single-frame as well as multi-frame GMM-based block quantization. For clean speech, the multi-frame GMM-based block quantizer achieved the best recognition at lower bitrates, exhibiting a negligible 1% degradation (word error rate of 2.5%) in recognition performance over the ASR accuracy at 800 bps and 5% degradation (word error rate of 7%) at 300 bps. Unlike vector quantization schemes, the multi-frame GMM-based block quantizer is scalable in bitrate and has a complexity that is independent of bitrate.

We also looked at the performance of vector quantization of MFCCs derived from noise corrupted speech at various SNR levels and compared this with the perceptually-weighted vector quantizer (PWVQ). Rather than quantizing MFCCs, the PWVQ works with logarithmic filterbank energies (LFBEs). The non-linearly weighted distance measure allows for the shaping of quantization noise, putting more emphasis on spectral peaks so that they are quantized more finely. We showed that this scheme improves noise-robustness for medium SNRs (10–15 dB) over the vector quantization of MFCCs.

References

ETSI Standard ES 201 108 (2003). *Speech Processing, Transmission and Quality Aspects (STQ); Distributed Speech Recognition; Front-End Feature Extraction Algorithm; Compression Algorithms.* Tech. Rep. Standard ES 201 108 v1.1.3, European Telecommunications Standards Institute (ETSI).

Chen, W., and Smith, C.H. (1977). "Adaptive Coding of Monochrome and Color Images." IEEE Trans. Commun. **COM-25**(11): 1285–1292.

Davis, S.B., and Mermelstein, P. (1980). "Comparison of Parametric Representations of Monosyllabic Word Recognition in Continuously Spoken Sentences." IEEE Trans. Acoust. Speech Signal Process. **ASSP-28**(4): 357–366.

Dempster, A.P., Laird, N.M., and Rubin, D.B. (1977). "Maximum Likelihood from Incomplete Data via the EM Algorithm." J. Roy. Stat. Soc. **39**: 1–38.

Digalakis, V.V., Neumeyer, L.G., and Perakakis, M. (1999). "Quantization of Cepstral Parameters for Speech Recognition over the World Wide Web." IEEE J. Select. Areas Commun. **17**(1): 82–90.

Gersho, A., and Gray, R.M. (1992). *Vector Quantization and Signal Compression.* Kluwer Academic Publishers, Massachusetts.

Gray, R.M., and Neuhoff, D.L. (1998). "Quantization." IEEE Trans. Inform. Theory **44**(6): 2325–2383.

Hirsch, H.G., and Pearce, D. (2000). The Aurora Experimental Framework for the Performance Evaluation of Speech Recognition Systems Under Noisy Conditions. ISCA ITRW ASR2000, Paris, France.

Huang, J.J.Y., and Schultheiss, P.M. (1963). "Block Quantization of Correlated Gaussian Random Variables." IEEE Trans. Commun. **CS-11**: 289-296.

Juang, B.H., Rabiner, L.R., and Wilpon, J.G. (1987). "On the Use of Bandpass Liftering for Speech Recognition." IEEE Trans. Acoust. Speech Signal Process. **1**: 597-600.

Kiss, I., and Kapanen, P. (1999). Robust Feature Vector Compression Algorithm for Distributed Speech Recognition. Eur. Conf. Speech Commun. Technol..

Leonard, R.G. (1984). A Database for Speaker-Independent Digit Recognition. Proc. IEEE. Int. Conf. Acoust. Speech Signal Process.

Linde, Y., Buzo, A., and Gray, R.M. (1980). "An Algorithm for Vector Quantizer Design." IEEE Trans. Commun. **28**(1): 84–95.

Lloyd, S.P. (1982). "Least Square Quantization in PCM." IEEE Trans. Inform. Theory **IT-28**(2): 129–137.

Lookabaugh, T.D., and Gray, R.M. (1989). "High-Resolution Quantization Theory and the Vector Quantizer Advantage." IEEE Trans. Inform. Theory **35**(5): 1020–1033.

Makhoul, J., Roucos, S., and Gish, H. (1985). "Vector Quantization in Speech Coding." Proc. IEEE **73**: 1551–1588.

Max, J. (1960). "Quantizing for Minimum Distortion." IRE Trans. Inform. Theory **IT-6**: 7–12.

Paez, M.D., and Glisson, T.H. (1972). "Minimum Mean-Squared-Error Quantization in Speech PCM and DPCM System." IEEE Trans. Commun. **COM-20**: 225–230.

Paliwal, K.K. (1982). "On the Performance of the Quefrency-Weighted Cepstral Coefficients in Vowel Recognition." Speech Commun. **1**: 151–154.

Paliwal, K.K. (1999). Decorrelated and Liftered Filterbank Energies for Robust Speech Recognition. Eur. Conf. Speech Commun. Technol., Budapest, Hungary.

Paliwal, K.K., and Atal, B.S. (1993). "Efficient Vector Quantization of LPC Parameters at 24 Bits/Frame." IEEE Trans. Speech Audio Process. **1**(1): 3–14.

Paliwal, K.K., and So, S. (2005). "A Fractional Bit Encoding Technique for the GMM-Based Block Quantization of Images." Digital Signal Process. **15**(3): 435–446.

Rabiner, L., and Juang, B.H. (1993). Fundamentals of Speech Recognition. Prentice Hall, New Jersey.

Ramaswamy, G.N., and Gopalakrishnan, P.S. (1998). Compression of Acoustic Features for Speech Recognition in Network Environments. IEEE Int. Conf. Acoust. SpeechSignal Process.

Segall, A. (1976). "Bit Allocation and Encoding of Vector Sources." IEEE Trans. Inform. Theory **IT-22**(2): 162–169.

So, S., and Paliwal, K.K. (2005). Improved Noise-Robustness in Distributed Speech Recognition via Perceptually-Weighted Vector Quantisation of Filterbank Energies. Eur. Conf. Speech Commun. Technol., Lisbon, Portugal.

So, S., and Paliwal, K.K. (2006). "Scalable Distributed Speech Recognition Using Gaussian Mixture Model-Based Block Quantisation." Speech Commun. **48**: 746–758.

Srinivasamurthy, N., Ortega, A., and Narayanan, S. (2006). "Efficient Scalable Encoding for Distributed Speech Recognition." Speech Commun. **48**(8): 888–902.

Strope, B., and Alwan, A. (1997). "A Model of Dynamic Auditory Perception and its Application to Robust Word Recognition." IEEE Trans. Speech Audio Process. **5**(2): 451–464.

Subramaniam, A.D., and Rao, B.D. (2003). "PDF Optimized Parametric Vector Quantization of Speech Line Spectral Frequencies." IEEE Trans. Speech Audio Process. **11**(2): 130–142.

Tsao, C., and Gray, R.M. (1985). "Matrix Quantizer Design for LPC Speech using the Generalized Lloyd Algorithm." IEEE Trans. Acoust. Speech Signal Process. **33**: 537–545.

Wallace, G.K. (1991). "The JPEG Still Picture Compression Standard." Commun. ACM **34**(4): 30–44.

Young, S., Evermann, G., Hain, T., Kershaw, D., Moore, G., Odell, J., Ollason, D., Povey, D., Valtchev, V., and Woodland, P. (2002). The HTK Book (for HTK Version 3.2.1). Cambridge University Engineering Department.

Zhu, Q., and Alwan, A. (2000). On the Use of Variable Frame Rate Analysis in Speech Recognition. IEEE Int. Conf. Acoust. Speech Signal Process.

Zhu, Q., and Alwan, A. (2001). An Efficient and Scalable 2D DCT-Based Feature Coding Scheme for Remote Speech Recognition. IEEE Int. Conf. Acoust. Speech Signal Process.

8

Error Recovery: Channel Coding and Packetization

Bengt J. Borgström, Alexis Bernard and Abeer Alwan

Abstract. Distributed Speech Recognition (DSR) systems rely on efficient transmission of speech information from distributed clients to a centralized server. Wireless or network communication channels within DSR systems are typically noisy and bursty. Thus, DSR systems must utilize efficient Error Recovery (ER) schemes during transmission of speech information. Some ER strategies, referred to as forward error control (FEC), aim to create redundancy in the source coded bitstream to overcome the effect of channel errors, while others are designed to create spread or delay in the feature stream in order to overcome the effect of bursty channel errors. Furthermore, ER strategies may be designed as a combination of the previously described techniques. This chapter presents an array of error recovery techniques for remote speech recognition applications.

This chapter is organized as follows. First, channel characterization and modeling are discussed. Next, media-specific FEC is presented for packet erasure applications, followed by a discussion on media-independent FEC techniques for bit error applications, including general linear block codes, cyclic codes, and convolutional codes. The application of unequal error protection (UEP) strategies utilizing combinations of the aforementioned FEC methods is also presented. Finally, frame-based interleaving is discussed as an alternative to overcoming the effect of bursty channel erasures. The chapter concludes with examples of modern standards for channel coding strategies for distributed speech recognition (DSR).

8.1 Distributed Speech Recognition Systems

Throughout this chapter various error recovery and detection techniques are discussed. It is therefore necessary to present an overview of a complete experimental DSR system, including feature extraction, a noisy channel model, and an automatic speech recognition engine at the server end (Fig. 8.1).

The feature extraction and source coding algorithms implemented for this chapter are similar to those described by the *ETSI* standards (ETSI 2000), whereby split vector quantization (SVQ) is used to compress the first 13 Mel-Frequency Cepstral Coefficients (MFCCs) as well as the log-energy of the speech frame. The SVQ then allocates 8 bits to the vector-quantization of the log-energy and the 0th cepstral coefficient pair, and 6 bits to each of the following 6 pairs. The vector quantizers were trained using the K-means algorithm, and quantization was carried out via an exhaustive search.

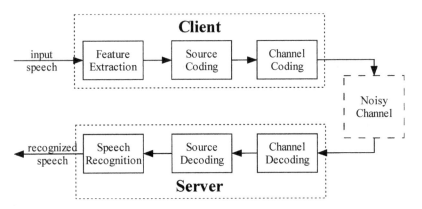

Fig. 8.1 Overview of the complete distributed speech recognition system

Two types of communication channels are studied, wireless circuit-switched and IP packet-switched. Although these channels are inherently different, they do share the characteristic that errors, whether they are flipped bits or packet erasures, tend to occur in bursts. Thus, similar models are used to simulate the effects of noisy channels in the wireless and IP network scenarios. Channel degradation and modeling are discussed in further detail in Sect. 8.2.

The server-end speech recognition engine is implemented using the HTK toolkit (Young et al. 2000). The training process uses an implementation of the forward-backward algorithm, and the recognition process uses an implementation of the Viterbi algorithm. The speech database used for experiments is the Aurora-2 database (Hirsch and Pearce 2000), which consists of connected digit strings, spoken by various male and female speakers. During the recognition process, 16-state word models were used, with each state comprised of 3 mixtures. 8,440 digit utterances were used for training, and 1,001 utterances were used for testing (500 males, 501 females for a total of 3,257 digits).

8.2 Characterization and Modeling of Communication Channels

Distributed speech recognition systems face the challenge of processing signal noise induced by communication channels. Such systems transmit extracted speech features from distributed clients to the server, and typically operate at lower bitrates than traditional speech communication systems.

8.2.1 Signal Degradation Over Wireless Communication Channels

Signal degradations caused by wireless channels are highly dependent on the specific physical properties of the environment between the transmitter and receiver, and it is a difficult task to accurately generalize performance results of communication over a wireless channel (Sklar 1997; Bai and Atiquzzaman 2003).

There are three general phenomena that affect the propagation of radio waves in wireless communication systems:

1. Reflection: Radio waves are reflected off smooth surfaces which are large in comparison to the wavelength of the wave.
2. Diffraction: Radio waves propagate around relatively large impenetrable objects when there exists no line of sight (LOS) between the transmitter and receiver. This effect is also referred to as "shadowing", since radio waves are able to travel from the transmitter to the receiver even when shadowed by a large object.
3. Scattering: Radio waves are disrupted by objects in the transmission path whose size is smaller than the wavelength of the radio wave.

These propagation phenomena cause fluctuations of the amplitude, phase, and angle of incidence of wireless signals, resulting in multipath propagation from the transmitter to the receiver. Multipath propagation leads to fading characteristics in the received signal. Large-scale fading occurs when the mobile receiver moves to/from the transmitter over large distances. Small-scale fading occurs when the distance between the receiver and transmitter changes in small increments. Thus, the speed of the mobile client has a great effect on the resulting channel behavior. When no dominant path exists, the statistics for the signal envelope can be described by a Rayleigh distribution, and when a dominant LOS path exists, the statistics of the signal envelope can be described by a Ricean distribution.

Since wireless communication systems are generally built upon circuit-switched networks, corrupted data occur as bit errors in the modulated bitstream. Furthermore, due to the fading nature of wireless channels, bit errors tend to occur in bursts. The probability of occurrence and expected duration of bit error bursts are dependent upon the time varying channel signal-to-noise ratio (SNR).

8.2.2 Signal Degradation Over IP Networks

IP networks rely on packet-switching, wherein packet loss and delay are caused mainly by congestion at the routers, and individual bit errors rarely occur (Tan et al. 2005). Specifically, packet losses may appear if the input flow of data is higher than the processing capacity of the switching logic, or if the processing capacity of the switching logic is higher than the output flow speed (Kurose and Rose 2003).

Similar to the occurrence of bit-error bursts in wireless networks, packet losses in packet-switched IP networks tend to occur in bursts (Jiao et al. 2002; Bolot 2003). Probability of occurrence and expected duration of the packet loss bursts are dependent on the congestion of the network.

8.2.3 Modeling Bursty Communication Channels

A common method for simulating a bursty channel in a communication system is the two-state *Gilbert-Elliot (GE) model* (Elliot 1963), which has been widely used for DSR studies. The GE model includes a good state, S_g, which incurs no loss, and a bad state, S_b, which assumes some probability of loss, as is illustrated in Fig. 8.2. Here, p represents the probability of transitioning from S_g to S_b, whereas q represents the probability of transitioning from S_b to S_g. To completely characterize the model, the parameter h, which represents the probability of loss while in state S_b, is used.

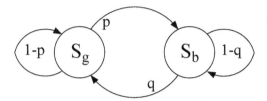

Fig. 8.2 The Gilbert-Elliot Model for simulating bursty channels: S_g and S_b represent the goodand bad states, respectively, and p and q represent transitional probabilities

8.2.3.1 Bit-Level Channel Models

Throughout this chapter, specific bit-level channel conditions are used for simulations and the probability of transition from S_g to S_b is set to $p = 0.002$. The noise incurred in S_b is additive white noise at a level of 2 dB, corresponding to an error probability of $P_{err}|S_b = 0.157$, whereas the noise incurred in S_g is at a level of 20 dB, corresponding to an error probability of $P_{err}|S_g \approx 0.00$. Thus, the remaining variable is the transitional probability q. The conditions to be used, similar to those described in (Han et al. 2004), are shown in Table 8.1.

Table 8.1 Channel conditions used for bit-level simulations

Channel condition	p	q	BER (%)
Clean	0.000	1.000	0.00
1	0.002	0.019	1.50
2	0.002	0.009	2.87
3	0.002	0.0057	4.10
4	0.002	0.003	6.32
5	0.002	0.002	7.90
6	0.002	0.0013	9.58
7	0.002	0.0009	10.90

8.2.3.2 Packet-Level Channel Models

An efficient method to simulate errors introduced by IP networks is to apply the GE model on the packet level, and thus the GE model iterates at each packet. Furthermore, the loss incurred in S_b represents a packet erasure.

Throughout this chapter, specific packet-level channel conditions will be used for simulations. These conditions, as described in Han et al. (2004), are shown in Table 8.2. Note that for simulating packet-level channels, the probability of packet loss while in S_b is assumed to be 1.0. In Fig. 8.2, the average burst duration can be determined as $1/q$, and the percentage of packets lost can be calculated as $100[p/(p+q)]\%$.

Table 8.2 Channel conditions used for packet-level simulations

Channel condition	p	q	Packets lost (%)	Average burst duration
Clean	0.000	1.000	0.0	0.00 frames
1	0.005	0.853	0.6	1.17 frames
2	0.066	0.670	9.0	1.49 frames
3	0.200	0.500	28.6	2.00 frames
4	0.250	0.400	38.5	2.50 frames
5	0.300	0.300	50.0	3.33 frames
6	0.244	0.200	55.0	5.00 frames

8.3 Media-Specific FEC

Media-specific FEC involves insertion of additional copies of speech features into the DSR datastream prior to transmission. The additional copies, referred to as replicas, are coarsely quantized in order to reduce the additional required bandwidth. Furthermore, due to the bursty nature of both wireless and IP-network transmission, speech feature replicas are inserted into the source-coded bitstream at certain frame intervals away from the original features.

In (Peinado et al. 2005a) the authors introduce a media-specific FEC method compatible with the *ETSI* DSR source coding standards (ETSI 2000). The media-specific FEC algorithm creates speech feature replicas by using B_{vq}-bit vector quantization applied to the entire 14-element feature vector of each speech frame. Additionally, the VQ replicas are inserted into the bitstream at intervals of T_{fec} frames away from the original speech feature, with the aim of overcoming the effects of clustered packet erasures characteristic of bursty channels. In general, optimal performance of media-specific FEC can be expected if T_{fec} is chosen to be at least as long as the expected burst duration. Figure 8.3 illustrates an example of media-specific FEC packetization for $T_{fec} = 4$. Here, the source-coded frames are shown along the top, and the FEC replicas are shown along the bottom. Note the separation between coded frames and corresponding vector quantized replicas. Additionally, Fig. 8.4 shows word-accuracy results obtained for various values of B_{vq}, for $T_{fec} = 6$ frames, and for various channel conditions described in Table 8.2.

As can be concluded from Fig. 8.4, media-specific FEC provides improved performance for DSR systems in the case of bursty channels, as compared with transmission without VQ replicas. However, the recognized speech at the server can be

	packet #4		packet #5		packet #6		packet #7	
Features	7	8	9	10	11	12	13	14
VQ Replicas	3	12	5	14	7	16	9	18

Fig. 8.3 Example of Media-Specific FEC packetization using vector-quantized replicas with $T_{fec} = 4$ (Based on Peinado et al. 2005a)

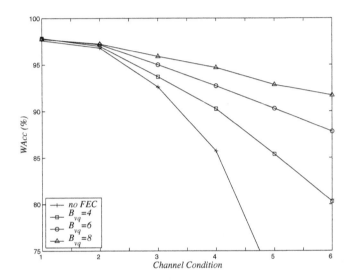

Fig. 8.4 Word-accuracy results obtained using the Aurora-2 database with Media-Specific FEC using B_{vq}-bit VQ replicas and packetization with $T_{fec} = 6$

delayed by up to T_{fec} frames, which may introduce problems for delay-sensitive applications.

8.4 Media-Independent FEC

Media-independent FEC techniques are applied within DSR systems with the aim of correcting transmission errors or predicting reliability of transmitted speech features, especially with wireless transmission. These techniques include the use of linear block codes, cyclic codes, or convolutional codes. It has been shown in (Bernard and Alwan 2002a, Bernard 2002) that packet losses or erasures degrade the word-accuracy perform-ance of DSR much less than incorrectly decoded packets. Specifically, it is shown that while channel errors have a disastrous effect on recognition accuracy, the recognizer is able withstand up to 15% of randomly inserted *channel erasures* with negligible loss of accuracy. Therefore, it may at many times be enough for the overall DSR system to simply detect packet errors, as opposed to attempting to decode unreliable data. Thereafter, lost speech features can be estimated at the server through various Error Concealment (EC) methods, which are discussed in Chap. 9.

8.4.1 Combining FEC with Error Concealment Methods

There exist numerous error concealment methods to deal with packet erasures in DSR systems which estimate lost speech features before speech recognition. The simplest of such methods include frame dropping and frame repetition. These methods involve low complexity, although they may produce poor recognition accuracy as error burst durations increase (Bernard and Alwan 2002a). Other algorithms to estimate lost speech features include various interpolation techniques, such as linear interpolation or polynomial interpolation, which provide better performance for longer error bursts (Peinado and Segura 2006).

More successful EC methods, however, are based on minimum-mean-square-error (MMSE) algorithms. In Peindao et al. (2005b), the authors successfully apply the Forward-Backward MMSE (FB-MMSE) algorithm to determine the maximum likelihood lost or erroneous observations given the most recent correct observation and the nearest correct future observation. However, due to the high complexity of the FB-MMSE algorithm (Peinado et al. 2005b) introduce simplified versions which greatly decrease the computational load without significantly reducing the performance.

There also exist recognizer-based EC methods which involve soft-feature Viterbi decoding at the server (Bernard and Alwan 2001), known as *weighted viterbi decoding*. In such techniques, channel decoding is performed to determine a measure of reliability of the current received packet. The reliability measure is then passed to the recognition engine, which applies corresponding weights to speech features during the Viterbi algorithm within the recognition process.

8.4.2 Linear Block Codes

For wireless communication channels, the transmitted bitstream generally becomes degraded due to reasons discussed in Sect. 8.2. An option for providing the detection or correction of transmission errors is through the use of linear block codes. Linear block codes are especially attractive for the problem of error-robust wireless communication due to their low delay, complexity, and overhead.

The aim of block codes is to provide $m = n - k$ redundancy bits to a block of k dataword bits, resulting in a block of n codeword bits prior to transmission. Such a code is referred to as a (n,k) block code.

Let $D = \{d_1, d_2, \ldots, d_k\}$ represent the set of all possible k-dimensional *datawords*, where $d_i \in \Re^{1 \times k}$. Also, let $C = \{c_1, c_2, \ldots, c_n\}$ represent the set of all possible n-dimensional *codewords*, where $c_i \in \Re^{1 \times n}$. Thus, block codes can be represented in matrix form as $c_i = d_i G$ where the matrix $G \in \Re^{k \times n}$ is referred to as the generator matrix.

Furthermore, a code is defined as systematic if each dataword is contained within its corresponding codeword (Blahut 2004). For systematic codes, the *generator matrix* must be in the form

$$G = [I_k \ P], \tag{8.1}$$

where $P \in \mathfrak{R}^{k \times (n-k)}$ is referred to as the *parity matrix*, and I_k is the $k \times k$ identity matrix. The *parity-check matrix*, $H \in \mathfrak{R}^{n-(k \times n)}$, which is used in the decoding process, is then defined as:

$$H = [-P^T \ I_{n-k}]. \tag{8.2}$$

An important parameter to measure the effectiveness of a specific linear block code is the minimum distance, d_{min}. The minimum distance of a code with codeword set C represents the minimum Hamming distance between any two codewords $c_i, c_j \in C$, for $i \neq j$. It can be shown that the d_{min} of a systematic code can be determined by finding the minimum Hamming weight of any nonzero codeword in C:

$$d_{min} = \min_{c \in C, c \neq 0} w(c), \tag{8.3}$$

where the function $w(\cdot)$ represents the Hamming weight, i.e. the number of 1's in a given binary vector.

A linear code with a minimum distance of d_{min} is able to detect no more than $d_{min} - 1$ errors, or is able to correct no more than $\left\lfloor \frac{1}{2}(d_{min} - 1) \right\rfloor$ errors.

Table 8.3 Systematic linear block codes for channel coding of speech features (From Bernard and Alwan 2002a, © 2002 IEEE)

(n,k)	m	P^T	d_{min}
(12,10)	2	[1,1,1,2,2,2,3,3,3,3]	2
(12,9)	3	[1,2,3,3,4,5,5,6,7]	2
(12,8)	4	[3,5,6,9,A,D,E,F]	3
(10,8)	2	[1,1,1,2,2,3,3,3]	2
(12,7)	5	[07,0B,0D,0E,13,15,19]	4
(10,7)	3	[1,2,3,4,5,6,7]	2

In Bernard and Alwan (2002a), the authors provide analysis of the performance of systematic linear block codes for the application of DSR over wireless channels. In this work, good codes are determined through exhaustive searches by maximizing d_{min} for various values of n and k, and minimizing the corresponding weights. Table 8.3 shows the resulting optimal codes. Note that the parity matrix P is given in hexadecimal form.

8.4.2.1 Hard vs. Soft Decoding

Linear block code theory discussed in the previous section builds on arithmetic over Galois Fields (Blahut 2004), most commonly over $GF(2)$, i.e. binary data with modulo-2 operations. However, the input data stream at the receiver is in the form of demodulated bits with superimposed channel noise. That is, the received data vector is in the form $y = x + n$, where x is the transmitted codeword and n is the channel noise vector. Binary decisions must be made for each received noisy bit, $y(i)$, before channel decoding of the block can be performed, resulting in the approximated bit-stream $\tilde{y}$, where $\tilde{y}(i) \in \{0,1\}$. There exists two classical methods of decoding noisy data into a binary bitstream: hard-decision decoding and soft-decision decoding, which will be denoted as $\tilde{y}_h$ and $\tilde{y}_s$, respectively.

Hard-decision decoding maps each received noisy data vector to an approximated transmitted bitstream by the following relationship:

$$\tilde{y}_h(i) = \arg\min_{b \in \{0,1\}} d_E(b, y(i)), \tag{8.4}$$

where $d_E(\cdot, \cdot)$ represents the Euclidean distance, and b corresponds to possible bit values.

As can be interpreted from Eq. 8.4, hard-decision decoding simply entails rounding to the nearest modulated bit. The resulting approximated bitstream, $\tilde{y}_h$, is then used to find the estimated transmitted codeword by minimizing the Hamming distance between itself and all possible true codewords. Thus, the detected codeword chosen for channel decoding, denoted as $\tilde{y}_h^{opt}$, is determined as:

$$\tilde{y}_h^{opt} = \arg\min_{d_i \in D} d_H(\tilde{y}_h, d_i), \tag{8.5}$$

where $d_H(\cdot, \cdot)$ represents the Hamming distance (Bernard and Alwan 2002a).

Since the mapping function described by Eq. 8.5 is not one-to-one, hard-decision decoding can lead to scenarios in which distinct transmitted codewords may be approximated as the same received codeword. That is, there may exist x_i and x_j, for $i \neq j$, such that $d_H(\tilde{y}_h^{opt}, x_i) = d_H(\tilde{y}_h^{opt}, x_j)$. In such situations, the decoder can detect an error, but cannot correct the error, since multiple distinct codewords could have been transmitted.

Figure 8.5a shows an example of hard-decision decoding. In this example, a (2,1) linear code with a generator matrix $G = [1,1]$ was used to transmit the dataword $d = [1]$. The next most likely dataword in this example is $[-1]$. In Fig. 8.5a, the region labeled **CD** corresponds to the region in which the received noisy data vector y would have been correctly decoded. Conversely, **UE** (undetected error) corresponds to the region in which y would have been incorrectly decoded. Furthermore, the regions labeled **ED** (error detected) refer to the regions in which y could not have been decoded, and an error would have been declared.

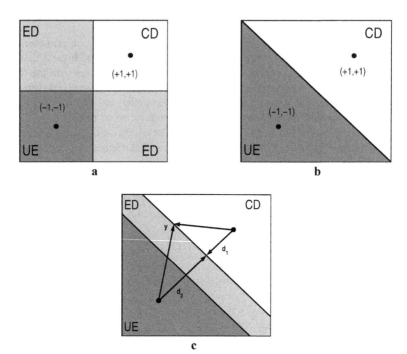

Fig. 8.5 Examples of (**a**) hard-decision decoding, (**b**) soft-decision decoding, and (**c**) λ-soft-decision decoding: This scenario represents a transmitted dataword of $d = [1]$, with $G = [1,1]$. Also, CE represents the region corresponding to a corrected error, UE corresponds to the region of an uncorrected error, and DE represents the region corresponding to a detected error (From Bernard and Alwan 2002a, © 2002 IEEE)

An alternative to the previously discussed hard-decision decoding is *soft-decision decoding*, which minimizes the Euclidean distance between the received noisy data vector and all possible codewords. The approximated bitstream obtained through soft-decision decoding, denoted by $\tilde{y}_s$, can be determined as:

$$\tilde{y}_s^{\text{opt}} = \arg \min_{d_i \in D} d_E^2(d_i, y). \qquad (8.6)$$

Figure 8.5b illustrates an example of soft-decision decoding for the same channel coding scheme explained for Fig. 8.5a. Once again, **CD** denotes the region corresponding to a correctly decoded transmitted dataword, and **UE** denotes the region corresponding to an incorrectly decoded transmitted dataword. Note that for soft-decision decoding, all received data vectors must be decoded, and there is no region for error detection (ED).

For AWGN channels, soft-decision decoding typically shows a 2 dB gain relative to hard-decision decoding in terms of the bit error rate (BER). However, for the specific application of DSR, the authors of (Bernard and Alwan 2002a) show the negative effect of incorrectly decoded transmission errors on word recognition results.

8.4.2.2 λ-Soft Decoding

The adaptive λ-soft decoding algorithm presented in Bernard and Alwan (2002a) provides the error-correcting advantage of soft-decision decoding with the error detecting capabilities of hard-decision decoding by recreating an "error detection" region based on the confidence in the decoding operation.

In order to accept a soft-decision decoded codeword as correct, the λ-soft decoding algorithm compares the likelihood of that transmitted codeword relative to the next most likely codeword. Let y represent the received codeword, and let x_1 and x_2 represent the first and second most probable transmitted codewords. The ratio of likelihoods of x_1 and x_2 can be expressed as:

$$\frac{P(y \mid x = x_1)}{P(y \mid x = x_2)} = \exp\left(\frac{d_E(y, x_2) - d_E(y, x_1)}{N_o}\right) = \exp\left(\frac{D^2}{N_o} \cdot \frac{d_2 - d_1}{D}\right), \quad (8.7)$$

where d_1 and d_2 represent the distances from the orthogonal projection of the received codevector y to the line joining x_1 and x_2, and D represents the distance between x_1 and x_2. Note that $d_2 + d_1 = D$. Also, N_0 is a constant related to the noise level of the channel.

The important factor in Eq. 8.7 is $\lambda = \dfrac{d_2 - d_1}{D}$, since it relates the relative proximity of the received codeword to the two possible transmitted codewords. Note that if $\lambda \approx 0$, the codevectors x_1 and x_2 are almost equiprobable, the soft-decision decoded codeword should be rejected, and an error should be declared. Conversely, if $\lambda \approx 1$, the soft-decision decoded codeword is highly likely. Thus, the region corresponding to error detection (**DE**) grows as λ increases.

For a system implementing the λ-soft decoding algorithm, parameter λ_0 is predetermined to perform soft-decision decoding with error detection capability. For received codewords resulting in $\lambda \geq \lambda_0$, soft-decision decoding is utilized, and for $\lambda < \lambda_0$, an error is declared. Recognition accuracy results for hard, soft, and λ-soft decoding are shown in Table 8.4. Here, the λ-soft parameter is set as $\lambda_0 = 0.16$, and the linear block codes used are described in Table 8.3. Note that λ-soft decoding significantly outperforms both hard and soft decoding by reducing the word error rate.

Table 8.4 Word accuracy results using hard-, soft-, and λ-soft decoding over Rayleigh fading channels (From Bernard and Alwan 2002a, © 2002 IEEE)

Code (n,k)	SNR (dB)	Hard (%)	Soft (%)	λ-Soft (%)
(10,10)	19.96	94.71	94.71	98.32
(10,9)	13.87	97.31	96.35	98.12
(10,8)	10.69	94.47	95.03	97.82
(11,8)	8.80	87.24	95.62	97.43
(12,8)	6.29	67.25	93.17	97.04
(12,7)	4.53	40.48	91.31	95.88

8.4.3 Cyclic Codes

Cyclic codes are a subclass of linear block codes, for which a shift of any codeword is also a codeword (Blahut 2004). That is, if C is a cyclic code, then whenever $c = [c_0,c_1,...,c_{n-1}]^T \in C$, this guarantees that $\hat{c} = [c_{n-1},c_0,...,c_{n-2}]^T \in C$. A convenient way to represent cyclic codes is in polynomial form. Thus, the *codeword* c can be described as:

$$c = c(x) = c_0 + c_1 x + \cdots + c_{n-1}x^{n-1} = \sum_{j=0}^{n-1} c_j x^j. \tag{8.8}$$

The *generator matrix* can also be written in polynomial form as:

$$g(x) = \sum_{j=0}^{n-k} g_j x^j. \tag{8.9}$$

Since a cyclic code is specific to its generator polynomial, a cyclic code can therefore be defined as all multiples of the generator polynomial $g(x)$ by a polynomial of degree $k-1$ or less. Thus, for any polynomial $a_i(x)$ of degree $d \le k-1$, the corresponding codeword polynomial is given by:

$$c_i(x) = g(x)a_i(x). \tag{8.10}$$

Cyclic codes are often used for DSR applications. Specifically, three types of cyclic codes are very useful: *cyclic redundancy codes* (CRCs), *Reed-Solomon* (RS) codes, and *Bose-Chaudhuri-Hocquenghem* (BCH) codes.

8.4.4 Convolutional Codes

A convolutional code is a channel coding technique which encodes the input bit stream as a linear combination of the current bit and past data in mod-2 arithmetic, and the encoder can be interpreted as a system of binary shift registers and mod-2 addition components. Similar to linear block codes, convolutional codes are characterized by their protection rate: a rate-k/n encoder processes n output bits for every k input bits. A major difference between linear block codes and convolutional codes is the fact that convolutional codes contain memory.

Although there exist various ways to decode convolutional coded data, the most common is the Viterbi algorithm (Viterbi 1971), due to its relatively low computational load. The goal of the Viterbi algorithm is to find the most likely transmitted bitstream by minimizing the total error between the entire received noisy datastream and potential transmitted bitstreams. The Viterbi algorithm accomplishes this minimization by iterating through a trellis, for which the number of states and possible paths within the trellis are determined by the structure of the given encoder (Wesel 2003).

Viterbi decoding of convolutional codes can be performed on hard or soft data. That is, the minimization discussed can be carried out on the hard-decision decoded bitstream, which can be determined from the received datastream through Eq. 8.5, or it can be carried out directly on the received soft data. Typically, soft-decision decoding outperforms hard-decision decoding by approximately 2 dB.

Convolutional codes also provide added flexibility to adjust the final protection rate through puncturing. Puncturing entails deleting certain outputs bits according to a specific pattern to obtain a lower final bitrate. For example, a rate-1/2 encoder can be punctured by deleting every 4th bit, and the resulting code will be rate-2/3. By puncturing bits in convolutional codes, the channel coding rate can be varied instantaneously, allowing for *rate compatible punctured codes* (RCPC) (Bernard et al. 2002b). Word recognition results for error protection schemes based on convolution codes are shown in Fig. 8.6.

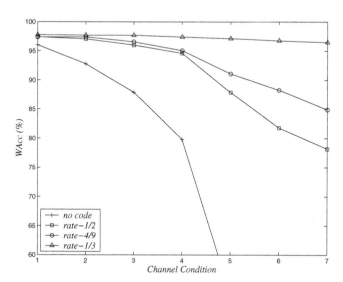

Fig. 8.6 Word recognition results tested on the Aurora-2 database for error protection using convolutional codes (Simulation details are given in Sect. 8.1.)

8.5 Unequal Error Protection

Performance of media-independent FEC codes, with respect to the level of redundancy inserted, can be enhanced by providing for each bit in the bitstream the adequate level of protection against channel errors. The bits comprising the transmitted datastream of DSR systems do not have equal effect on the word recognition performance of the overall system (Boulis et al. 2002; Weerackody et al. 2002). For example, in the case of scalar quantization of extracted speech features, the most significant bits (MSBs) generally provide more information for the recognizer than the least significant bits (LSBs). Furthermore, in some source coding algorithms, various speech features affect the system performance differently. For example, in ETSI (2000), the log-energy and 1st MFCC are considered to play a greater role in speech recognition than higher order MFCCs.

In bandwidth-restrictive systems, it may therefore be beneficial to use Unequal Error Protection algorithms to provide more protection for certain bits, while providing less protection for others. Cyclic codes provide an efficient tool for constructing UEP schemes, due to their flexibility regarding protection rate.

For example, in Boulis et al. (2002), the authors present an UEP scheme for DSR systems utilizing shortened RS codes. The proposed source coding scheme extracts the first 5 MFCCs, and quantizes the 1st, 2nd, and 4th coefficients using 6 bits, while allocating 4 bits each to the 3rd and 5th coefficient. Also, 20 adjacent windowed frames are packetized, and corresponding bits are grouped across time to form 20 bit symbols. Furthermore, these symbols are allocated to different streams, each of which is protected by a separate RS code. Table 8.5 illustrates the UEP scheme proposed in Boulis et al. (2002), where integers represent the MFCC corresponding to the given symbol, and F denotes an error correction symbol. In this example, streams 1 and 2 are protected with (12,4) RS codes, stream 3 is protected with a (12,6) code, and stream 4 is transmitted without protection.

Convolutional codes can be used for UEP schemes to provide various amounts of error protection for bits within a data block. In Weerackody et al. (2002), the authors propose two different UEP schemes for DSR applications based on convolutional codes. The source coding used in the study involves the extraction of the frame energy and the first 11 cepstral coefficients, either obtained through Mel Filterbank analysis or through Linear Prediction analysis.

Table 8.5 Unequal error protection scheme utilizing cyclic codes: integers represent the MFCC corresponding to the given symbol, and F denotes an error correction symbol (From Boulis et al. 2002, © 2002 IEEE)

Stream	Symbols (1-12)											
1	1	2	3	4	F	F	F	F	F	F	F	F
2	5	1	1	3	F	F	F	F	F	F	F	F
3	5	1	2	2	3	4	F	F	F	F	F	F
4	4	5	4	1	1	4	3	4	2	2	2	2

The source coder allocates 6 bits each to the frame energy and the first 5 cepstral coefficients, while allocating 4 bits each to the remaining coefficients, resulting in a payload bitrate of 4.8 kbps.

The UEP algorithms presented in Weerackody et al. (2002) groups the total 60 bits of each block into 4 different levels of importance. Each level is then encoded into a separate bitstream using a distinct channel coding scheme, and the total bitrate for each case is 9.6 kbps. Let $e^i(n)$ represent the ith bit of the frame energy value at block n, and let $c^i_j(n)$ represent the ith bit of the jth cepstral coefficient for block n. Tables 8.6 illustrates the bit allocation and channel coding scheme involved in the UEP scheme proposed in Weerackody et al. (2002), referred to as UEP_1. The corresponding results reported for Rayleigh fading channels are provided in Table 8.7.

Table 8.6 Unequal error protection scheme UEP_1 utilizing convolutional codes (From Weerackody et al. 2002, © 2002 IEEE)

Level	Speech feature bits	Error protection
1	$e^0(n),e^1(n),c^0_1(n),c^0_2(n),c^0_3(n),c^0_4(n),c^0_5(n)$	rate-1/2 Convolutional code
2	$e^2(n),c^1_1(n),c^1_2(n),c^1_3(n),c^1_4(n),c^1_5(n)$	rate-1/2 convolutional code
3	$e^3(n),e^4(n),c^2_1(n),c^3_1(n),c^2_2(n),c^3_2(n),\ldots$ $c^0_6(n),c^1_6(n),c^0_7(n),c^1_7(n),\ldots,c^0_{11}(n),c^1_{11}(n)$	rate-1/2 convolutional code + puncturing
4	$e^5(n),c^4_1(n),c^5_1(n),c^4_5(n),c^5_5(n),$ $c^2_6(n),c^3_6(n),c^2_7(n),c^3_7(n),\ldots,c^2_{11}(n),c^3_{11}(n)$	No code

Table 8.7 Word accuracy results for unequal error protection scheme utilizing convolutional codes: "dec. type" refers to the type of data used, i.e. hard-decision decoded data (*Hard*) or soft-decision decoded data (*Soft*) (From Weerackody et al. 2002, © 2002 IEEE)

UEP scheme	Dec. type	SNR			
		15 dB	10 dB	7 dB	5 dB
UEP_1	Hard	92.6%	89.8%	82.7%	71.7%
UEP_1	Soft	92.7%	91.3%	87.1%	80.1%

8.6 Frame Interleaving

It is known that for speech recognition purposes, errors that occur in groups are more degrading to word-accuracy performance than errors which occur randomly. Frame interleaving is a technique aimed at countering the effects of such bursty channels, through the addition of delay but no redundancy.

Frame interleaving aims to reorder speech frame packets within the transmitted bitstream so that frames which are adjacent with respect to the original speech signal are separated within the transmitted signal. Thus, grouped errors in the interleaved signal will result in scattered errors in the de-interleaved signal.

Two parameters commonly used to measure the effectiveness of an interleaver are the *delay*, δ, and the *spread*, *s*. The delay of an interleaver is defined as the maximum delay that any packet experiences before being transmitted. That is, for the interleaving function described by $j = \pi(i)$, where j is the original packet position and i is the resulting packet position for transmission, the delay can be determined as:

$$\delta = \max_i \left(\pi^{-1}(i) - i\right) \tag{8.11}$$

An interleaver is said to have a spread of *s* if for any packet positions, p_1 and p_2, within the original ordering:

$$\left|\pi(p_1) - \pi(p_2)\right| < s \;\Rightarrow\; \left|p_1 - p_2\right| \geq s. \tag{8.12}$$

In James and Milner (2004), three classes of interleavers are discussed for the use of DSR: optimal spread block interleavers, convolutional interleavers, and decorrelated block interleavers. They are successfully shown to provide improved results for DSR systems.

8.6.1 Optimal Spread Block Interleavers

A defining parameter of a block interleaver is the degree *d*. A block interleaver of degree *d* operates on a set of d^2 packets by reordering them prior to transmission. A pair of interleaving functions, $\pi_1(\cdot)$ and $\pi_2(\cdot)$, of degree *d* are considered optimal with respect their spread if, and only if:

$$\pi_1(i \cdot d + j) = (d - 1 - j)d + i, \tag{8.13}$$

and

$$\pi_2(i \cdot d + j) = j \cdot d + (d - 1 - i), \tag{8.14}$$

for $0 \leq i, j \leq d - 1$. Note that $\pi_1(\cdot)$ and $\pi_2(\cdot)$ are inverse functions, i.e. $\pi_1(\pi_2(i)) = i$. Furthermore, it is shown in (James et al. 2004) that optimal spread block interleaver function pairs can be interpreted as 90° clockwise and counter-clockwise rotations of $d \times d$ packet matrices. Figure 8.7 illustrates the $d = 3$ case.

It can also be concluded that for optimal spread block interleavers,

$$\delta_{opt} = d^2 - d, \tag{8.15}$$

and

$$s_{opt} = d. \tag{8.16}$$

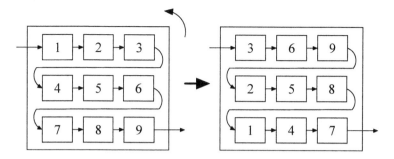

Fig. 8.7 Example of an optimal spread block interleaver with $d = 3$. Note that the interleaving function can be interpreted as a 90° counter-clockwise rotation of the $d \times d$ block

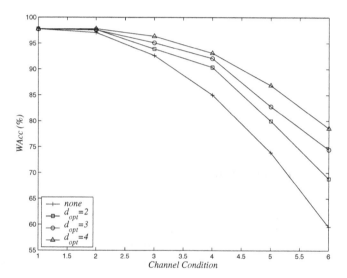

Fig. 8.8 Word recognition results tested on the Aurora-2 database using optimal spread block interleavers and decorrelated block interleavers of various degrees: d_{opt} refers to the degree of the optimal spread interleaver (Simulation details are given in Sect. 8.1)

The performances of optimal spread block interleavers of various degrees are shown in Fig. 8.8.

8.6.2 Convolutional Interleavers

A convolutional interleaver can be interpreted as a multirate device operating on a stream of packets. It involves an input multiplexer, followed by a group of shift registers in parallel, and finally an output multiplexer. A convolutional interleaver of degree $d = 4$ is illustrated in Fig. 8.9.

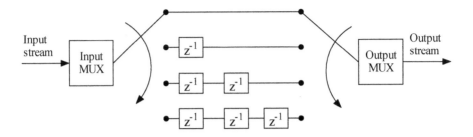

Fig. 8.9 Example of a convolutional interleaver with degree $d = 4$

The interleaving function of the convolutional interleaver with degree d is given by:

$$\pi_{conv}(i) = i - (i \bmod d).$$ (8.17)

It is also shown in (James et al. 2004) that the maximum delay and spread are given by:

$$\delta_{conv} = d^2 - d,$$ (8.18)

and

$$s_{conv} = d - 1.$$ (8.19)

8.6.3 Decorrelated Block Interleavers

The optimal spread block interleaver and convolutional interleaver previously discussed utilize the maximum delay and spread parameters to measure the effectiveness of specific interleaving functions. However, there exist other parameters by which the performance of an interleaving function can be measured.

For example, in James et al. (2004), the authors introduce a decorrelation measurement given by:

$$D = \sum_{i=1}^{d^2}\sum_{j=1}^{d^2} \frac{|\pi(i) - \pi(j)|}{|i - j|}.$$ (8.20)

The decorrelation measurement D shows the ability of an interleaving function to spread the input feature stream. In order to illustrate this, 2000 interleaving functions with degree $d = 4$ were randomly created, and the corresponding decorrelation values were determined. Additionally, the chosen interleavers were tested on a bursty packet-level channel with parameters $p = 0.75$ and $q = 0.75$. Note that the given channel parameters produce an average burst duration of $\bar{d}_b = 4$ without interleaving. The average burst lengths of the de-interleaved feature streams were then plotted against the corresponding decorrelation values in Fig. 8.10. The final decorrelated block interleaving function chosen for each degree is determined through an exhaustive search of randomly created permutations by minimizing D.

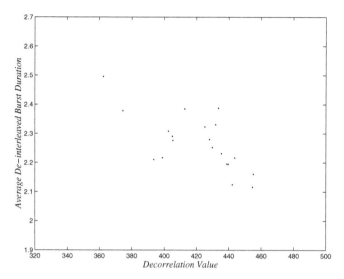

Fig. 8.10 Average burst durations of de-interleaved feature streams as a function of the corresponding decorrelation values

As can be concluded from Fig. 8.10, there is a strong correlation between the D values of interleaver functions and the resulting burst durations.

8.7 Examples of Modern Error Recovery Standards

There exist a number of standards for DSR and speech recognition systems currently in use, for which a main contributer is the *European Telecommunications Standards Institute (ETSI)*. *ETSI* has developed error protection and packetization standards for both low bitrate DSR systems (ETSI 2000), and for full rate NSR systems (ETSI 1998). The channel coding schemes described by these systems are summarized in the following sections.

8.7.1 ETSI DSR Standard (ETSI 2000)

The ETSI DSR standard describes a low bitrate speech recognition system intended for DSR applications. The compression algorithm implemented by the given standard extracts the frame energy, as well as the first 13 MFCCs, for every window of speech, at a windowing frequency of 100 Hz. A total of 44 bits is allocated to the extracted speech data through a split vector quantization (SVQ) system, resulting in a data bitrate of 4.4 kbps.

The *packetization* scheme in (ETSI 2000) includes a multiframe format in order to reduce the overhead required for synchronization and header information. Thus, speech feature data from 24 adjacent speech frames are grouped with one synchronization sequence and one header field.

The information contained in the header field is considered to be critical for decoding purposes, and thus a great deal of error protection is allocated to this data. The header field is protected by an extended (31, 16) systematic cyclic code, with the addition of an overall parity check bit. The *cyclic code* has a minimum distance of $d_{min} = 8$, and can therefore support correction of up to 3 channel errors or detection of up to 7 channel errors.

The 24 frames included with the header field and synchronization sequence are grouped into 88 bit pairs. Each pair of frames is then protected with a 4-bit CRC code. The resulting error protection and packetization scheme requires a data rate of 4.8 kbps.

8.7.2 ETSI GSM/EFR Standard (ETSI 1998)

The *ETSI* GSM Enhanced Full Rate (EFR) system, described by (ETSI 1998), is intended for speech transmission, and thus operates at a higher rate than the system described by (ETSI 2000). The GSM/EFR speech coding algorithm compresses 20 ms windows of speech signal into 244-bit blocks of data, which corresponds to a preliminary data rate of 12.2 kbps. The channel coding algorithm presented by (ETSI 1998) is a *UEP* scheme involving *CRC cyclic codes, convolutional codes*, and interleaving.

The first step in the channel coding and packetization scheme described for the GSM/EFR system is an expansion of the 244-bit blocks into 260-bit blocks, which results in 16 redundancy bits. Of these 16 redundancy bits, 8 are repetition bits, and 8 correspond to a CRC code.

Each 260-bit expanded data block is then grouped into two classes of bits. The first 50 bits of class 1, referred to as class 1a, are protected with three parity check bits generated with a shortened cyclic code. Both class 1a and class 1b bits are protected with a rate-1/2 convolutional code and interleaved with a bit-level interleaving function. Class 2 bits are transmitted without protection, and thus the resulting bitrate of the GSM/EFR system after channel coding and packetization becomes 28.4 kbps. The UEP scheme described in (ETSI 1998) is illustrated in Table 8.8.

Table 8.8 UEP Scheme for channel coding in ETSI GSM/EFR system (Based on ETSI 1998)

Class	Original bits	Channel coded bits	Error protection
1a	50	106	Cyclic code + rate-1/2 convolutional code
1b	132	264	rate-1/2 convolutional code
2	78	78	No code

8.8 Summary

This chapter focused on error recovery (ER) methods for transmission of speech features over error-prone channels in remote speech recognition applications. Such systems tend to be bandwidth-restrictive and often delay-sensitive, and thus channel coding schemes must be efficient. We discussed both FEC techniques, which add redundancy to the source coded data stream prior to transmission, as well as frame-based interleaving, which creates spread in the data stream prior to transmission to overcome the effect of bursty errors. Table 8.9 summarizes the discussed ER methods.

The individual ER techniques discussed in this chapter, and summarized in Table 8.9, serve as useful tools in providing protection against channel noise or packet erasures within DSR systems. However, channel coding strategies can be enhanced by combining these techniques and offering more error protection to those bits in the bitstream which provide more utility to the overall system, as in UEP schemes. This chapter concludes with example standards of channel coding for modern DSR and speech recognition systems.

Table 8.9 Summary of discussed error recovery techniques

FEC technique	Error pattern application	Advantages	Disadvantages
Media-specific FEC: coarsely quantized feature replicas are inserted into the datastream with the aim of reconstructing lost packets			
Media-specific FEC (Sect. 8.3)	Packet erasures	Low complexity	Introduces delay
Media-independent FEC: redundancy is added to blocks of data in order to detect or correctly decode corrupted data in the presence of bit-level errors			
Linear block codes (Sect. 8.4.2)	Bit errors or additive noise	Low delay, low overhead	Complexity increases for soft and λ-soft decoding
Cyclic codes (Sect. 8.4.3)	Bit errors or additive noise	Low delay, efficient for decoding bursty channels	Restrictive in terms of block length
Convolutional codes (Sect. 8.4.4)	Bit errors or additive noise	Low delay, flexible in terms of block length	Complexity increases for soft-decision decoding
Frame interleaving: frames are interleaved prior to transmission so that bursty packet losses in the transmitted stream may become scattered packet losses in the de-interleaved stream			
Optimal spread block interleavers (Sect. 8.6.1)	Packet erasures	No required additional bandwidth	Introduces delay
Convolutional interleavers (Sect. 8.6.2)	Packet erasures	No required additional bandwidth	Introduces delay
Decorrelated interleavers (Sect. 8.6.3)	Packet erasures	No required additional bandwidth	Introduces delay, requires extensive and complex training

Acknowledgments

This work was supported in part by the NSF and by a Radcliffe Institute Fellowship to Abeer Alwan.

References

Bai, H., and Atiquzzaman, M. (2003). Error Modeling Schemes for Fading Channels in Wireless Communications: A Survey. *IEEE Communications Surveys and Tutorials*, Fourth Quarter, vol. 5, no. 2, pp. 2–9.

Bernard, A., and Alwan, A. (2001). Joint Channel Decoding- Viterbi Recognition for Wireless Applications. *Proceedings of Europspeech*, vol. 3, pp. 2703–2706.

Bernard, A., and Alwan, A. (2002a). Low-Bitrate Distributed Speech Recognition for Packet-Based and Wireless Communication. *IEEE Transactions on Speech and Audio Processing*, vol. 10, no. 8, pp. 570–579.

Bernard, A., Liu, X., Wesel, R., and Alwan, A. (2002b). Speech Transmission Using Rate-Compatible Trellis Codes and Embedded Source Coding. *IEEE Transactions on Communications*, vol. 50, no. 2, pp. 309–320.

Bernard, A. (2002). *Source and Channel Coding for Speech Transmission and Remote Speech Recognition*, PhD Thesis, University of California, Los Angeles.

Blahut, R.E. (2004). *Algebraic Codes for Data Transmission*. Cambridge University Press, Cambridge, UK.

Bolot, J.-C. (2003). End-to-End Packet Delay and Loss Behavior in the Internet. *ACM Sigcomm*, pp. 289–298.

Boulis, C., Ostendorf, M., Riskin, E. A., and Otterson, S. (2002). Graceful Degradation of Speech Recognition Performance Over Packet-Erasure Networks. *IEEE Transactions on Speech and Audio Processing*. vol. 10, no. 8, pp. 580–590.

Elliot, E. (1963). Estimates of Error Rates for Codes on Bursty Noise Channels. *Bell Systems Technical Journal*, vol. 42, no. 9.

ETSI EN 300 909 v7.3.1 (1998). Digital Cellular Telecommunications System (Phase 2+)(GSM); Channel Coding.

ETSI ES 201 108 v1.1.2 (2000). Distributed Speech Recognition; Front-end Feature Extraction Algorithm; Compression Algorithms.

Han, K.J., Srinivasamurthy, N., and Narayanan, S. (2004). Robust Speech Recognition over Packet Networks: An Overview. *Proceedings of ICSLP*, vol. 3, pp. 1791–1794.

Hirsch, H.G., and Pearce, D. (2000). The AURORA Experimental Framework for the Performance Evaluation of Speech Recognition Systems under Noisy Condition. In *Proceedings of ISCA ITRW ASR 2000*.

James, A.B., and Milner, B.P. (2004). An Analysis of Interleavers for Robust Speech Recognition in Burst-Like Packet Loss. In *Proceedings of ICASSP*, vol. 1, pp. 853–856.

Jiao, C., Schwiebert, L., and X, B. (2002). On Modeling the Packet Error Statistics in Bursty Channels. *IEEE LCN*, pp. 534–538.

Leon-Garcia, A. (2007). *Probability and Random Processes for Electrical Engineers*. Prentice-Hall.

Peinado, A.M., Gomez, A.M., Sanchez, V., Perez-Cordoba, J.L., and Rubio, A.J. (2005a). Packet Loss Concealment Based on VQ Replicas and MMSE Estimation Applied to Distributed Speech Recognition. *Proceedings of ICASSP*. vol. 1, pp. 329–330.

Peindo, A.M., Sanchez, V., Perez-Cordoba, J.L., and Rubio, A.J. (2005b). Efficient MMSE-Based Channel Error Mitigation Techniques. Application to Distributed Speech Recognition Over Wireless Channels. *IEEE Transactions on Wireless Communications,* vol. 4, no. 1, pp.14–19.

Peindo, A.M., and Segura, J.C. (2006). *Speech Recognition Over Digital Channels.* Wiley, New York, West Sussex, England.

Sklar, B. (1997). Rayleigh Fading Channels in Mobile Digital Communication Systems Part 1: Characterization. IEEE Communications Magazine, pp. 90–100.

Tan, Z.-H., Dalsgaard, P., and Lindberg, B. (2005). Automatic Speech Recognition over Error-Prone Wireless Networks. *Speech Communication*, vol. 47, pp. 220–242.

Weerackody, V., Reichl, W., and Potamianos, A. (2002). An Error-Protected Speech Recognition System for Wireless Communications. *IEEE Transactions on Wireless Communications*, vol. 1, no. 2, pp. 282–291.

Viterbi, A. (1971). Convolutional Codes and Their Performance in Communication Systems. *IEEE Transactions on Communications*, vol. 19, no. 5, part 1, pp. 751–772.

Wesel, R.D. (2003). Convolutional Codes, from *Encyclopedia of Telecommunications*. Wiley, New York.

Young, S., Kershaw, D., Odell, J., Ollason, D., Valtchev, V., and Woodland, P. (2000). *The HTK Book*. Microsoft Corporation.

9

Error Concealment

Reinhold Haeb-Umbach and Valentin Ion

Abstract. In distributed and network speech recognition the actual recognition task is not carried out on the user's terminal but rather on a remote server in the network. While there are good reasons for doing so, a disadvantage of this client-server architecture is clearly that the communication medium may introduce errors, which then impairs speech recognition accuracy. Even sophisticated channel coding cannot completely prevent the occurrence of residual bit errors in the case of temporarily adverse channel conditions, and in packet-oriented transmission packets of data may arrive too late for the given real-time constraints and have to be declared lost. The goal of error concealment is to reduce the detrimental effect that such errors may induce on the recipient of the transmitted speech signal by exploiting residual redundancy in the bit stream at the source coder output. In classical speech transmission a human is the recipient, and erroneous data are reconstructed so as to reduce the subjectively annoying effect of corrupted bits or lost packets. Here, however, a statistical classifier is at the receiving end, which can benefit from knowledge about the quality of the reconstruction. In this book chapter we show how the classical Bayesian decision rule needs to be modified to account for uncertain features, and illustrate how the required feature posterior density can be estimated in the case of distributed speech recognition. Some other techniques for error concealment can be related to this approach. Experimental results are given for both a small and a medium vocabulary recognition task and both for a channel exhibiting bit errors and a packet erasure channel.

9.1 Introduction

In a client-server speech recognition system the client, e.g. a cellular phone, captures the speech signal, codes it and sends it via a digital communication link to the remote recognition server. At the server side, the received signal is decoded and forwarded to the speech recognition engine, which outputs the decoded word string. Depending on the type of data transmitted, one distinguishes between *distributed* (DSR) and *network speech recognition* (NSR). In DSR speech recognition features, such as Mel-Frequency Cepstral Coefficients (MFCC), are computed, coded and transmitted (Pearce 2000), while in NSR a typical speech codec, such as the adaptive multi-rate (AMR) codec is employed (Fingscheidt et al. 2002).

Compared to a realization of the recognizer on the client, the client-server architecture has many obvious advantages, such as ease of maintainability of the application data on the server and avoidance of resource-intensive tasks on the client. However, the price to pay is an additional processing delay due to transmission and the potential corruption of the digitized speech data due to channel-induced errors. Here we are concerned with the latter and show how error concealment techniques help mitigate the negative effects of transmission errors on the speech recognition accuracy.

Two channel models exhibiting different error types are considered in the following: a channel characterized by bit errors and a *packet erasure channel*. Channel degradations at the bit level are for example typical of cellular circuit-switched transmission, where noise, multi-path fading and interference from neighboring stations are frequent error causes. Packet loss is a typical phenomenon of packet-based transmission of data with real-time constraints over the internet. The combination of both error types is an approximate model for communications over a wireless packet network, or communications that involve both a wireless and a (packet-based) wireline link (Lahouti and Khandani 2007).

To mitigate transmission errors researchers have proposed several different approaches. One category is comprised of methods that perform error or packet-loss concealment techniques at the receiving end. Another class of techniques requires certain coordination with the transmitter side, e.g. forward error correction or diversity schemes based on multiple description coding. A third category requires a certain degree of support from the network, such as using packets with different priorities. The schemes rely on the network to drop the packets with low priority during congestion periods. Currently, this support, however, may only be available in proprietary networks and in the next generation of the Internet Protocol (IPv6) (RFC 2460, 1998).

In this contribution we restrict ourselves to purely receiver (server) based techniques which leave the transmitter (client) side untouched, since they have the striking advantage that they are fully compatible with the current European Telecommunications Standards Institute (ETSI) standards for distributed speech recognition (ETSI Standard ES 202 050, 2002; ETSI Standard ES 201 108, 2003a) and can be readily applied in current networks. Actually, the frame repetition scheme proposed in the ETSI standard is an example of such an error concealment method.

The term *error concealment* denotes techniques which aim to reduce or even eliminate the effect of uncorrected transmission errors on the quality as perceived by the consumer of the transmitted data. For data transmission with no latency constraints a virtually error-free transmission can be achieved by a combination of forward and backward error correction. This no longer holds for speech, audio or video transmissions, which typically have to adhere to real-time constraints.

The detrimental effect of transmission errors can be concealed by exploiting residual redundancy still present at the output of the (in the Shannon sense) imperfect source coder. One might argue that, since low-bit rate source coding has been an issue since the early days of digital speech transmission, it is unlikely to find enough residual redundancy in the output bit stream of the speech coder to be exploited for error concealment. But even for the low-rate codes used in GSM successful error concealment based on exploiting the non-uniform bit pattern probabilities and the correlation between successive frames has been demonstrated (Fingscheidt et al. 2007).

Error concealment has been studied extensively in the field of mobile communications and, more recently, for voice or other real-time data transmission over the internet protocol (IP). In cellular systems standards such as GSM error concealment algorithms are proposed as non-mandatory recommendations (GSM 06.11 Recommendation, 1992), and very sophisticated techniques have been developed in recent years (Vary and Martin 2006). In Voice-over-IP packet loss is a frequent phenomenon, which is addressed, among others, by replacing the missing segments of speech with estimates constructed form previous or future available speech segments. For example, a waveform substitution algorithm based on pitch detection has been proposed for G.711 pulse code modulation speech coding standard (ITU-T Recommendation G.711 Appendix I, 1999). Packet loss concealment methods for code excited linear prediction (CELP)-based coders often replace the missing parameters with the corresponding parameters of the previous frame (Cox et al. 1989) and use scaled-down gains. Methods that interpolate between previous and future frames can also be employed.

Similar techniques have been proposed for distributed speech recognition (DSR), where speech recognition related parameters, such as MFCCs (Davis and Mermelstein 1980) are computed in the user's terminal and then transmitted to the remote speech recognition engine (Tan et al. 2005). Feature reconstruction techniques range from quite simple methods such as substitution (with silence, noise or source-data), repetition or interpolation (Boulis et al. 2002; Milner and Semnani 2000) to more elaborated schemes, such as repetition on a subvector level (Tan et al. 2004) and *minimum mean square error* (MMSE)-based reconstruction which models *interframe correlation* by a first-order Markov model (Peinado et al. 2003; James et al. 2004; Haeb-Umbach and Ion 2004). However, in DSR we can do even considerbly more.

In a DSR scenario we would like to alleviate the effect that transmission errors have on the consumer of the data, the automatic speech recognition (ASR) decoder. Unlike a human recipient, the recognizer not only benefits from a good reconstruction of lost or corrupted data but also from knowledge about the quality of the reconstruction. The ASR decoder is then modified such that features deemed unreliable are deemphasized (Bernard and Alwan 2001, 2002) or completely excluded from consideration in the recognizer (Weerackody et al. 2002; Endo et al. 2003). However, it is not an easy task to identify corrupt features or even quantify the degree of corruption, at least on a channel exhibiting bit errors. While in Haeb-Umbach and Ion (2004) the availability of a soft-output channel decoder was assumed, in Ion and Haeb-Umbach (2005) a technique was proposed which estimates bit error probabilities based on a priori knowledge of plausible bit patterns.

Actually, the close connection between feature reconstruction and modification of the decoding engine becomes apparent once the problem of speech recognition in the presence of unreliable feature vectors is cast in a Bayesian framework. Here, results form noisy speech recognition can be borrowed, where so-called *Uncertainty Decoding* has been investigated already for a couple of years (Morris et al. 1998; Morris et al. 2001; Arrowood and Clements 2002; Droppo et al. 2002; Kristjansson and Frey 2002). Let the speech be corrupted by additive noise or by transmission errors, in either case the original clean or uncorrupted speech feature vector is not observable, but rather a distorted version of it. Traditionally, the goal of *speech feature enhancement* is to obtain a *point estimate* of the clean speech feature, such as the MMSE estimate. This estimate is then "plugged into" the Bayes decision rule and used in the ASR decoder as if it were the true clean speech feature.

However, one can do better if one takes the reliability of the estimate into account. In one formulation of uncertainty decoding the probability density function of the corrupted speech feature vector, conditioned on the unobservable clean speech feature vector, is computed and averaged over the observation probability of the clean speech (Liao and Gales 2004). In another formulation the front-end delivers uncertain observations, expressed as a posteriori density of the clean speech feature vector, given the observed noisy vector. It is well known, that the mean of the posterior is exactly the MMSE estimate. Its variance is a measure of the uncertainty about this estimate. In the case of jointly Gaussian random variables, it is even equal to the variance of the estimation error. This frame-level uncertainty can be incorporated in the decoding process by using a modified Bayesian decision rule, where integration over the uncertainty in the feature space is carried out. Under certain assumptions this can be accomplished by a simple modification of the means and variances of the observation probabilities.

In the context of distributed speech recognition the concept of uncertainty decoding has been proposed for the first time in Haeb-Umbach and Ion (2004). Here, inter-frame correlation has been identified as a major knowledge source which helps in reconstructing lost or corrupted features.

This volume chapter is organized as follows. In the following section we present the probabilistic framework of speech recognition in the presence of corrupted observations. In Sect. 9.3 this concept is applied to distributed speech recognition, where we consider channels characterized by either bit errors or packet loss. Experimental results, both for a small and a medium vocabulary recognition task, are given in Sect. 9.4, followed by some conclusions drawn in Sect. 9.5.

9.2 Speech Recognition in the Presence of Corrupted Features

9.2.1 Modified Observation Probability

The *Bayesian decision rule* is at the heart of statistical speech recognition. Given the sequence of T (uncorrupted) feature vectors $\mathbf{x}_1^T = (x_1, \dots, x_T)$ extracted from an utterance, the goal is to find the sequence of words $\hat{\mathbf{W}}$ from of a given vocabulary, which maximizes the probability $P(\mathbf{W} \mid \mathbf{x}_1^T)$. Using the Bayesian theorem for conditional probabilities this can be expressed more conveniently as maximizing the product between observation probability $p(\mathbf{x}_1^T \mid \mathbf{W})$ and word sequence probability $P(\mathbf{W})$:

$$\hat{\mathbf{W}} = \arg\max_{\mathbf{W}} \{ p(\mathbf{x}_1^T \mid \mathbf{W}) \cdot P(\mathbf{W}) \}. \tag{9.1}$$

Introducing the hidden state sequence $s_1^T = (s_1, s_2, ..., s_T)$ we obtain

$$p(x_1^T \mid \mathbf{w}) = \sum_{s_1^T} p(x_1^T, s_1^T \mid \mathbf{w}) = \sum_{s_1^T} p(x_1^T \mid s_1^T) P(s_1^T),$$

(9.2)

where the sum is over all state sequences within $\mathbf{w}$. As there is exactly one word sequence corresponding to a state sequence, the condition on $\mathbf{w}$ can be left out.

A common assumption employed in speech recognition is the so-called *conditional independence* assumption, which states that x_t is conditionally independent of neighboring feature vectors, given the HMM state s_t:

$$p(x_1^T \mid s_1^T) = \prod_{t=1}^T p(x_t \mid x_1^{t-1}, s_t) = \prod_{t=1}^T p(x_t \mid s_t).$$

(9.3)

Using this in Eq. 9.9.2 we obtain

$$p(x_1^T \mid \mathbf{w}) = \sum_{s_1^T} \prod_{t=1}^T p(x_t \mid s_t) P(s_1^T),$$

(9.4)

Often we are unable to observe the uncorrupted feature vector sequence x_1^T. We observe a corrupted sequence $y_1^T = (y_1, ... y_T)$, which may differ from x_1^T. In DSR, transmission errors are the reason for this difference. The speech recognition problem thus amounts to solving

$$\hat{\mathbf{W}} = \arg\max_{\mathbf{w}} \{ p(y_1^T \mid \mathbf{w}) \cdot P(\mathbf{w}) \}.$$

(9.1)

In solving this we need to find an efficient way to compute $p(y_1^T \mid s_1^T)$. To this end we introduce the (hidden) uncorrupted feature sequence:

$$p(y_1^T \mid s_1^T) = \int_{x_1^T} p(y_1^T \mid x_1^T) p(x_1^T \mid s_1^T) dx_1^T$$

(9.2)

Using Eq. 9.9.3 and noting that

$$p(y_1^T \mid x_1^T) = \prod_{t=1}^T p(y_t \mid x_t)$$

(9.3)

we obtain

$$p(y_1^T \mid s_1^T) = \int_{x_1^T} \prod_{t=1}^T p(y_t \mid x_t) p(x_t \mid s_t) dx_1^T = \prod_{t=1}^T \int_{x_t} p(y_t \mid x_t) p(x_t \mid s_t) dx_t$$

(9.4)

i.e. it is possible to interchange the product and the integral since the terms inside the integral only depend on t.

Often it is more convenient to express $p(\mathbf{y}_t \mid \mathbf{x}_t)$ via a posterior probability

$$p(\mathbf{y}_t \mid \mathbf{x}_t) = \frac{p(\mathbf{x}_t \mid \mathbf{y}_t)p(\mathbf{y}_t)}{p(\mathbf{x}_t)} \tag{9.5}$$

If inter-frame correlation among the feature vector sequence is to be taken into account, $p(\mathbf{x}_t \mid \mathbf{y}_t)$ has to be replaced by $p(\mathbf{x}_t \mid \mathbf{y}_1^T)$, i.e. the *a posteriori density* of the clean feature sequence, given all observed corrupted features. This posterior is, from an estimation theory point of view, the complete solution to the problem of estimating the clean feature vector, given all observations. In Sect. 9.3 we will show how this posterior can be efficiently estimated in a distributed speech recognition scenario.

Since we are eventually only interested in the word (state) sequence which maximizes Eq. 9.9.4, the probability of the noisy features $p(\mathbf{y}_t)$ can be disregarded. Further, replacing $p(\mathbf{x}_t \mid \mathbf{y}_t)$ by $p(\mathbf{x}_t \mid \mathbf{y}_1^T)$ in Eq. 9.9.5 and using it in Eq. 9.9.4 we arrive at

$$p(\mathbf{y}_1^T \mid s_1^T) = \prod_{t=1}^{T} \int_{\mathbf{x}_t} \frac{p(\mathbf{x}_t \mid \mathbf{y}_1^T)}{p(\mathbf{x}_t)} p(\mathbf{x}_t \mid s_t) d\mathbf{x}_t. \tag{9.6}$$

Replacing $p(\mathbf{x}_1^T \mid s_1^T)$ by $p(\mathbf{y}_1^T \mid s_1^T)$ in Eq. 9.9.2 and using Eq. 9.9.6 we finally arrive at

$$p(\mathbf{y}_1^T \mid \mathbf{W}) = \sum_{s_1^T} \prod_{t=1}^{T} \int_{\mathbf{x}_t} \frac{p(\mathbf{x}_t \mid \mathbf{y}_1^T)}{p(\mathbf{x}_t)} p(\mathbf{x}_t \mid s_t) d\mathbf{x}_t \cdot P(s_1^T). \tag{9.7}$$

The only difference to the standard ASR decoder is that the observation probability $p(\mathbf{x}_t \mid s_t)$ has to be replaced by the modified observation probability:

$$p(\mathbf{x}_t \mid s_t) \rightarrow \int_{\mathbf{x}_t} \frac{p(\mathbf{x}_t \mid \mathbf{y}_1^T)}{p(\mathbf{x}_t)} p(\mathbf{x}_t \mid s_t) d\mathbf{x}_t. \tag{9.8}$$

It is instructive to consider the extreme cases of an error-free transmission and a completely unreliable transmission. In case of an error-free transmission there is $\mathbf{y}_t = \mathbf{x}_t$, and the a posteriori density $p(\mathbf{x}_t \mid \mathbf{y}_1^T)$ reduces to a Dirac delta-impulse. As a result, the modified observation probability, Eq. 9.9.8, reduces to the standard observation probability (the denominator is then a constant and can be neglected as it does not influence the maximization in Eq. 9.9.1).

In the other extreme case the channel does not transmit any information, which can be expressed by $p(\mathbf{x}_t \mid \mathbf{y}_1^T) = p(\mathbf{x}_t)$ for all $t = 1,...,T$. In this case the modified observation probability evaluates to one and Eq. 9.9.1 reduces to $\hat{\mathbf{W}} = \arg\max_{\mathbf{w}}\{p(\mathbf{W})\}$. As the observed features are uninformative, the recognizer has to rely solely on the prior word probabilities.

The key element of the novel decoding rule is the posterior density $p(\mathbf{x}_t \mid \mathbf{y}_1^T)$. The processing of the corrupted features in front of the recognizer has to produce a posterior density instead of a point estimate. It is well-known from estimation theory, that the posterior density comprises all information about the parameter to be estimated, here $\mathbf{x}_t$, that is available from the observations, here $\mathbf{y}_1^T$. Optimal point estimates, such as MMSE or *maximum a posteriori* (MAP) can be obtained as the mean or mode of this density. Further, the (co)variance of the posterior is a measure of reliability of the point estimate. For this reason the posterior has sometimes been called *soft feature* (Haeb-Umbach and Ion 2004).

Related decoding rules can be found e.g. in Morris et al. (1998, 2001), Arrowood and Clements (2002), Droppo et al. (2002), Kristjansson and Frey (2002), Liao and Gales (2004). However, in most cases past and future observed feature vectors are not taken into account for the estimation of the posterior density of the current uncorrupted feature vector, i.e. $p(\mathbf{x}_t \mid \mathbf{y}_1^T)$ is replaced by $p(\mathbf{x}_t \mid \mathbf{y}_t)$. In doing so interframe correlation is neglected for the posterior estimation. In Sect. 9.3, however, we will show that inter-frame correlation is a powerful knowledge source to be utilized for transmission error-robust speech recognition.

9.2.2 Gaussian Approximation

Still, the modified observation probability given in Eq. 9.9.8 looks intimidating. The computation of the observation probabilities is the single most time consuming processing step in speech recognition. Replacing the evaluation of a mixture density by the numerical evaluation of an integral may increase the computational burden beyond the limits of practical interest. Fortunately, the integral can be solved analytically, if we make the following assumptions:

1. The observation probability is a Gaussian mixture density:

$$p(\mathbf{x}_t \mid s_t) = \sum_{m=1}^{M} c_{s,m} N(\mathbf{x}_t; \boldsymbol{\mu}_{s,m}, \boldsymbol{\Sigma}_{s,m}) \tag{9.9}$$

2. The *a priori density* of the uncorrupted feature vector can be modeled by a Gaussian density:

$$p(\mathbf{x}_t) = N(\mathbf{x}_t; \boldsymbol{\mu}_{\mathbf{x}}, \boldsymbol{\Sigma}_{\mathbf{x}}) \tag{9.10}$$

3. The a posteriori density of the uncorrupted feature vector, given the sequence of received feature vectors, can be approximated by a Gaussian density:

$$p(\mathbf{x}_t \mid \mathbf{y}_1^T) \approx p_N(\mathbf{x}_t \mid \mathbf{y}_1^T) = N(\mathbf{x}_t; \boldsymbol{\mu}_{\mathbf{x}_t|\mathbf{y}}, \boldsymbol{\Sigma}_{\mathbf{x}_t|\mathbf{y}}) \tag{9.11}$$

Further we assume that all Gaussians, Eqs. 9.9.9–9.9.11, have diagonal covariance matrices. Since the individual elements of a diagonal-covariance Gaussian are independent, the densities can then be factorized over the feature vector elements. Let $\mu_{s,m}$, μ_x and $\mu_{x_t|y}$ denote the means and $\sigma_{x,m}^2$, σ_x^2 and $\sigma_{x_t|y}^2$ the corresponding variances of the Gaussians of an individual vector component of the observation, prior and posterior density, respectively. Then the integral present in Eq. 9.9.8 can be solved analytically (Droppo et al. 2002; Ion and Haeb-Umbach 2006c), where for each dimension we obtain the following:

$$\int \sum_{m=1}^{M} c_{s,m} N\left(x_t ; \mu_{s,m}, \sigma_{s,m}^2\right) \frac{N\left(x_t ; \mu_{x_t|y}, \sigma_{x_t|y}^2\right)}{N\left(x_t ; \mu_x, \sigma_x^2\right)} dx_t = \sum_{m=1}^{M} c_{s,m} A N\left(\mu_e ; \mu_{s,m}, \sigma_{s,m}^2 + \sigma_e^2\right) \quad (9.12)$$

where

$$\frac{\mu_e}{\sigma_e^2} = \frac{\mu_{x_t|y}}{\sigma_{x_t|y}^2} - \frac{\mu_x}{\sigma_x^2}$$

$$\frac{1}{\sigma_e^2} = \frac{1}{\sigma_{x_t|y}^2} - \frac{1}{\sigma_x^2} \qquad (9.13)$$

$$A = \frac{N\left(0; \mu_{x_t|y}, \sigma_{x_t|y}^2\right)}{N\left(0; \mu_x, \sigma_x^2\right) N\left(0; \mu_e, \sigma_e^2\right)}$$

if $\sigma_x^2 > \sigma_{x_t|y}^2$. Equation 9.9.12 states that the variance of the original observation probability of the uncorrupted features is to be increased by σ_e^2 and that it is to be evaluated at μ_e and weighted by A.

The assumption of Eq. 9.9.9 is the standard model for observation probabilities. Further, the prior density of the feature vector $p(x_t)$ can be reasonably well approximated by a Gaussian density. The most critical assumption seems to be Eq. 9.9.11. We often observed a multi-modal shape of the posterior density. However, the Gaussian approximation was adopted due to computational complexity reasons.

9.3 Feature Posterior Estimation in a DSR Framework

The decoding rule derived in the last section requires knowledge of $p\left(x_t \mid y_1^T\right)$, the a posteriori density of the transmitted feature vector, given all received feature vectors. In this section we show how this term can be estimated in the case of distributed speech recognition, where coded MFCCs are transmitted over an error-prone channel. We first describe the ETSI DSR standard to the extent necessary for understanding the subsequent derivation. Subsection 9.3.2 quantifies the redundancy present in the output bit stream of the source coder. The two channel models under consideration are explained in Subsects. 9.3.3, and 9.3.4 shows how the feature posterior density can be computed from a priori and "transmission probabilities".

This section is concluded by relating other approaches for error concealment to the one presented here.

9.3.1 ETSI DSR Standards

The ETSI distributed speech recognition standards define two feature extraction schemes, standard front end and advanced front end processing, together with the source coding, packet construction, and the backend source decoding scheme (ETSI Standard ES 202 050, 2002; ETSI Standard ES 201 108, 2003a). For the purpose of error concealment we need to consider the source coder in more detail.

A *source coder* is a mapping of the N-dimensional Euclidian space into a finite index set J of 2^M elements. It consists of two components: the *quantizer* and the *index generator*. The quantizer maps the N-dimensional parameter vector $\mathbf{x}$ to a N-dimensional codeword (*centroid*) $\mathbf{c}$ in the finite codebook C. This codeword represents all vectors falling in this quantization cell. The index generator then maps this codeword $\mathbf{c}$ to an index (bit pattern) $\mathbf{b}$ in an index set J.

The source coder of the ETSI DSR standard employs a *split vector quantizer* (VQ) for the quantization of the static MFCC parameters. The input to the quantizer is the $N = 14$ dimensional parameter vector, consisting of the thirteen-dimensional MFCC feature vector and as a fourteenth component the logarithmic frame energy ($\log E$). The parameter vector is split into seven *subvectors*, each of dimension two, which are quantized with bit-rates (6,6,6,6,6,5,8) bits, respectively. Including one bit for voice-activity information this sums to 44 bits per frame. Before transmission two quantized frames are grouped together creating a frame pair. A 4-bit cyclic redundancy check (CRC) is calculated for each frame pair, resulting in a total of 92 bits per frame pair.

In our notation we will not distinguish between individual subvectors in the following, since the same operations are performed for all subvectors. We even do not make a distinction between the complete vector and any of the subvectors in our notation. Which interpretation is used should become clear from the context.

9.3.2 Source Coder Redundancy

The key to error concealment is the exploit the residual redundancy present in the source coder output bit stream. Let $\mathbf{x}_t$ denote any of the real-valued MFCC subvectors at time frame t produced by the front-end. The source coder quantizes the subvector to a codeword $\mathbf{c}_t$ and maps the codeword to a bit pattern $\mathbf{b}_t = (b_t(0),...b_t(M-1))$ of M bits, which is transmitted over an equivalent discrete-time channel.

Table 9.1 gives the entropies $H(\mathbf{b}_t)$ and mutual information $I(\mathbf{b}_t; \mathbf{b}_{t-1})$ of the individual subvectors. The values have been obtained on the training set of the Aurora 2 database (Hirsch and Pearce 2000) using the ETSI advanced feature extraction front-end. Here, subvector 1 denotes the bit pattern corresponding to the first and second mel-frequency cepstral coefficient, subvector 2 the third and fourth, and so on. Subvector 7 comprises the zero-th cepstral coefficient and $\log E$. M is the number of bits used to code a subvector, i.e. the length of the bit pattern $\mathbf{b}_t$. Comparing M with the entropy $H(\mathbf{b}_t)$ of the bit pattern, one can observe that for all subvectors the two values are fairly close to each other. This indicates that the bit pattern has almost a uniform distribution. Not much redundancy is left within a subvector which could be utilized for error concealment.

Table 9.1 Entropies and mutual information among the subvectors produced by the ETSI advanced DSR front-end (measured on Aurora 2 training database)

Subvector	1	2	3	4	5	6	7
M	6	6	6	6	6	5	8
$H(\mathbf{b}_t)$	5.8	5.8	5.8	5.8	5.8	4.8	7.7
$I(\mathbf{b}_t;\mathbf{b}_{t-1})$	2.6	2.1	1.6	1.4	1.2	1.0	3.4
$I(\mathbf{b}_t;(\mathbf{b}_{t-1},\Delta\mathbf{b}_{t-1},\Delta^2\mathbf{b}_{t-1}))$	3.0	2.4	1.9	1.7	1.5	1.3	4.5

The mutual information $I(\mathbf{b}_t;\mathbf{b}_{t-1})$ indicates how much information about the current bit pattern $\mathbf{b}_t$ is already present in the previous $\mathbf{b}_{t-1}$. The larger the mutual information the better a bit pattern following in time can be predicted from the one of the previous frame. Obviously, strong inter-frame correlation exists.

The last line of the table gives the mutual information between the current bit pattern and the bit pattern $(\mathbf{b}_{t-1},\Delta\mathbf{b}_{t-1},\Delta^2\mathbf{b}_{t-1})$ of the previous frame, which consists of the coded static MFCC components $\mathbf{b}_{t-1}$ and the coded dynamic features. For this experiment a $D_1 = 3$ bit vector quantizer was used for the delta (velocity) and just a $D_2 = 1$ bit quantizer for the delta-delta (acceleration) parameter. Obviously, the dynamic parameters of the previous frame provide additional knowledge about the static parameters of the current frame, since the measured mutual information is larger than the one observed between $\mathbf{b}_t$ and $\mathbf{b}_{t-1}$. This comes to no surprise, as the dynamic features capture the trend present in the feature trajectory.

Obviously, the key to successful error concealment is the exploitation of the strong inter-frame correlation of MFCC feature vectors. In specifying the inter-frame correlation models of different complexity may be chosen. A good compromise between modeling accuracy and complexity is to assume that the source vector sequence $\mathbf{b}_t$, $t = 1,2,...$ is a homogeneous first-order Markov process, whose "transition probabilities" $P(\mathbf{b}_t^{(i)} \mid \mathbf{b}_t^{(j)})$, $i,j = 1,...,2^M$ are independent of time. With this model, however, long-term dependencies e.g. on the phone level cannot be captured.

9.3.3 Channel Models

Let us now consider the transmission model of Fig. 9.1. At the channel output a bit pattern $\mathbf{y}_t = (y_t(0),...y_t(M-1))$ is observed. Due to transmission errors $\mathbf{y}_t$ and $\mathbf{b}_t$

are not identical. Please note that $\mathbf{y}_t$ is a discrete random variable here, while we assumed $\mathbf{y}_t$ to be a continuous random variable in Sect. 9.2. We prefer this abuse of notation to more easily link the DSR case considered in this section to the more general theory presented in Sect. 9.2.

In the following we use a superscript if we want to denote a specific bit pattern, i.e. $\mathbf{b}_t^{(i)}$ indicates the bit pattern corresponding to the i-th codebook centroid $\mathbf{c}_t^{(i)}$, $i \in \{1,...,2^M\}$.

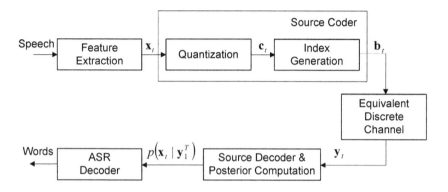

Fig. 9.1 Block diagram of distributed speech recognition system

We consider two channel models:

a) Time-variant *binary symmetric channel* (TV-BSC)

The TV-BSC is an equivalent discrete channel which models the effects of additive white Gaussian noise on the transmitted bit sequence. While one usually assumes constant bit error probability in a BSC, we want to allow here the bit error probability p_t to be time-variant. This model can be used to characterize wireless circuit-switched transmission, where the bit error rate varies, e.g. due to time-variant multi-path fading.

As the channel is assumed to be memoryless, the probability of a received bit pattern given the sent can be expressed as

$$P\left(\mathbf{y}_t \mid \mathbf{b}_t^{(i)}\right) = \prod_{m=0}^{M-1} P\left(y_t(m) \mid b_t^{(i)}(m)\right) \tag{9.18}$$

where

$$P\left(y_t(m) \mid b_t^{(i)}(m)\right) = \begin{cases} 1 - p_t(m) & \quad y_t(m) = b_t^{(i)}(m) \\ p_t(m) & \quad y_t(m) \neq b_t^{(i)}(m) \end{cases} \text{if} \tag{9.19}$$

Here, $p_t(m)$ is the (instantaneous) bit error probability of the m-th bit of the t-th bit pattern. This probability can either be obtained from a soft-output channel decoder or can be estimated from consistency checks applied to the received bits (Ion and Haeb-Umbach 2006a).

b) Packet erasure channel

In this channel model, a data packet is either completely lost or received without any bit error. It models the random loss of data packets, e.g. due to network congestion. Most real communication channels exhibit packet losses occurring in bursts. Such channels can be modeled by a 2-state Markov chain, known as *Gilbert model*, see Fig. 9.2. In the figure p is the probability that the next packet is lost, provided the previous one has arrived; q is the probability that the next packet is not lost, given that the previous one was lost. The parameter q can be seen as controlling the burstiness of packet losses. This channel model is often described in terms of the *mean loss probability* $mlp = p/(p+q)$, the average probability of loosing a packet, and *conditional loss probability* $clp = 1-q$, i.e. the probability of loosing a packet, conditioned on the event that the previous packet was lost.

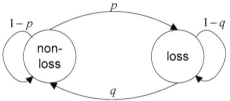

Fig. 9.2 Gilbert model

It is important to model the bursty nature of packet losses. It was shown that the word error rate of a DSR system depends strongly on the burstiness of the channel: Frame losses of up to 50% hardly have an effect on the word error rate, provided the average burst length is one packet (i.e. one frame pair), while the word error rate dramatically increases for longer average burst lengths (Gómez et al. 2007).

For a packet erasure channel model the probability of the received bit pattern, given the sent, is as follows:

$$P\left(\mathbf{y}_t \mid \mathbf{b}_t^{(i)}\right) = \begin{cases} \delta\left(\mathbf{y}_t - \mathbf{b}_t^{(i)}\right) & \text{packet received} \\ \dfrac{1}{2^M} & \textit{if} \quad \text{packet lost} \end{cases} \qquad (9.20)$$

Here $\delta(\cdot)$ denotes the Kronecker delta impulse.

Note that in practice often a combination of both error types is present. Communications that involve both a wireless and a packet-based wireline link may exhibit both packet losses and bit errors. Packets with bit errors are discarded by the User Datagram Protocol (UDP). While this is reasonable for many payloads, for DSR or speech transmission it would make more sense to deliver packets with bit errors, as it allows for more effective error concealment. UDP-Lite (RFC 3828, 2004) is a transport protocol that allows the application to receive partially corrupted packets.

9.3.4 Estimation of Feature Posterior

At the receiving end we are given the sequence $\mathbf{y}_1^T = \mathbf{y}_1,...\mathbf{y}_T$, and our goal is to carry out speech recognition by employing the modified observation probability given by Eq. 9.9.8.

To this end we need to compute the a posteriori probability density $p(\mathbf{x}_t \mid \mathbf{y}_1^T)$. Figure 9.3 illustrates the different processing steps. Note, that the input to the ASR decoder is no longer a feature vector, but a probability density function.

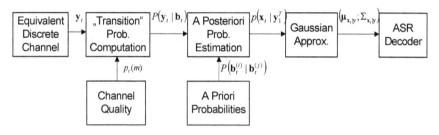

Fig. 9.3 Block diagram of posterior estimation and uncertainty decoding

Introducing the hidden (unobservable) sent bit pattern, we can express the posterior density as follows:

$$p(\mathbf{x}_t \mid \mathbf{y}_1^T) = \sum_{i=1}^{2^M} p(\mathbf{x}_t \mid \mathbf{b}_t^{(i)}) P(\mathbf{b}_t^{(i)} \mid \mathbf{y}_1^T) \tag{9.14}$$

The computation of the posterior probability $P(\mathbf{b}_t^{(i)} \mid \mathbf{y}_1^T)$ can be accomplished using the *Forward-Backward (FB) algorithm* (Bahl et al. 1974; Peinado et al. 2003):

$$P(\mathbf{b}_t^{(i)} \mid \mathbf{y}_1^T) = \frac{\alpha_t^{(i)} \beta_t^{(i)}}{\sum_{j=0}^{2^M-1} \alpha_t^{(j)} \beta_t^{(j)}} \tag{9.15}$$

where

$$\begin{aligned} \alpha_t^{(i)} &= P(\mathbf{b}_t^{(i)}, \mathbf{y}_1, \mathbf{y}_2,...\mathbf{y}_t) \\ \beta_t^{(i)} &= P(\mathbf{y}_{t+1}, \mathbf{y}_{t+2},...\mathbf{y}_T \mid \mathbf{b}_t^{(i)}) \end{aligned} \tag{9.16}$$

Both $\alpha_t^{(i)}$ and $\beta_t^{(i)}$ are computed recursively.

Using the FB algorithm the a posteriori density can be computed for either of the two channel models outlined in Sect. 9.3.3 and either of the two source models considered in Sect. 9.3.2. In the case of a packet erasure channel a very efficient realization of the recursions can be found exploiting the property of Eq. 9.9.20 (Ion and Haeb-Umbach 2006b).

Although the dynamic vector components are not transmitted, error concealment can benefit from the superior prediction quality of a source model including static and dynamic vector components. For the source model which models the sequence of bit patterns corresponding to the static MFCC vectors only as a first-order Markov model, there are 2^M bit patterns $\mathbf{b}_t^{(i)}$, $i \in \{1,...,2^M\}$ per subvector, and the inter-frame correlation is captured by a $2^M \times 2^M$ matrix, whose $(i, j)^{th}$ element is $P(\mathbf{b}_t^{(i)} \mid \mathbf{b}_t^{(j)})$. On the other hand, for the source model which considers a feature vector including dynamic components, inter-frame correlation is captured by a $2^{M+D_1+D_2} \times 2^{M+D_1+D_2}$ matrix, where M, D_1, and D_2 are the number of bits used to code the subvector of static, first-order and second-order differential coefficients. The matrices are estimated beforehand on clean training data. Since only the bits corresponding to the static MFCC vector are actually transmitted, the "transition probability" is independent of the bits corresponding to the dynamic part of the feature vector: $P(\mathbf{y}_t \mid \mathbf{b}_t, \Delta \mathbf{b}_t, \Delta^2 \mathbf{b}_t) = P(\mathbf{y}_t \mid \mathbf{b}_t)$. In Sect. 9.4 we compare the two source models w.r.t. speech recognition accuracy obtained on an error-prone channel. To simplify notation we will assume the source model of static components only in the remainder of this section.

Note that the FB algorithm needs to be performed only inside isolated erroneous regions (error bursts), i.e. when $P(\mathbf{y}_t \mid \mathbf{b}_t^{(i)})$ is not a Delta impulse. Then the FB recursions are initialized using the last uncorrupted feature vector before and the first uncorrupted feature vector after the error burst. Detecting the presence of an uncorrupted feature vector is trivial in the case of a packet erase channel, but it is not that trivial in the case of a time-variant BSC. In the latter case erroneous bit patterns can be detected based on consistency checks among subsequent bit patterns and on cyclic redundancy check failure (Ion and Haeb-Umbach 2006a).

The other term needed in Eq. 9.9.14, $p(\mathbf{x}_t \mid \mathbf{b}_t^{(i)})$, is the probability density function (pdf) of the feature vector, given the i-th centroid. This VQ cell-conditioned pdf is modeled as a Gaussian $p(\mathbf{x}_t \mid \mathbf{b}_t^{(i)}) = N(\mathbf{x}_t; \mathbf{c}_t^{(i)}, \Sigma_t^{(i)})$, where $\mathbf{c}_t^{(i)}$ is the VQ centroid corresponding to $\mathbf{b}_t^{(i)}$. The within-cell covariance matrix $\Sigma_t^{(i)}$ can be estimated on the training data.

In order to simplify subsequent processing, the feature posterior, Eq. 9.9.14, is approximated a Gaussian density $p_N(\mathbf{x}_t \mid \mathbf{y}_1^T) = N(\mathbf{x}_t; \mu_{\mathbf{x}_t|\mathbf{y}}, \Sigma_{\mathbf{x}_t|\mathbf{y}})$, see Eq. 9.9.11. The parameters $\mu_{\mathbf{x}_t|\mathbf{y}}, \Sigma_{\mathbf{x}_t|\mathbf{y}}$ of this Gaussian can be obtained by finding that Gaussian which has the smallest *Kullback-Leibler divergence* to the original non-Gaussian posterior $p(\mathbf{x}_t \mid \mathbf{y}_1^T)$. This results in the following estimates:

$$\mu_{\mathbf{x}_t|\mathbf{y}} = \sum_{i=1}^{2^M} P(\mathbf{b}_t^{(i)} \mid \mathbf{y}_1^T) \mathbf{c}_t^{(i)} \qquad (9.24)$$

$$\Sigma_{\mathbf{x}_t|\mathbf{y}} = \sum_{i=1}^{2^M} P\!\left(\mathbf{b}_t^{(i)} \mid \mathbf{y}_1^T\right)\!\left(\!\left(\mathbf{c}_t^{(i)} - \mathbf{\mu}_{\mathbf{x}_t|\mathbf{y}}\right)\!\left(\mathbf{c}_t^{(i)} - \mathbf{\mu}_{\mathbf{x}_t|\mathbf{y}}\right)^T + \Sigma_t^{(i)}\right) \tag{9.17}$$

This result makes intuitively sense: The mean $\mathbf{\mu}_{\mathbf{x}_t|\mathbf{y}}$ of the Gaussian is equal to the mean of the original posterior, and the covariance is the sum of the between-VQ-cell covariance and the within-VQ-cell covariance. For high resolution, i.e. sufficiently large M, as is e.g. the case for the vector quantizer used in the ETSI DSR standard, the within-cell variance is negligibly small, such that Eq. 9.9.17 simplifies to

$$\Sigma_{\mathbf{x}_t|\mathbf{y}} \approx \sum_{i=0}^{2^M-1} P\!\left(\mathbf{b}_t^{(i)} \mid \mathbf{y}_1^T\right)\!\left(\mathbf{c}_t^{(i)} - \mathbf{\mu}_{\mathbf{x}_t|\mathbf{y}}\right)\!\left(\mathbf{c}_t^{(i)} - \mathbf{\mu}_{\mathbf{x}_t|\mathbf{y}}\right)^T. \tag{9.18}$$

The posterior probability is the complete solution to the problem of estimating the uncorrupted features from the corrupted ones. The mean of the posterior given in (9.24) is the MMSE estimate of the feature vector $\mathbf{x}_t$. If one were only interested in the reconstruction of the uncorrupted feature vector, one could, for example, use this estimate. The maximum of the posterior is the maximum a posteriori estimate of the feature vector, another estimate commonly used in various estimation problems. The covariance matrix of the posterior, Eq. 9.9.17, is a measure of reliability of the reconstructed features. If the parameter to be estimated and the observation are jointly Gaussian, it equals the covariance matrix of the MMSE estimation error.

9.3.5 Related Work

Several server based error mitigation schemes proposed for distributed speech recognition can be related to the framework presented in this article.

Peinado et al. (2003) employ the MMSE estimate, Eq. 9.9.24, to reconstruct corrupted feature vectors on a channel exhibiting bit errors. A crucial issue, however, is the determination of the instantaneous bit error probability $p_t(m)$ needed in Eq. 9.9.19. It may either be obtained from the soft-output of the channel and SNR estimation (Peinado et al. 2003) or a soft-output channel decoder (Haeb-Umbach and Ion 2004). If the soft-output is not available the bit error probability can be estimated from consistency checks applied to the received bit patterns (Ion and Haeb-Umbach 2006a).

Marginalization reformulates the classification to perform recognition based on the reliable features alone (Endo et al. 2003). On a packet erasure channel there is a straightforward association of packet loss with unreliable data. However on a channel characterized by bit errors it is difficult to decide whether a feature is reliable or not, even if the instantaneous bit error probability of all bits making up the representation of the feature is available. In Endo et al. (2003) a threshold was experimentally determined. If the bit error probability was larger than the threshold the corresponding feature was marginalized.

Marginalization can be obtained in the presented framework, if the (simpler) feature posterior $p(\mathbf{x}_t \mid \mathbf{y}_t)$, which is only conditioned on the received data corresponding to the current frame, is used instead of $p(\mathbf{x}_t \mid \mathbf{y}_1^T)$ in the modified observation probability of Eq. 9.9.8. If a feature is declared lost, then $p(\mathbf{x}_t \mid \mathbf{y}_t) = p(\mathbf{x}_t)$. Using this in Eq. 9.9.8, the integral evaluates to one, i.e. the corresponding frame is marginalized.

The binary reliability measure used in marginalization can be replaced by a continuous confidence measure γ, taking values between zero and one. *Weighted Viterbi* (WV) decoding takes into account the confidence about a feature vector by raising the observation probability to the power of γ (Bernard and Alwan 2001). Obviously, for the correctly received feature vectors there is $\gamma = 1$, and no changes to the observation probability occur. For a lost feature vector the maximum uncertainty is expressed by $\gamma = 0$, resulting in an observation probability evaluating to one and being independent of the state. Thus with binary weighting WV is equivalent to marginalization. However, raising the observation probability to some power γ anywhere between zero and one lacks a probabilistic interpretation. Moreover, determining an optimal value for γ is not an easy task. The methods proposed to determine the confidence measure γ are rather empirical, and the optimal value depends on the recognition task (Cardenal-López et al. 2006).

The effect of raising the observation probability to some power between zero and one is to deemphasize the contribution of this frame to the ASR decision. The same effect is achieved with the observation probability of Eq. 9.9.8 proposed in this paper, if the feature posterior is not a Dirac delta impulse.

9.4 Performance Evaluations

In this section we present experimental results for distributed speech recognition employing the proposed error concealment techniques. We first describe the experimental setup and then give speech recognition results for the two channel models outlined in Sect. 9.3.3 and for two recognition tasks, a small vocabulary and a medium vocabulary task.

9.4.1 Experimental Setup

We consider a setup which is compatible to the ETSI standards for DSR. The whole front-end processing, consisting of feature extraction, source coding and packetization is carried out according to the ETSI advanced front-end (ETSI 2002) standard.

As an example for a channel exhibiting bit errors the GSM data channel was considered. A realistic simulation of the GSM physical layer processing was carried out including channel coding/decoding, interleaving/deinterleaving, modulation/demodulation. The channel coding was TCH/F4.8 described in (ETSI 2003b) which uses convolutional coding at a rate $r = 1/3$. The channel decoding employed the FB algorithm (Bahl et al.

1974) which is able to provide the instantaneous bit error probability $p_t(m)$. We preferred this full channel simulation, since if we had used merely GSM error patterns, the instantaneous bit error rate would not have been available.

We have chosen a channel model approximating a "typical urban" profile specified by COST 207 (COST 1989). The model is characterized by 12 propagation paths, delay spread of 1.03 µs and Rayleigh fading. The terminal was assumed to be moving at 50 km/h. Various Carrier-to-Interference (C/I) power ratios were simulated, ranging from 10 dB to 2.5 dB. Note that C/I=2.5 dB is a very poor channel, where the bit error rate is as high as 3.6%.

For the packet erasure channel we adopted the Gilbert model to model that packet losses occur in bursts. In the literature often four channel conditions are evaluated, with C1 corresponding to mildly bad and C4 to very poor channel conditions. Table 9.2 gives the conditional and mean loss probabilities of the four conditions (Boulis et al. 2002). In our simulations we transmitted one frame pair per packet.

Table 9.2 Packet erasure channel test conditions

Condition	C1	C2	C3	C4
clp	0.147	0.33	0.50	0.60
mlp	0.006	0.09	0.286	0.385

Different error concealment techniques were applied at the receiving (server) side and compared in terms of achieved word error rate obtained on two databases.

The small vocabulary task is the clean test set of the Aurora 2 database, which consists of 4004 utterances from 52 male and 52 female speakers distributed over four subsets. The sampling rate is 8 kHz. The acoustic models used in the recognizer were those described in (Hirsch and Pearce 2000): 16 states per word, 3 Gaussians per state.

The medium vocabulary task is the Wall Street Journal WSJ0 5k Nov. '92 evaluation test set (Paul and Baker 1992) comprising 330 utterances of 4 male and 4 female speakers, summing up to 40 min of speech. Here, the sampling rate is 16 kHz. Recognition experiments were carried out on this test set using a closed vocabulary bigram language model. The acoustic model consisted of 3437 tied states. The parameters of the 10-component mixture densities were trained on the SI-84 set of the WSJ corpus using the HTK toolkit (Young et al. 2004).

9.4.2 Results on GSM Data Channel

Figure 9.4 gives an illustrative example of the reconstruction achieved by employing the a posteriori density. The figure shows how the feature $\log E$ is reconstructed in the presence of bit errors during transmission. The continuous solid line labeled x_t is the sent ("true") value of the parameter over the frame index t. $\mu_{x_t|y}$ is the MMSE estimate, and $\mu_{x_t|\mathbf{y}} \pm \sigma_{x_t|\mathbf{y}}$ the MMSE estimate plus/minus one standard deviation of the a posteriori density. The interval given in this way can be interpreted as confidence interval for the MMSE estimate. The curve NFR shows the reconstruction by *nearest frame repetition*, which is the error concealment strategy proposed in the ETSI standard. The grey areas show intervals in which transmission errors occurred. We used two grey scales to distinguish between regions where transmission errors occurred in the bit pattern carrying the $\log E$ component (dark grey) and regions where the bit patterns corresponding to other subvectors of the same frame were affected by errors (light grey). It can be seen that the $\log E$ component is not affected by transmission errors in other subvectors. This can be attributed to the fact that the a posteriori computation operates on a per-subvector basis. Uncorrupted parts are forwarded to the recognizer without modification. A subvector-based error concealment, such as this or the one proposed by Tan et al. (2004) is superior to a frame-based scheme, such as NFR, where a complete frame is modified, even if only one subvector is degraded by transmission errors. But even if the illustrated $\log E$ component is affected by transmission errors, much better feature reconstruction is achieved with the proposed method compared to NFR.

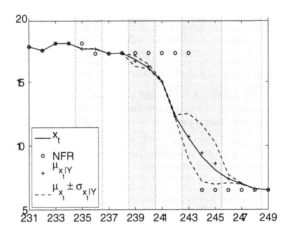

Fig. 9.4 Example of feature reconstruction. The figure shows the trajectory of the $\log E$ feature over time (labeled x_t) and its reconstructions, either by nearest frame repetition (NFR) or by the proposed scheme. The shaded areas indicate regions where bit errors occurred during transmission, either in the $\log E$ component (dark grey) or another component of the feature vector (light grey)

Figures 9.5 and 9.6 present word error rates for different Carrier-to-Interference (C/I) power ratios for the Aurora 2 and WSJ0 database, respectively. In these figures, the performance of the proposed scheme, termed *uncertainty decoding* (UD), is compared with *marginalization* (M), *nearest frame repetition* (NFR) and *Weighted Viterbi* decoding (WV). For WV, the confidence γ was computed as in Potamianos and Weerackody (2001), however using the instantaneous bit error probability from the channel decoder. For UD, we employed the source model based on the correlation of static features only. It can be seen, that UD outperforms all other schemes. Speech recognition accuracy is hardly affected for C/I-values as low as 2.5 dB. Figure 9.5 also shows the bit error rate (BER) at the output of the channel decoder. It is interesting to note that BER increases by almost three orders of magnitude when C/I is reduced from 10 dB to 2.5 dB, while the word error rate achieved by UD is only mildly affected. This underscores that uncertainty decoding makes the ASR decoder very robust towards degraded channel conditions.

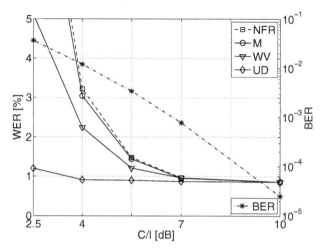

Fig. 9.5 Word error rates for transmission over GSM TCH/F4.8 channel using different error concealment schemes; Aurora 2 task. The dash-dotted line indicates the bit error rate (BER) at the output of the channel decoder

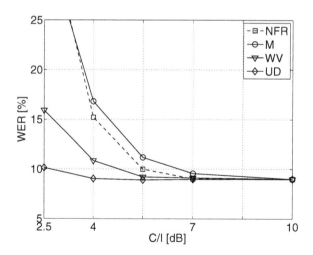

Fig. 9.6 Word error rates for transmission over GSM TCH/F4.8 channel using different error concealment schemes; WSJ0 task

As the two are closely related, the frame error rate increases similarly to BER, from 0.08% at C/I = 10 dB to 60% at C/I = 2.5 dB. As a consequence marginalization and nearest frame repetition, which operate on a vector rather than a subvector basis, perform poorly.

9.4.3 Results on Packet Erasure Channel

For the experiments on the packet erasure channel we used the channel conditions C1 to C4, specified in Table 9.2. Figures 9.7 and 9.8 display the word error rates of different error concealment techniques for the Aurora 2 and WSJ0 task, respectively. In the figures we included a condition C0 as a reference, which corresponds to an error-free transmission. Results are presented for two variants of the proposed scheme: *uncertainty decoding* employing an a priori model of the source which captures correlation among the static MFCCs alone (UD) and the one utilizing the correlation among the full (static and dynamic) feature vector (UD-dyn). It can be seen, that UD outperforms *marginalization* (M) and *nearest frame repetition* (NFR). The performance of *Weighted Viterbi* (WV) decoding comes close to UD. For the WV curve the lost features were reconstructed by NFR, and their confidence $\gamma(t)$ was chosen dependant on the relative position (τ) within an error burst. It equals one at the start and end of the burst and decreases exponentially according to $\gamma(t_{start} + \tau) = \gamma(t_{end} + \tau) = \alpha^\tau$ towards the middle (Cardenal-López et al. 2006). Here, t_{start} and t_{end} denote the starting and ending time of the error burst. The optimal value of α was experimentally found to be $\alpha = 0.7$ for this task.

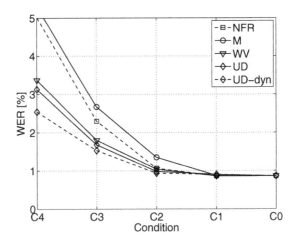

Fig. 9.7 Word error rates for packet erasure channel using different error concealment schemes; Aurora 2 task

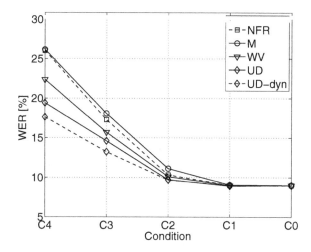

Fig. 9.8 Word error rates for packet erasure channel using different error concealment schemes; WSJ0 task

9.5 Conclusion

Error concealment is concerned with mitigating the detrimental effect that transmission errors may have on the recipient of the signal by exploiting residual redundancy in the bit stream of the source coder output. In distributed speech recognition (DSR) the recipient is the ASR decoder, which, unlike a human listener, can take advantage of both the optimally reconstructed transmitted data and information about the reliability of the reconstruction. The Bayes decision rule therefore has to be reformulated to account for a corrupted or unreliable feature vector sequence. This results, under certain assumptions, in just a modification of the observation probability computation, while the structure of the decoder, which is based on the Viterbi search, remains unchanged. Crucial to the performance of this modified decoding rule is the accuracy of the a posteriori probability density estimate of the uncorrupted feature vector, given all the received corrupted ones. For DSR we were able to find an efficient estimation method, both for channels characterized by bit errors and channels exhibiting packet losses. The key was to exploit the high inter-frame correlation of MFCC feature vectors. Using these techniques high recognition accuracy can be maintained over a wide range of channel conditions.

It should be noted that server-based error concealment techniques, as the ones described in this contribution, are fully compatible with the ETSI standards for distributed speech recognition.

Acknowledgments

This work was supported by Deutsche Forschungsgemeinschaft under contract numbers HA 3455/2-1 and HA 3455/2-3.

References

Arrowood, J.A. and Clements, M.A. (2002). Using observation uncertainty in HMM decoding. In *Proc. ICSLP*, Denver, Colorado.
Bahl, L., Cocke, J., Jelinek, F. and Raviv, J. (1974). Optimal decoding of linear codes for minimizing symbol error rate, *IEEE Trans. Inf. Theory*, vol. 10, pp. 284–287.
Bernard, A. and Alwan, A. (2001). Joint channel decoding—Viterbi recognition for wireless applications. In *Proc. Eurospeech*, Aalborg, Denmark.
Bernard, A. and Alwan, A. (2002). Low-bitrate distributed speech recognition for packet-based and wireless communication. *IEEE Trans. Speech and Audio Process.*, vol. 10, no. 8, Nov., 2002.
Boulis, C., Ostendorf, M., Riskin, E.A. and Otterson, S. (2002). Graceful degradation of speech recognition performance over packet-erasure networks. *IEEE Trans. on Speech and Audio Processing*, vol. 10, no. 8, Nov. pp. 580–590.
Cardenal-López, A., García-Mateo, C. and Docío-Fernández, L. (2006). Weighted Viterbi decoding strategies for distributed speech recognition over IP networks, Speech Communication, vol. 48, no. 11, Nov., pp. 1422–1434.
COST 207 (1989). *Digital land mobile radio communication—Final report*. Office for official publications of the European Communities, Luxembourg.
Cox, R.V., Kleijn, W.B. and Kroon, P. (1989). Robust CELP coders for noisy backgrounds and noisy channels. In *Proc. IEEE Int. Conf. Acoust. Speech Signal Process.*, 1989, pp. 739–742.
Davis, S.B. and Mermelstein P. (1980). Comparison of parametric representations for monosyllabic word recognition in continuously spoken sentences. *IEEE Trans. on Acoust. Speech and Signal Process.*, vol. 28, pp. 357–366.
Droppo, J., Acero, A. and Deng, L. (2002). Uncertainty decoding with Splice for noise robust speech recognition. In *Proc. IEEE Int. Conf. Acoust. Speech Signal Process.*, Orlando, Florida.
Endo, T., Kuroiwa, S. and Nakamura, S. (2003). Missing feature theory applied to robust speech recognition over IP networks. In *Proc. Eurospeech*, Geneva, Switzerland.
ETSI Standard ES 202 050 (2002). Speech processing, transmission and quality aspects (STQ); distributed speech recognition; advanced front-end feature extraction algorithm; compression algorithms. v1.1.1, Oct.
ETSI Standard ES 201 108 (2003a). Speech processing, transmission and quality aspects (STQ);distributed speech recognition; front-end feature extraction algorithm; compression algorithms. v1.1.3, Sep.
ETSI Standard TS 100 909 v8.7.1 (2003b). *Digital cellular telecommunications system (phase 2+); channel coding.* (3GPP TS 05.03 version 8.7.0; Release 1999).
Fingscheidt, T., Aalburg, S., Stan, T. and Beaugeant, C. (2002). Network-based versus distributed speech recognition in adaptive multi-rate wireless systems. In *Proc. Int. Conf. on Spoken Language Proc.*, Denver.
Fingscheidt, T. and Vary, P. (2001). Softbit speech decoding: A new approach to error concealment. *IEEE Trans. Speech and Audio Proc.*, vol. 9, no. 3, March, pp. 1–11.

Gómez, A.M., Peinado, A.M., Sánchez, V. and Rubio, J. (2007). On the Ramsey class of interleavers for robust speech recognition in burst-like packet loss, *IEEE Trans. Audio Speech and Lang. Process.*, vol. 15, no. 4, May, pp. 1496–1499.

GSM 06.11 Recommendation (1992). Substitution and muting of lost frames for full rate speech traffic channels. ETSI TC-SMG.

Haeb-Umbach, R. and Ion, V. (2004). Soft features for improved distributed speech recognition over wireless networks. In *Proc. ICSLP*, Jeju, Korea.

Hirsch, H.G. and Pearce, D. (2000). The Aurora experimental framework for the performance evaluation ofspeech recognition systems undernoisy conditions. In *Proc. ISCA ITRW Workshop ASR2000*, Paris, France, pp. 181–188.

Ion, V. and Haeb-Umbach, R. (2005). A unified probabilistic approach to error concealment for distributed speech recognition. In *Proc. Interspeech*, Lisbon.

Ion, V. and Haeb-Umbach, R. (2006a). Uncertainty decoding for distributed speech recognition over error-prone networks, *Speech Communication* 48, pp. 1435–1446.

Ion, V. and Haeb-Umbach, R. (2006b). An inexpensive packet loss compensation scheme for distributed speech recognition based on soft-features. In *Proc. IEEE Int. Conf. Acoust. Speech Signal Process.*, Toulouse, France.

Ion, V. and Haeb-Umbach, R. (2006c). Improved source modeling and predictive classification for channel robust speech recognition. In *Proc. Interspeech*, Pittsburgh.

ITU-T Recommendation G.711 Appendix I (1999). A high quality low-complexity algorithm for packet loss concealment with G.711.

James, A.B., Gomez, A. and Milner, B.P. (2004). A comparison of packet loss compensation methods and interleaving for speech recognition in burst-like packet loss. In *Proc. ICSLP*, Jeju, Korea.

Kristjansson, T.T. and Frey, B.J. (2002). Accounting for uncertainty in observations: A new paradigm for robust speech recognition. In *Proc. IEEE Int. Conf. Acoust. Speech Signal Process.*, Orlando, Florida.

Lahouti, F. and Khandani, A.K. (2007). Soft reconstruction of speech in the presence of noise and packet loss. *IEEE Trans. Audio Speech and Lang. Proc.*, vol. 15, no. 1, Jan., pp. 44–56.

Liao, H. and Gales, M.J.F. (2004). *Uncertainty decoding for noise robust automatic speech recognition*. Technical Report TR.499, Cambridge University Engineering Department.

Milner, B. and Semnani, S. (2000). Robust speech recognition over IP networks. In *Proc. Int. Conf. Acoust. Speech Signal Process.*, Istanbul, Turkey.

Morris, A., Cooke, M. and Green, P. (1998). Some solutions to the missing feature problem in data classification, with application to noise-robust ASR. In *Proc. Int. Conf. Acoust. Speech Signal Process.*, Seattle.

Morris, A., Barker, J. and Bourlard, H. (2001). From missing data to maybe useful data: Soft data modeling for noise robust ASR. In *Proc. WISP*, vol. 6.

Paul, D. and Baker, J. (1992). *The design for the Wall Street Journal-based CSR corpus*. DARPA Technical Report.

Pearce, D. (2000). Enabling new speech driven services for mobile devices: An overview of the ETSI standards activities for distributed speech recognition front-ends. In Proc. *Voice Input/Output Soc. Speech Applications Conference*, May.

Peinado, A.M., Sanchez, V., Perez-Cordoba, J.L. and de la Torre, A. (2003). HMM-based channel error mitigation and its application to distributed speech recognition. *Speech Communication*, 41, pp. 549–561.

Potamianos, A. and Weerackody, V. (2001). Soft-feature decoding for speech recognition over wireless channels. In *Proc. IEEE Int. Conf. Acoust. Speech Signal Process.*, Salt Lake City, Utah.

RFC 2460 (1998). *Internet Protocol, Version 6 (IPv6) Specification*, http://www.ietf.org/rfc/rfc2460.txt, Internet Engineering Task Force, Dec.

RFC 3828 (2004). *The Lightweight User Datagram Protocol (UDP-Lite)*, http://www.ietf.org/rfc/rfc3828.txt, Internet Engineering Task Force, July.

Tan, Z.-H., Dalsgaard, P. and Lindberg, B. (2004). A subvector-based error concealment algorithm for speech recognition over mobile networks. In *Proc. IEEE Int. Conf. Acoust. Speech Signal Process.*, Montreal, Quebec, Canada.

Tan, Z.H., Dalsgaard, P. and Lindberg, B. (2005). Automatic speech recognition over error-prone wireless networks, *Speech Communication*, vol. 47, no. 1–2, Sep.–Oct., pp 220–242.

Vary, P. and Martin, R. (2006). Digital Speech Transmission—Enhancement, Coding and Error Concealment. John Wiley, New York.

Weerackody, V., Reichl, W. and Potamianos, A. (2002). An error-protected speech recognition system for wireless communications. *IEEE Trans. on Wireless Communications*, vol. 1, no. 2, April, pp. 282–291.

Young, S.J. et al. (2004). *HTK: Hidden Markov Model Toolkit V3.2.1 Reference Manual*. Cambridge University Speech Group, Cambridge, U.K.

Part III

Embedded Speech Recognition

10

Algorithm Optimizations: Low Computational Complexity

Miroslav Novak

Abstract. Advances in ASR are driven by both scientific achievements in the field and the availability of more powerful hardware. While very powerful CPUs allow us to use ever more complex algorithms in server-based large vocabulary ASR systems (e.g. in telephony applications), the capability of embedded platforms will always lag behind. Nevertheless as the popularity of ASR application grows, we can expect an increasing demand for functionality on embedded platforms as well. For example, replacing simple command and control grammar-based applications by natural language understanding (NLU) systems leads to increased vocabulary sizes and thus the need for greater CPU performance. In this chapter we present an overview of ASR decoder design options with an emphasis on techniques which are suitable for embedded platforms. One needs to keep in mind that there is no one-size-fits-all solution; specific algorithmic improvements may only be best applied to highly restricted applications or scenarios. The optimal solution can usually be achieved by making choices with respect to algorithms aimed at maximizing specific benefits for a particular platform and task.

10.1 Introduction

While systems for dealing with large vocabulary recognition in real-time ASR have been available for many years, deployment on platforms with limited resources presents new challenges with respect to many aspects of the overall ASR system design. There is an obvious tendency to utilize any new beneficial technique over the wide spectrum of ASR applications and platforms. But in numerous respects embedded platforms lag behind the state of the art in speech recognition with respect to general CPU architectures. The limitations of the embedded platforms affecting this state of affairs includes: lower CPU clock speeds, limited or missing ability to process floating point operations, and restricted memory capacity both with respect to speed and size having implications to all levels of caching. Those constraints are bad enough, but added to this is the relatively restricted capability of tools/platforms tailored specifically to embedded systems development, which results in many important language features (e.g., templates in the C++ case) not being fully supported.

Hence any new algorithm targeting embedded platforms must therefore be evaluated in light of the above restrictions. Possible actions that one needs to consider taking may include finding more efficient implementation of a given algorithm, finding and applying an even more efficient algorithm with the same or similar results, or resorting to the use of approximations applied to the algorithm or simply reducing the overall model complexity, while at the same time striving to ensure an acceptable trade-off between accuracy and efficiency.

10.2 Common Limitations of Embedded Platforms

In comparison to a typical workstation used for large vocabulary ASR, the embedded platforms are limited both in terms of CPU power and memory. While the workstation market for desktop and server applications is dominated by a single platform (i.e. 32-bit or more recently 64-bit Intel), the situation for embedded devices is much more diverse (a variety of chips ranging from 8 to 32 bit architectures are fairly common), which makes the development of one-size-fits-all implementation all the more difficult if not completely impossible.

10.2.1 Memory Limitations

Memory (not only the amount available but, more importantly, its speed) seems to be the most pronounced limiting factor. It is well known that a fast CPU is not very helpful when it is used in combination with slow memory – especially when a very small memory *cache* is used.

There is a popular belief that in the implementation of many algorithms, speed can be improved by using more memory. Caution is advised with embedded systems however because the use of more memory can often lead to a higher cache miss rate and an overall degradation in performance. Algorithmic improvements developed on a large computer do not always translate to improvements on systems with limited resources.

Most embedded CPUs have small instruction and data caches (typically in the 4-32 kB range). Also, slow RAM is often used due to constraints on the hardware cost and power consumption. Simple CPUs sometimes even have no data cache. Instead they use a block of fast scratchpad memory and content must be managed by the application software and DMA transfers. Porting more complex code to such architectures can be extremely difficult.

Embedded systems that use memory management (i.e. logical to physical address translation) usually have only simple memory management units (MMUs),which have only a limited number of Translation Lookaside Buffer (TLB) entries (typically 16-64). Moreover, they rely on software handling of TLB faults by the operating system that cost hundreds of CPU cycles per fault. As most systems use a memory page size of 4 kB, the memory size that can be handled without TLB faults is only 64-256 kB. Any task accessing a large amount of memory will be slowed down by expensive handling of TLB fault interrupts.

One of the impacts of slow memory access is the high cost of heap allocation routines, often observed on embedded systems. In addition, memory allocation failures are much more common than on desktop or server systems due to limits of the memory size and runtime environment.

These memory limitations must be kept in mind during the implementation of any ASR algorithm. One goal is to maximize the locality of memory access by organizing data structures so that concurrently accessed elements are always close to each other. Another goal is to maximize the use of any model data when it is accessed. A typical task encountered in ASR algorithms requires some type of evaluation of a data stream against model data in memory. This model can occupy a significant amount of memory so it is beneficial to process several samples of the data stream when a particular part of the model is accessed and loaded into the memory cache. A typical example of such an approach involves the evaluation of a Gaussian mixture component on a sequence of several feature vectors as opposed to the evaluation of all components during each time frame.

10.2.2 CPU Limitations

Typical embedded CPUs often lack hardware support for floating point operations. Although software emulation is usually available, it is often prohibitively expensive. Hence, many algorithms need to be redesigned in order to work efficiently in situations involving integer arithmetic and for specific word lengths (32 or 16 bit). The ASR front end requires several multiplication steps applied to each feature vector so proper scaling is essential. To avoid the overhead of dynamic scaling, the scaling factors are typically precomputed offline. To maximize use of the limited dynamic range, a separate scaling factor can be estimated for each step.

Proper application of architecture knowledge can significantly improve the execution efficiency of the code. Many embedded CPUs are RISC load-store architectures, often with limited number of general-purpose registers (GPRs). This constrains the number of local variables that can be efficiently accessed. For example, on a CPUs with 16 GPRs (e.g. ARM), usually only about 5 local variables can be stored in registers. Other local variables must be located on the stack and using them in operations is more expensive. This can have a serious performance impact, for example on code performing various iterative computations.

Modern CPUs use sophisticated branching prediction methods, which are usually not available on embedded CPUs. For example, insertion of a conditional statement inside of a loop has an adverse effect on the instructions *cache* and can significantly slow down execution of the code, even if it were intended to avoid unnecessary computations. In such situations performing more operations can be faster, as long as the code does not contain branches.

10.3 Overview of an ASR System

A block diagram of an ASR decoder is shown in Fig. 10.1. It contains three major components:

- *Feature extraction (front end)*
- *Observation model (labeler)*
- Search

While this is a very simple diagram which does not show the complex interactions in ever more complex modern ASR designs, these fundamental blocks can be found in basically any decoder implementation.

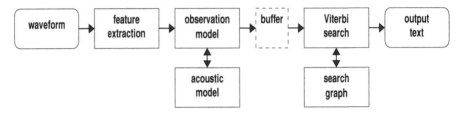

Fig. 10.1 ASR block diagram

The waveform is converted into a sequence of feature vectors by the *front end*. Each feature vector represents a short segment of the utterance corresponding to an acoustic event which is assumed to be stationary. These vectors are statistically modeled as a set of acoustic classes (*acoustic model*), each having its own probability density function. These classes typically model phoneme parts in a specific acoustic context (e.g. tri-phones). The *observation model* provides the means to efficiently compute probabilities of a particular feature vector. These are called "emission" or "output" probabilities because they are used as state output probabilities of *Hidden Markov Model* (HMM) states in the *search graph* network representing all possible utterances. The role of the search module is to find the best matching path through this network. Often it is required to find the N- best matching paths, or a word *lattice* representing alternative hypotheses in a compact form.

10.4 Front End

The role of the front end is to convert the input waveform into a sequence of acoustic feature vectors. Each feature vector describes a portion of the waveform, called frame (typically 10–15 ms). FFT (Fast Fourier Transform) followed by MEL band filtering or similar technique is applied on each frame to produce a feature vector. The vector elements are converted to cepstral coefficients by computing their logarithms and applying Discrete Cosine Transformation (DCT). The DCT has a decorrelating effect facilitating use of diagonal *Gaussian* models in the observation model. Dynamic features are captured either by computing double-delta feature vectors or by splicing several vectors together and applying a linear discrimination analysis (LDA) transformation. The transformation more expensive but usually leads to higher accuracy and the vector dimension can be reduced with a minimal impact on overall accuracy.

The final dimension of the feature vector typically varies from 13 to 60. Choice of the dimensionality entails making a tradeoff between CPU cost and recognition accuracy. Eventually, a transformation for runtime adaptation (usually unsupervised) can be applied to compensate for channel and/or speaker mismatch (Feature space Maximum Likelihood Linear regression *fMLLR*) (Gales 1997).

The computational cost of the front end is relatively small in comparison to the rest of the decoder. There are two main operations utilized by the front end:

- FFT computation
- Matrix multiplication

Efficient implementations of these algorithms for specific platforms have been extensively covered in the literature and a more detailed description in beyond the scope of this chapter.

If feature space adaptation is employed, it is preferable to use algorithms that do not require matrix inversion (which can cause instability), particularly in integerized implementations. (Balakrishnan 2003) proposes stochastic gradient based estimation, which also significantly reduces the computational cost from $O(n^3)$ to $O(n^2)$ and memory use from $O(n^3)$ to $O(n^2)$, where n is the vector dimension, in comparison to the originally proposed method (Gales 1997).

10.5 Observation Model

10.5.1 Model Organization

The observation model computes the likelihoods of acoustic model classes (typically context dependent sub-phonetic units). The CPU cost of the observation likelihood computation is often the dominating one. The models are almost exclusively based on *Gaussian Mixture Models* (GMM) of some form (untied or tied) such that each acoustic class is modeled by one Gaussian mixture. In a tied system, the Gaussian components can be shared across classes.

Proper design of the observation model is essential for the performance of the ASR system. The main design criteria include:

- Number of acoustic classes
- Number of components in each mixture
- Size of the phonetic *context*

The theoretical upper limit on the number of Gaussian components is given by the amount of training data available for reliable estimation of their parameters (means, variances and weights). In practice, this limit is determined by the amount of both CPU resources available for the Gaussian likelihood computation and the memory for their storage. Lowering the number of Gaussians to reduce the CPU cost is effective up to a certain degree. If the observation model is less accurate, the search tends to get wider (i.e. more states becoming active) and thus slower.

The Bayesian Information Criterion (BIC) has been successfully used to assign the number of Gaussian components to each state (Deligne et al. 2002). The objective function for GMM estimation is extended by a factor which imposes a penalty for the number of parameters in the final model. This way, new Gaussians are added during the model building only if they significantly contribute to the likelihood increase.

Further improvement in accuracy can be achieved by using gender dependent models. To avoid the memory cost of having two models, (Olsen and Dharanipragada 2003) propose a method which uses only one set of Gaussians but adjusts the mixture weights adaptively in accordance with the automatically detected gender of the speaker.

The size of phonetic *context* information used to create context dependent models is also a very important factor. The wider the context, the better is the coarticulation modeling at the expense of a more complex model. The effect is twofold: the number of acoustic classes increases and the complexity of the search increases. The latter is particularly evident when the phonetic context is modeled across word boundaries. Often the phonetic context is modeled only within each word and not across word boundaries to reduce the complexity of the resulting search graph. Use of word internal context modeling also significantly simplifies the construction of the graph.

The dynamic range of the observation likelihood is very large. To have the capability to process the likelihood in integer arithmetic, logarithms are employed. Computation of logarithms is very expensive but can be avoided by using the best Gaussian approximation where only the component within the mixture of a particular context-dependent state with the highest likelihood is used. The likelihood of one diagonal Gaussian mixture model λ with M components for a given D-dimensional feature vector x can be computed as:

$$\log(L(x \mid \lambda)) = \log(\sum_{i=1}^{M} p_i \frac{1}{(2\pi)^D \sqrt{\prod_{j=1}^{D} \sigma_{ji}}} \exp(-\frac{1}{2} \sum_{j=1}^{D} \frac{(x_j - \mu_{ji})^2}{\sigma_{ji}}))$$

$$\approx \max_{i=1}^{M}(K_i - \frac{1}{2} \sum_{j=1}^{D} \frac{(x_j - \mu_{ji})^2}{\sigma_{ji}})$$

(10.1)

10.5.2 Efficient Computation Strategies

Regardless of the particular scheme, the number of Gaussians in the model needed for good accuracy will likely be much greater than the number that can be evaluated at every time frame. Since there are only a limited number of states active in any given time frame, it is clear that not all Gaussians are always needed. There exist two basic approaches to reducing Gaussian evaluation:

- On-demand computation
- Gaussian selection

In the on-demand scheme only those mixtures corresponding to active states in the search are computed. This approach does not scale well. As the search space grows, the number of active mixtures grows as well. As the number of components in a mixture grows, the on-demand computation is even less effective since only a few components contribute significantly to the mixture likelihood.

Another disadvantage of the on-demand approach is that the likelihoods must be computed synchronously with the search, i.e. it is more difficult to efficiently evaluate the likelihood of one state several frames ahead (some form of a fast match is needed).

In the selection-based schemes, the goal is to evaluate only a limited number of Gaussian components chosen either completely independently of the active states or in some other manner which will achieve reasonable active state coverage. Popular techniques are those which create a shortlist of components; some form of an inexpensive metric is used to partition the feature space and to create an active shortlist(s) given the feature vector. Bocchieri (1993) uses vector Quantization (VQ). Ortmanns et al. (1997) compare space partitioning schemes, projection search, Hamming distance, and VQ pre-selection. Fritsch and Rogina (1996) propose the Bucket Box Intersection method, which partitions the space into rectangular regions, each defined as a bounding box around the Gaussian component within which the likelihood is greater than a fixed threshold.

Novak et al. (2002) reduce the memory overhead associated with the Gaussian selection implementation by using non-overlapping shortlist and an n-ary search tree. These non-overlapping lists can be represented much more efficiently in memory. The structure of the tree can be seen in Fig. 10.2. At each level of the tree, one Gaussian is used to represent each shortlist at the next level. The Gaussians in the top tree level are evaluated first. Only the top N scoring Gaussians are then processed. To reduce the cost for the top N search, the likelihood of the best Gaussian is found and all Gaussians with likelihood less the maximum minus a certain threshold are discarded first. Each Gaussian represents a shortlist in the next tree level. These shortlists are combined and evaluated. Subsequently, the top N list is created and expanded to the next level until the final level is reached. In this level, a contribution of each Gaussian component is added to its corresponding HMM state likelihood. In the selection scheme, the state likelihood computation can be completely decoupled from the search and be performed on blocks of feature vectors. This can significantly improve the memory *cache* utilization.

There are also hybrid schemes Saon et al. (2005) combining the advantages of the on-demand and selection schemes. The higher levels of the tree are evaluated as in the selection scheme for several frames at a time and the shortlists are stored. During the search, state likelihoods are computed on demand by using the precomputed shortlists.

Bahl et al. (1994) use an alternative method for state likelihood computation based on *rank* likelihoods. Rather than using the likelihood of the Gaussian mixture directly, the probability of the rank of the mixture can be used instead. The observation model distribution represents the probability that a given state will have a certain rank when all states are sorted by their likelihoods. The concept of rank distribution was originally introduced in an asynchronous stack decoder implementation and its main purpose was to normalize the likelihood range. It has been shown that its use has advantages in the Viterbi decoder as well. A likelihood value from the tail of the rank distribution provides a robust estimate for states which are not evaluated in the Gaussian selection scheme. While the direct use of mixture likelihoods yields a slightly lower error rate when all or a large portion of Gaussians are evaluated, the rank-based likelihoods are more robust against underestimation when only a small fraction is computed. In such cases the use of rank likelihoods leads to a smaller accuracy loss due to the limited number of evaluated Gaussians.

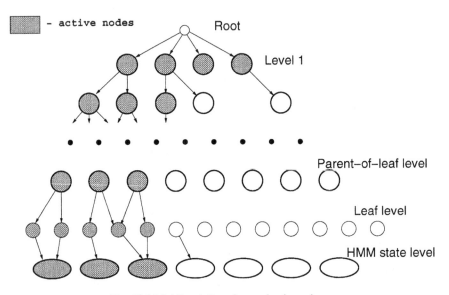

Fig. 10.2 Multilevel Gaussian evaluation scheme

The difference in performance can be seen in Table 10.1. String error rates are measured on a grammar containing 30 thousand stock names using an acoustic model with a total of 150k Gaussians. Each line shows the error rate when only the top N Gaussians were used in each frame. It can be seen that while the mixture likelihoods eventually lead to better accuracy, the *rank*-based system is much less affected when the number of evaluated Gaussians is significantly reduced.

Table 10.1 String error rate comparison of rank and mixture likelihood

Number of evaluated Gaussians	Rank likelihoods	Mixture likelihoods
1000	8.51	9.13
1500	8.21	8.43
2000	8.19	8.10
3000	8.10	8.05

The use of a rank distribution also has an effect on the state transition probabilities. While many believe that *transition probabilities* do not play a significant role due to the huge dynamic range of mixture likelihoods in comparison to the range of

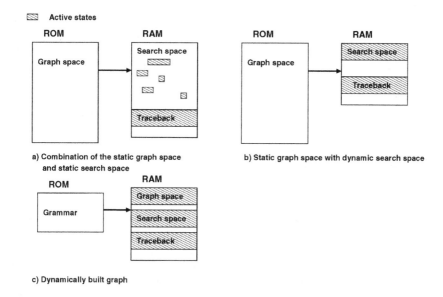

a) Combination of the static graph space and static search space

b) Static graph space with dynamic search space

c) Dynamically built graph

Fig. 10.3 Search memory organization

transition probabilities, the rank distribution brings the dynamic range of the observation likelihoods closer to that of transition probabilities. It has been observed that the use of transition probabilities can improve the recognition accuracy, particularly when the language model is weak, and in a presence of noise.

10.6 Search

Search is often the most computationally expensive part of the decoder, depending on the vocabulary and *language model* size. While the search implementations predominantly use *Viterbi* search, there are many variations in the implementation details. An excellent overview of search techniques is presented by Aubert (2002).

Several comparisons (Kanthak et al. 2000; Dolfing 2002) have shown that the use of *finite state transducers* (FST) (Mohri et al. 2002) to produce a minimized static graph is the most computationally efficient design for Viterbi based decoders on unrestricted platforms. The *search graph* is prepared offline and use of global minimization guaranties that the smallest possible graph is searched. Embedded system memory restrictions put a limit on the size of the usable acoustic and language models. Large models may require a dynamic scheme, e.g. (Ortmanns et al. 1998), but there is a significant computational overhead association with dynamic composition of the search graph.

10.6.1 Viterbi Search Implementation

The details of a Viterbi search implementation can vary significantly. It is difficult to objectively compare the various techniques found in the literature as the speed of the decoder results from the mutual interaction of many factors, such as type and speed of CPU, C compilers, amount and speed of memory, task complexity, etc.

We will focus our comparison mainly on memory organization, since it is a primary factor affecting the efficiency on embedded platforms. Memory usage in a Viterbi decoder can be divided into three classes:

- static representation of the HMM states—**graph space** (*search graph*),
- memory for active states, updated at each time frame—**search space**,
- **traceback** information (tokens).

All information related to the graph space can possibly be stored in read-only memory, while both the search space and traceback inherently require read-write memory to evaluate likelihoods for each frame and to propagate the *traceback* information. The search can be implemented in three distinct ways:

Combination of the static graph space and static search space Fig. 10.3a represents the most efficient implementation from the search point of view. Here all memory is allocated before the search starts with a one-to-one correspondence between the graph state and search state, i.e. the position of each state of the search space is fixed during the search. This arrangement minimizes overhead during the search. To improve locality of access on platforms with enough RAM to store the search graph, states of the search graph and search space can be placed next to each other. The static search space can be represented in a very compact form using local properties of the network, e.g. linear sequences of HMM states can be stored very efficiently without the need for explicit connections between states. The code for evaluation of these linear segments can be very compact and efficient; Deligne at al. (2002) found that faster decoding can be achieved by avoiding full *minimization* of the graph to preserve longer linear state sequences.

This method is very efficient for small tasks, but does not scale well to systems with large vocabularies. As the size of the network grows, a significant portion of the search space memory is wasted since only a small portion of it will be used. In this scenario, it is more difficult to take advantage of more complex memory reduction techniques such as graph factoring (Mohri at al. 2002).

Static graph space with dynamic search space Fig. 10.3b is a more memory efficient option. The amount of search space memory corresponds to the number of active states at each time frame. This also gives the option to limit the amount of state space memory, e.g. by top N active state pruning. There is some runtime and memory overhead associated with the mapping between the graph and search spaces which changes for each time frame. On the other hand the improved locality of the memory access due to a much smaller search space can significantly improve the search speed.

Graph factoring can be used to reduce the memory needed for the search graph, where reoccurring linear state sequences are referenced by a single index. Expansion of these indices back to state sequences creates some runtime overhead, so for the most efficient search space implementation (i.e. the dynamic assignment of the search memory to the corresponding states in the graph space), the graph representation needs to be simple, i.e. without factoring. The efficiency of this approach has been demonstrated on LVCSR in the context of the DARPA Switchboard 1xRT evaluation (Saon et al. 2003).

Dynamically built graph Fig. 10.3c offers the most memory efficient decoder, but there can be a significant cost associated with the graph construction. The advantage is obvious – only the part of the graph which is searched is constructed in memory, with possibly significant memory savings in comparison to static graph methods. Another advantage is that *grammar* can be quickly modified (e.g. adding new pronunciations). The drawback is a higher computational cost, which can be construed as having two main parts. The first part is associated with the overhead of the on-the-fly HMM network construction. Secondly, the less obvious cost is caused by the loss of the minimal property of the dynamically built graph. This leads to duplicate evaluations of some graph parts, which is eliminated by global *minimization* in the static graph approach.

The memory savings, i.e., the ratio between the expected number of states visited during the search and the static graph size, depends on many factors (e.g. vocabulary size, number of alternative pronunciations, specific structure of the grammar) and is not easily predictable. This is the technique of choice when the static graph size is too large or when it is not possible to build the full graph. On systems with limited memory this approach clearly offers benefits, but the dynamic graph building algorithm needs to be carefully designed to limit its overhead cost.

Recent methods for on-the-fly composition employ the *finite state transducer* framework (Caseiro and Trancose 2006). Techniques which try to combine advantages of the static and dynamic methods have been proposed. (Willet and Katagiri 2002) statically compile only the part of the graph corresponding to the unigram language model, and Dolfing (2002) proposes an incremental application of a factorized language model to the search graph.

10.6.1.1 Pruning

The most critical part of any search implementation is the pruning of active states. Proper application of the pruning algorithm always involves a tradeoff between the efficiency and admissibility of the search. There are two main approaches to pruning: *beam pruning* and *rank* pruning.

In the beam pruning approach, at each time frame the state with the highest likelihood is found and then only those states are kept whose likelihood lie within a certain threshold from this maximum. This is an inexpensive and very efficient method; there exist variations of this approach which utilize multiple thresholds, e.g. for word internal states and for inter-word transitions.

Rank based pruning can be used in addition the beam pruning on larger tasks by keeping only top N active state at each time frame. One disadvantage of the beam pruning is that the number of active states can vary significantly. The ambiguity is usually much higher at the beginning of an utterance, in particular on grammars representing large list of choices. Application of a uniform pruning threshold may cause the search to be too slow in the beginning and have too many search errors at the end of the utterance. In such a case, the search cost can be reduced by selecting the top N active states. This is clearly a more expensive algorithm and its effect should be evaluated specifically for each task.

10.6.1.2 Lattice Generation

The implementation of the Viterbi search becomes much more complex when alternative hypotheses to the best path are required. The alternative paths are typically represented as a word *lattice*. Word lattices can be used for many purposes including multiple-pass search algorithm, computation of a confidence measure and *N-best* list generation. Exact lattice generation that would consider all possible state sequences is too expensive for practical use, but approximations can be used with only a small degradation in the lattice quality (Schwartz and Austin 1993).

Finding the best path only requires storage of one traceback record for the best path at each word end in each time frame. The traceback record contains the score of the path, identification of the word and a pointer the previous record. For lattice generation, significantly more memory is needed to store the traceback records for the alternative paths as well. There are two alternative approaches to this problem. The first approach is illustrated in Fig. 10.4a on a segment of an HMM network. For simplicity, HMM states shown in the figure are either word end states (white circles) or internal states with path merges (black circles). In this approach, multiple hypotheses (shown as gray wide lines) are propagated through the HMM network segment towards the next word end. Their new traceback records are created for each hypothesis. These records are then used during the lattice generation to create links between word ends (dotted lines). This approach increases the cost of the search, since some HMM states are effectively evaluated multiple times.

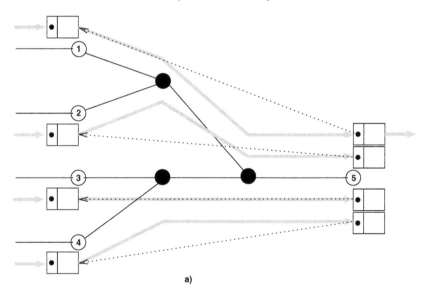

a)

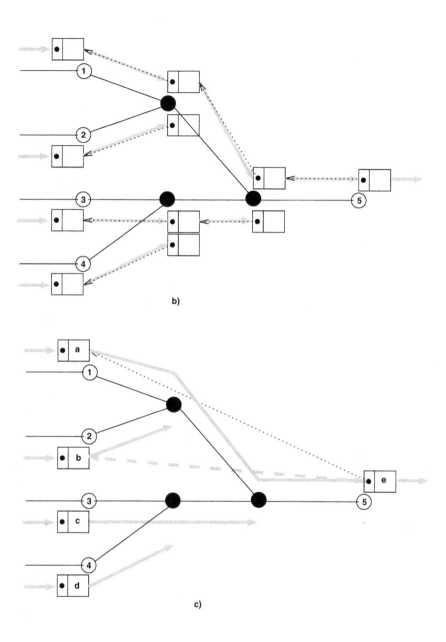

Fig 10.4 Lattice generation methods

In the second approach, shown in Fig. 10.4b, a new traceback record is created for each alternative path at each word internal merge state. Lattice links are created by concatenating connections between these records, considering all alternative records in each state.

This approach does not increase the cost of search but is more memory demanding. Application of global *minimization* on the search graph actually aggravates this problem, since it creates merges of word internal states. One possible way to reduce the runtime memory overhead is to use a non-minimal graph in which alternative paths can merge at word ends only.

Novak (2005) proposes a method that saves the runtime memory by adding auxiliary information to the search graph, which can be used to recover alternative paths even when traceback records are created at word ends only. It can be seen in Fig. 10.4c that only one lattice link can be created between states 1 and 5. Under the assumption that only internal states of the same words are merged, i.e. the state 5 represents an end state of a unique word, links of the alternative paths can still be recovered (with some approximation).

For example, a link between states 5 and 2 can be created since there is a valid record *b* in state 2. The score of this link is assumed to be the same except for the difference between language model scores of the path from state 1 to 5 and from state 2 to 5. To be able to calculate this difference, information about possible word links and their language model scores for each word end state is added to the static search graph during the offline graph compilation.

10.6.2 Search Graph Construction

In comparisons of the static graph scenario with that of dynamic decoders, the time needed to construct the graph is usually not considered as a run-time cost. But the assumption that the search graph never changes is clearly not practical. Having an efficient method for the search graph compilation is desirable for several reasons. Even if the static graph scenario is used, there may exist a need to quickly compile the graph using the limited resources of the particular platform before the search starts (e.g., when the grammar is constructed dynamically in response to a certain dialog state). The ability to customize the system by the user also requires that the graph be built locally.

The use of *finite state transducers* (Mohri et al. 2002) has become popular in the speech recognition community. It provides a solid theoretical framework for the operations needed to create a search graph. In general, the search graph is the result of a composition:

$$C \circ L \circ G \tag{10.2}$$

where G represents the language model as a finite state acceptor (FSA), L represents the pronunciation model and C converts the context independent phones to context dependent HMMs.

Language models fall into two distinct categories: *n-grams* and *grammars*. Each has different properties from the graph construction point of view. For n-gram models, the back-off model type is widely preferred for use in a Viterbi decoder. A straightforward approximate method can be used to construct the FSA representation of an n-gram back-off model using non-emitting (null) arcs representing the back-off transitions. If the back-off symbol is treated as a part of the vocabulary, G can be considered deterministic. It would be impractical – and even impossible – to determinize the *n-gram* model by removing the null arcs. Each state represents a unique history and the graph is usually close to being minimal.

Grammars represent a way to define a set of (possibly infinite) allowed sentences. Most of the systems use a formal syntax based on Context Free Grammars with regular expression construct extensions to enable more compact representations. By allowing only right recursion, regular languages can be generated and a corresponding FSA can be found. The task of a grammar compiler is to convert the grammar definition (e.g. written in BNF language) into a (weighted) WFSA. The first step of the compiler usually produces a nondeterministic WFSA. The next step is *determinization*, usually the most expensive part of the process (possibly with exponential cost). In some situations determinization can significantly increase the number of states and arcs. The final compilation step finds a minimal form of the deterministic WFSA. The complexity of *minimization* is $O(NA\ log(N))$ where N is the number of states and A is the average branching factor. For acyclic FSA much faster $(O(N+A))$ minimization algorithms exist.

Various schemes have been proposed for more efficient search graph compilation of grammars. A common idea is to avoid full expansion of the search graph and compile only those parts used at search time by exploiting properties of a particular grammar.

The concept of Recursive Transition Networks and a late binding scheme is used by Schalkwyk et al. (2003). The grammar is factored into several parts that are compiled into separate search graphs, and which are composed together at the runtime. This approach allows fast modification of the final search graph, e.g. by using a user specific address list as one of the parts. Zheng and Franco (2002) propose hierarchical non-deterministic grammar compilation to avoid graph size increases caused by the determinization step in some situations.

Instead of working with transducers, it is possible to use acceptors. An acceptor has only one label on each arc and can be constructed by taking a union of the input and output alphabets. Both determinization and minimization are simpler and faster when performed on finite state acceptors. By not using transducers, the full advantage of minimization by pushing the output labels is not utilized. However, if such minimization is used, the word labels are no longer attached to actual word ends. Keeping the placement of word labels at the word ends simplifies the decoder design, since the proper time alignment is preserved. Novak and Bergl (2004) uses a graph construction method that performs all operations (composition, *determinization, minimization* and weight pushing) in a single step, applied incrementally to each state of G (and for each context class for cross-word context models) one at a time. The resulting graph is fully minimized (as an acceptor) with no intermediate step requiring more memory than the final graph. In other words, during the compilation the graph never has more arcs or states than the final one. An important advantage of the incremental construction is that it does not require use of a general minimization procedure; rather, a much more efficient local acyclic graph minimization is used.

10.6.3 Fast Match

Search cost can be reduced by the use of a fast match. The idea is to use inexpensive models to look into the near future to decide which paths can be pruned. The effectiveness of the fast match increases with the accuracy of the approximate models and the amount of look-ahead time and decreases with the computation costs, so a trade-off needs to be found. In an asynchronous stack decoder (Gopalakrishnan et al. 1995) a fast match can be used very efficiently with a look-ahead of whole words. In a synchronous scheme, that is difficult to do. Hence, the most popular schemes, e.g. (Ortmanns et al. 1997), are usually limited to predicting the next active phone. In this instance, most of the time savings comes from the elimination of the observation likelihood computation for the states not selected by the fast match. In the Gaussian selection scheme, where the Gaussian computation is independent of the search, the short term fast match is less effective.

Long-term fast match can be used in certain tasks (Novak et al. 2003). When the utterances are relatively short and the grammar can be effectively expressed as a tree, then a two pass method can be used with inexpensive phonetic models used in the first pass to find the *N-best* list of paths to be rescored in a second pass.

10.6.4 Alternative Decoding Schemes

A decoding strategy based on building HMM networks covering the entire search space (either statically or dynamically) has limitations. Memory use is certainly an issue for static graphs, but even for dynamically build graphs the expansion of the active space may be too memory intensive. Direct incorporation of both the acoustic and the language models into the search graph may not be the most optimal approach in all situations. The context affecting the acoustic search (several phones) is typically much shorter than the context affecting the language model (several words). The acoustic search space can be much smaller (or significantly reduced) without the application of the language model.

This idea is utilized in multi-pass decoding schemes, where a more complex model is applied at each pass. For example, only the uni-gram language model is applied in the first pass and bi-gram and tri-gram parts are applied on the *N-best* list of paths found in the first pass. The multi-pass approach has several disadvantages, in particular the inherent latency, which makes it less attractive for use in interactive systems. To compute the top N choices, additional memory is needed to store significantly large amount of traceback information over that required by the single best path search. Finally, the two pass approach is not necessarily faster since the lower *n-gram* model is less discriminatory and more paths need to be explored by the search.

The time conditioned search (Ortmanns and Ney 2000) can be seen as a method which performs acoustic search independently of the language model in a single pass decoding scheme. Acoustic scores for words are evaluated for all possible starting times and then combined with the language model to find the best hypothesis. The advantage of this approach is that both the acoustic scores of each word and a particular start time are computed only once regardless of the number of language model contexts it appears in. A significant advantage was not found when a 3-gram model was used, as the cost of recombination with the language model reduced the efficiency of the method. One can expect that for much more complex language models the benefits would be more apparent.

Another alternative to the Viterbi search, an asynchronous decoding scheme, can be considered (Gopalakrishnan et al. 1995). This approach is best suited (in combinations with a fast match) for dynamic search cases where acoustic and language model decoupling is desired. Renals and Hochberg (1999) use synchronous acoustic search and asynchronous search for the language model part. In an asynchronous decoder, Novak and Picheny (2000) reported a significant speed improvement by reusing (with some approximations not affecting the accuracy) acoustic scores of words in multiple language model contexts. Unfortunately, the asynchronous decoder implementation is significantly more complex in comparison to the Viterbi search and less robust, i.e. it requires many more parameters which need to be carefully tuned.

10.7 Conclusion

We have presented several ASR algorithm implementations aimed at achieving high efficiency, which make them suitable for deployment on embedded platforms. This is a very wide area and we tried here to cover only the most commonly encountered issues. The reader should refer to the extensive list of reference for specific details. As one might expect, this area is constantly evolving. As more and more powerful embedded hardware devices become available, we can certainly expect more complex algorithms being deployed.

Acknowledgments

The author would like to acknowledge members of the Speech and Language Technologies department at IBM T.J. Watson Research Center for their help and suggestions in preparations of this chapter.

References

Aubert, X.L. (2002). An overview of decoding techniques for large vocabulary continuous speech recognition. *Computer Speech & Language*, vol. 16, no. 1, pp. 89–114.

Bahl, J.L.R., de Souza, P.V., Gopalakrishnan, P.S., Nahamoo, D. and Picheny, M. (1994). Robust methods for using context-dependent features and speech recognition models in a continuous speech recognizer. In *Proceedings of ICASSP*.

Balakrishnan, S. (2003). Fast incremental adaptation using maximum likelihood regression and stochastic gradient descent. In *Proceedings of Eurospeech*.

Bocchieri, E. (1993). Vector quantization for the efficient computation of continuous density likelihoods. In *Proceedings of ICSLP*, pp. 692–695.

Caseiro, D. and Trancose, I. (2006). A specialized on-the-fly algorithm for lexicon and language model composition. *IEEE Transactions on Audio Speech and Language Processing*, vol. 14, no. 4, pp. 1281–1291.

Deligne, S., Dharanipragada, S., Gopinath, R., Maison, B., Olsen, R. and Printz, H. (2002). A robust high accuracy speech recognition system for mobile applications. *IEEE Transactions on Speech and Audio Processing, Special issue on automatic speech recognition for mobile and portable devices*, 10(8), pp. 551–561.

Dolfing, H.J.G.A. (2002). A comparison of prefix tree and finite-state transducer search space modelings for large vocabulary speech recognition. In *Proceedings of ICSLP*, pp. 1305–1308.

Frichtsch, J. and Rogina, I. (1996). The bucket box intersection (BBI) algorithm for fast approximative evaluation of diagonal mixture gaussians. In *Proceedings of ICASSP*.

Gales, M.J.F. (1997). Maximum likelihood linear transformations from HMM-based speech recognition. *CUED Technical Report TR291*.

Gales, M.J.F., Knill, K.M. and Young, S.J. (1992). State-based Gaussian selection in large vocabulary continuous speech recognition using HMMs. *IEEE Transactions on Speech and Audio Processing*, vol. 7, no. 2, pp. 154–161.

Gopalakrishnan, P.S., Bahl, L.R. and Mercer, R.L. (1995). A tree search strategy for large vocabulary continuous speech recognition. In *Proceedings of ICASSP*, pp. 572–575.

Kanthak, S., Ney, H., Riley, M. and Mohri, M. (2000). A comparison of two LVR search optimization techniques. In *Proceedings of ICSLP*, pp. 1309–1312.

Mohri, M., Pereira, F. and Riley, M. (2002). Weighted finite-state transducers in speech recognition. *Computer Speech & Language*, vol. 16, no. 1, pp. 69–88.

Novak, M., and Picheny, M. (2000). Speed improvement of the tree-based time asynchronous search. In *Proceedings of ICSLP*, pp. 334–337.

Novak, M., Gopinath, R.A. and Sedivy, J. (2002). Efficient hierarchical labeler algorithm for Gaussian likelihoods computation in resource constrained speech recognition systems. *http://www.research.ibm.com/people/r/rameshg/novak-icassp2002.ps*.

Novak, M., Hampl, R., Krbec, P., Bergl, V. and Sedivy, J. (2003). Two-pass search strategy for large list recognition on embedded speech recognition platforms. In *Proceedings of ICASSP*, pp. 200–203.

Novak, M. (2005). Memory efficient approximative lattice generation for grammar based decoding. In *Proceedings of Eurospeech*, pp. 573–576.

Novak, M. and Bergl, V. (2004). Memory efficient decoding graph compilation with wide cross-word accoustic context. In *Proceedings of ICSPL*, pp. 281–284.

Olsen, P. and Dharanipragada, S. (2003). An efficient integrated gender detection scheme and time mediated averaging of gender dependent acoustic models. In *Proceedings of Eurospeech*, pp. 2509–2512.

Ortmanns, S., Firzlaff, T. and Ney, H. (1997a). Fast likelihood computation for continuous mixture densities in large vocabulary speech recognition. In *Proceedings of Eurospeech*, pp. 143–146.

Ortmanns, S., Ney, H., Eiden, S.A. and Coenen, N. (1997b). Look ahead techniques for fast beam search. In *Proceedings of ICASSP*, pp. 1783–1786.

Ortmanns, S., Eiden, S.A. and Ney, H. (1998). Improved lexical tree search for large vocabulary speech recognition. In *Proceedings of ICASSP*, pp. 817–820.

Ortmanns, S. and Ney, H. (2000). The time-conditioned approach in dynamic programming search for LVCSR. *IEEE Transactions on Speech and Audio Processing*, vol. 8, no. 6, pp. 676–687.

Renals, S. and Hochberg, M.M. (1999). Start-synchronous search for large vocabulary continuous speech recognition. *IEEE Transactions on Speech and Audio Processing*, vol. 7, no. 5, pp. 542–553.

Saon, G., Zweig, G., Kingsbury, B., Mangu L. and Chaudhari, U. (2003). An architecture for rapid decoding of large vocabulary conversational speech. In *Proceedings of Eurospeech*, pp. 1977–1980.

Saon, G., Zweig, G. and Povey, D. (2005). Anatomy of an extremely fast LVCSR decoder. In *Proceedings of Interspeech*, pp. 549–552.

Schalkwyk, J., Hetherington, L. and Story, E. (2003). Speech recognition with dynamic grammars using finite-state transducers. In *Proceedings of Eurospeech*, pp. 1969–1972.

Schwartz, R. and Austin, S. (1993). A comparison of several approximate algorithms for finding multiple (N-best) sentence hypotheses. In *Proceedings of ICASSP*.

Willet, D. and Katagiri, S. (2002). Recent advances in efficient decoding combining on-line transducer composition and smoothed language model incorporation. In *Proceedings of ICASSP*, pp. 713–716.

Zheng, J. and Franco, H. (2002). Fast hierarchical grammar optimization algorithm toward time and space efficiency. In *Proceedings of ICSLP*, pp. 393–396.

11

Algorithm Optimizations: Low Memory Footprint

Marcel Vasilache

Abstract. For speech recognition algorithms targeting mobile devices the memory footprint is a critical parameter. Although the memory consumption can be both static (long-term) and dynamic (run-time) in this chapter we focus mainly on the long-term memory requirements and, more specifically, on the techniques for acoustic model compression. As all compression methods, acoustic model compression is exploiting redundancies within the data as well as the limits for the parameter representation accuracy. Considering data redundancies specific for hidden Markov models (HMMs), parameter tying and state or density clustering algorithms are presented with cases like semicontinuous HMMs (SCHMMs) and subspace distribution clustered HMMs (SDCHMMs). Regarding parameter representation a simple scalar quantized representation is shown for the case of quantized HMMs (qHMMs). The effects on computational complexity are also reviewed for all the compression methods presented.

11.1 Introduction

In practical speech recognition systems, especially when targeting the mobile or embedded environment, complexity considerations play a major role when selecting the type of algorithms employed. Computational and storage complexity limits often require making performance compromises in order to meet the implementation constraints. Fortunately, there are numerous techniques which aim at minimizing the loss of performance with respect to the complexity savings.

In the following the focus will be on reducing the memory footprint of the acoustic models as they usually represent the most significant memory expenditure of the classifier. Since a large body of literature targeting this area exists and a comprehensive presentation of specific algorithms would require a book of its own, in this chapter we aim at offering an overview of the main design factors, a few selected methods and links to relevant references. With this goal we first proceed in revising the fundamentals of hidden Markov models (HMMs) based classification and stating the optimization problem. Following this, a few model selection criterions are presented. In Sect. 11.4 are presented the main levels of parameter tying for the continuous density HMMs and next, in Sect. 11.5, we illustrate the main options for parameter representation. We then examine some of the methods frequently used for model size reduction like quantized parameters HMMs in

Sect. 11.6 and subspace distribution clustering HMMs in Sect. 11.7. The computational complexity implications are briefly mentioned in Sect. 11.8. Finally, some practical implementation aspects are revealed and a few concluding statements are made.

11.2 Notations and Problem Statement

In speech recognition the input audio waveform is transformed into a sequence of observation vectors $o_{1:n} = o_1, o_2, ..., o_n$ which is often modeled as a 1st order Markov chain using *hidden Markov models* (HMM).

An HMM (Rabiner 1989; Jelinek 1998) consists of:

- a set of states

$$S = \{s_i \mid i \in \overline{1, N}\}$$

- the initial probability distribution for the states

$$\pi = \{\pi_i \mid i \in \overline{1, N}, \pi_i = P(s(0) = s_i)\}$$

- a matrix of state transition probabilities

$$A = \{a_{i,j} \mid i \in \overline{1, N}, j \in \overline{1, N}, a_{i,j} = P(s(t+1) = s_j \mid s(t) = s_i)\}^1$$

- a set of state dependent probability distributions or probability density functions (*pdf*) for observation vectors

$$B = \{b_i(o) \mid i \in \overline{1, N}, b_i(o) = P(o \mid s_i)\}$$

More compactly, the parameters for such a model can be grouped into a set

$$\lambda = \{\pi, A, B\}.$$

HMMs allow us to effectively compute

$$P(o_{1:n} \mid \lambda)$$

which is the probability that an observation sequence $o_{1:n}$ was generated by the model λ. In addition, the set of parameters λ can be optimized such that the previous probabilities are maximized for observation sequences of selected acoustic classes.

For a majority of the current practical systems the observation vectors form a continuous space therefore the set B consists of probability density functions and the resulting HMMs are named continuous density hidden Markov models (*CDHMMs*). For a simpler parameter estimation the functions B are formed as

---

[1] $s(t)$ being a function denoting the temporal state sequence.

mixtures of log-concave or elliptically symmetric densities, very frequently Gaussians or Laplacians using the following formula

$$b_i(o) = \sum_{n=1}^{M} c_{in} G(o, \mu_{in}, \sigma_{in}^2) \qquad (11.1)$$

where G denotes, for instance, the Gaussian *pdf* parameterized by the mean vector μ and variance vector σ^2 and the mixture coefficients $c_{in} \geq 0$ satisfy the normalizing condition $\sum_{n=1}^{M} c_{in} = 1$.

When having a set of words from a given vocabulary

$$W = \{w \mid w \in Vocab\}$$

the recognition problem consists in finding the word with the maximum a-posteriori probability given the current observations

$$w_{rec} = \arg\max_{w \in W} P(w \mid o_{1:n}) = \arg\max_{w \in W} P(o_{1:n} \mid w) P(w).$$

To compute the probabilities above, a set of HMMs can be used, each one representing one word and having its parameters λ_w optimized after a training process. With this, the recognizer's job consists in the evaluation of the expression below

$$w_{rec} = \arg\max_{w \in W} P(o_{1:n} \mid \lambda_w) P(w).$$

Acoustic model compression aims, in essence, at maximizing the recognition performance when the memory complexity is upper bounded by practical implementation limits.

With an HMM based speech recognizer the memory complexity is directly dependent on the total *parameter space*

$$\Lambda = \{\lambda_w = \{\pi_w, A_w, B_w\} \mid w \in Vocab\}.$$

Consider we have a set of modeling options indexed by m with the corresponding parameter spaces denoted by Λ_m and memory cost and performance functions denoted with Γ and Θ, respectively. The indexing m covers only the model structure and it is not dependent on the parameter values. For optimality, in general, the parameters for each m have to be chosen as

$$\Lambda_m^* = \arg\max_{\Lambda_m} \Theta(m, \Lambda_m)$$

This results in finding the optimum model set m^* within the memory constraints *Mem* as

$$m^* = \arg \max_{\{m | \Gamma(m) \leq Mem\}} \Theta(m, \Lambda_m^*)$$

The direct, exhaustive search approach for solving this under the assumption that Γ does not depend on actual parameter values[2] can be summarized as:

Step 1. Generate all possible model sets within the limit $\Gamma(m) \leq Mem$.

Step 2. For each of them find the optimal parameter values, Λ_m^*.

Step 3. Pick the best set.

Enumerating all possible model structures at step 1 can be a very difficult task. The combinatorial explosion due to the multiple distinct modeling parameters (feature vector dimension, number of states, mixture sizes, sparsity of transition matrices, ...) quickly results into an intractable size for the search space. Even if the modeling search space is severely pruned, the optimization at step 2 is very expensive for nearly all practical cases. Each of the typical objective functions: maximum likelihood (Rabiner and Juang 1986, 1993; Huang et al. 1990), minimum classification error (Juang et al. 1997; Katagiri et al. 1998), maximum mutual infor-mation (Bahl et al. 1986; Normandin et al. 1994), largest classification margin (Hui et al. 2006), require expensive optimization procedures usually performed on very large training databases.

In practice, the usual alternatives are:

1. Gradually grow a model structure following a set of transformation rules until either its performance does not improve or the memory limits are reached. The typical transformations consist of model, state or mixture density splitting.
2. Start with a model set which has good classification performance but it is over the imposed memory limit. Apply then a set of compression transformations which, while minimally degrading the objective function Ω, allow for the complexity to fit within the imposed limit. In this case the key operations are parameter pruning, tying, clustering and quantization.

Most frequently, from a simple set of models increasingly more complex ones are created by iterating type 1 transforms until the performance saturates or a modeling performance/complexity criterion is maximized. At this point, the type 2 transforms are used to bring the models within more acceptable complexity limits.

An illustration is provided in Fig. 11.1 where all the possible model con-figurations are bounded by an optimal performance/complexity curve. The required maximal complexity limit is marked by the vertical dotted line. The ideal configuration is the highest performing model on the left side of the limit line which,

---

[2] If Γ depends on the parameter values but a lower bound for it exists, then generat-ing all models which have this lower bound below the imposed memory limit is sufficient to guarantee that the optimal model is included in the search space.

in this case, is the one closest to the intesection point (marked with a black circle). With arrows are shown model evolutions when procedures of type 1 or 2 are used. Also visible is a typical sequence of complexity and performance growing transforms followed by a complexity reduction stage.

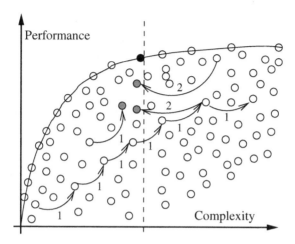

Fig. 11.1 Model structures in the performance/complexity space

11.3 Model Complexity Control

As introduced in the previous section, the total parameter set is

$$\Lambda = \{\lambda_w = \{\pi_w, A_w, B_w\} \mid w \in Vocab\}$$

Controlling this set involves the selection of model sizes (number of states), model topology (i.e., A matrices), and the degree of accuracy in modelling the state *pdf*s (B functions). For most practical situations the dominant number of parameters is formed by the B functions for which the total number of states and number of mixture densities are the key elements.

A fundamental problem in pattern recognition is having the parameter set of "adequate" complexity given the classification task at hand. To address this objective one can use either direct or indirect methods.

The direct method consists in selecting a representative *validation set* for the targeted use case and then monitoring the classification performance on it for increasingly complex models in order to find the complexity point from where the performance is no longer improved.

Since a direct method can be costly to implement the indirect approaches aim at estimating how a model behaves on unseen data given its performance on the training set. In the following, we denote by O the concatenation of all observation vectors in the training set, by W_{seq} the corresponding sequence of words of the correct transcription and we index by m all the model structures under evaluation.

11.3.1 Akaike's Information Criterion

For addressing the issue of model selection, *Akaike's Information Criterion* (AIC) was proposed in a pioneering work (Akaike 1973, 1974). AIC has the form shown in Eq. 11.2 below and it has to be maximized by the optimal model.

$$\log P(O \mid \Lambda_m^*, W_{seq}) - N(\Lambda_m) \tag{11.2}$$

Here with $N(\Lambda_m)$ is denoted the number of parameters for model m and with Λ_m^* the maximum likelihood parameters.

This criterion was derived starting with the relative entropy (Kullback-Leibler divergence) between the true *pdf* and the modelled one and linking it to the maximum likelihood. Akaike found that the maximum likelihood value is a biased estimate of the model dependent part of the relative entropy and that a bias correction term in the form of the number of parameters must be included. With such correction minimizing the expected divergence more closely amounts to maximizing Eq. 11.2 hence resulting in a much simpler model selection rule.

11.3.2 Bayesian Information Criterion

In a Bayesian framework the optimal model is the one maximizing the *evidence integral* over the parameter space for the training data. The best model is hence found as in the following formula

$$\arg\max_m P(m) \int P(O \mid \Lambda_m, W_{seq}) P(W_{seq}) P(\Lambda_m) d\Lambda_m$$

In practice it is not feasible to use the previous expression therefore approximation schemes are derived.

A commonly used 1st order approximation near the maximum likelihood parameters (Λ^*) is the *Bayesian Information Criterion* (BIC) (Schwartz 1978). Under the assumption of uninformative priors for the models this transforms the previous formula into the maximization of the expression below

$$\log P(O \mid \Lambda_m^*, W_{seq}) - \frac{\tau}{2} N(\Lambda_m) \log N(O)$$

where $N(O)$ denotes the size of the training data. Here τ is a tuning parameter which allows the original value of 1 to be better adapted to the specific task (Chou and Reichl 1999). Other practical examples of using this criterion can be seen in Chen and Gopalakrishnan (1998) and Mak (2004).

11.3.3 Second Order Approximation

In the previous cases the model parameters are equally treated irrespective of their impact on the likelihood function. A 2nd order Laplace approximation of the evidence integral can be made with an assumption of a local Gaussian curvature for the likelihood function at the maximum point in the parameter space. This results in maximizing

$$\log P(O \mid \Lambda_m^*, W_{seq}) + \frac{N(\Lambda_m)}{2} \log 2\pi - \frac{1}{2} \log \left| -\nabla^2_{\Lambda_m = \Lambda_m^*} \log P(O \mid \Lambda_m, W_{seq}) \right|$$

where in the final term is the determinant of the Hessian matrix for the log-likelihood function computed at the ML point.

This criterion, however, is far more demanding in practical use since for large systems the Hessian becomes intractable and approximations are needed (Roberts et al. 1998)

11.3.4 Other Measures

A different perspective over the problem based on information and coding theory, is offered by Minimum Description Length (MDL) (Barron et al. 1998) and Minimum Message Length (MML) (Wallance and Boulton 1968). Although using different premises, form a practical perspective, all these model selection criteria can be viewed as penalizing the likelihood with a method specific term which has linear or near linear variation on the number of model parameters. In Yang and Barron (1998) a multitude of such measures is presented while excellent historical perspectives with a closer examination of the various criteria can be found in Lanterman (2001) and Burnham and Anderson (2002, 2004). More recent approaches based on discriminative or predictive methods which are directly targeting the speech recognition domain can also be found in Padmanabhan and Ban (2000), Chien and Furui (2005), and Liu and Gales (2007).

11.4 Parameter Tying

The main conclusion concerning the measures introduced in the previous section is that when having model sets giving identical performance on the training set, for better performance on unseen data, it is best to select the set with the minimum number of parameters. In addition, a reduced number of parameters allows in most cases more reliable estimates of their values, especially when the training data does not cover extensively the multitude of acoustic events which the models are expected to encounter.

Parameter tying is an effective approach for reducing the number of model parameters with immediate gains in terms of memory and computational complexity savings. Due to this, the subject has received a good deal of attention (Huang and Jack 1989;Young 1992).

Parameter tying can be implemented on several levels as briefly described next and illustrated in Fig. 11.2. However, this is not a complete coverage of the tied structures as also other possibilities exist [e.g., for Gaussian mixtures tying the covariance matrices with particular cases such as global variance or semi-tied covariances Gales (1999)].

11.4.1 Model Level

If we consider the totality of words in the recognition vocabulary as forming the parameter space for the classifier, the basic *model level tying* consists in building

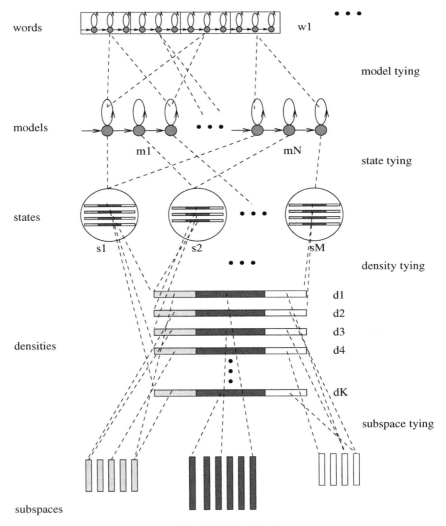

Fig. 11.2 Tying levels

each word as a concatenation of smaller units shared for the whole recognition lexicon. These units are often a set of allophones or even syllables (Ganapathiraju et al. 2001) which allow for the complete representation of the words in the recognition lexicon.

Even with the relatively smaller number of allophones, further tying is sometimes demanded, especially in the context of multilingual speech recognition where phonetic similarities across languages can be exploited (Harju et al. 2001). Although the tying decisions can be phonetically motivated, data-driven methods or combinations are also being used (Vihola et al. 2002).

11.4.2 State Level

The next level of tying involves individual HMM states. This type of tying is often done when a large number of HMM units are used for capturing contextual information. The typical cases are biphones or triphones models in large vocabulary speech recognition systems for which the middle states are shared among the models corresponding to the same allophone.

The selection of the tying structure can be done top-down when using phonetical rules and decision trees or bottom-up with data driven clustering procedures (Nock et al. 1997; Young et al. 1994; Junqua and Vassallo 1996).

11.4.3 Density Level

Semicontinuous HMMs (*SCHMM*s), also known as tied-mixture HMMs, implement tying at the level of the mixture density of the state emission functions (Huang and Jack 1989; Huang 1992). In SCHMMs all states share the same set of densities. For Gaussian densities the state *pdf* has nearly the same form as in Eq. 11.1 with the difference that now the mixture components are indexed over a global codebook and only the mixture weights c_{in} are still state dependent.

$$b_i(o) = \sum_{n=1}^{M_{glob}} c_{in} G(o, \mu_n, \sigma_n^2)$$
(11.3)

With this structure and by allowing a sparse representation of the mixture weights we can observe that, in fact, SCHMMs offer a generalization for the mixture based CDHMM. Following this a more general framemork for mixture tying is created as presented in Digalakis et al. (1996) and Willett and Rigoll (1997) where algorithms for automatic sharing the mixture components among states are proposed.

11.4.4 Subspaces

When going below the density level the natural approach is to examine the models from the perspective of the feature space. As this is usually of moderately high dimensions, splitting it into disjoint orthogonal subspaces allows for a new level of tying where each density is split into several components as given by its projections into the selected subspaces. The number of densities in each subspace can be significantly reduced using a clustering procedure which replaces the initial densities with the corresponding cluster representatives. For each subspace a density codebook is therefore formed and a tying structure induced. The likelihood for a full space density is a product of likelihoods of its corresponding subspace densities. This model structure is named Subspace Distribution Clustering HMM (*SDCHMM*) (Mak 1998; Mak and Bocchieri 2001b).

SCHMMs also have an extension to multiple subspaces as a parallel concept to *multi-stream HMMs* (Rabiner and Juang 1993). In multi-stream HMMs the state emission score is given by

$$b_i(o) = \prod_{k=1}^{K} b_{ik}(o)^{w_k}$$

where we considered K streams (subspaces) and the stream weights w_k are all positive.[3] For SCHMMs, in each subspace the functions $b_{ik}(o)$ have a similar form as in Eq. 11.3 where the densities are all shared from a subspace specific codebook. It is now visible that, even with unity stream weights, SDCHMMs will be equivalent to SCHMMs only when each substream's *pdf* consists of a single density. In all other cases the difference consists in having mixture weight parameters on all subspaces for SCHMMs, while only allowing for a single set, at state level, in the case of SDCHMMs.

At the limit, when the subspaces are of unitary dimension and if considering only the mean values of the distributions, a feature level tying can be obtained, as presented in Takahashi and Sagayama (1995b) The additional effect of tying variance values is examined in Takahashi and Sagayama (1995a). If both mean and variances are scalar quantized then the quantizers can also be seen to introduce an implicit tying as in Vasilache (2000). However, in this latter case, the tying is not explicit since parameter changes (e.g., due to model adaptation) do not preserve the original tying structure in the model updates.

11.4.5 Clustering

Of fundamental importance for pattern recognition in general, data clustering is playing a significant role in the selection of the tied structures presented earlier. Aiming at creating optimal partitions for a set of objects, the clustering can be done with hierarchical or partitional types of algorithms.

The hierarchical algorithms which are further divided into divisive (top-down) and agglomerative (bottom-up) methods create the partition with a succession of splits, respectively, unions of the clusters until a termination criterion is valid. Typically, in the first case we start with all the elements placed into a single class while in the second case we start with each element forming a class of its own.

---

[3] The unity summation condition is frequently relaxed since the value of this sum is effectively balancing the impact on recognition scores of the A matrix of transition probabilities against the state score values B.

In partitional algorithms the number of clusters and initial cluster memberships are given at start. These types of algorithms are then changing the element memberships until an optimality criterion is reached. The *K-means algorithm* is representative for this case (Duda et al. 2001, Chap. 10).

When evaluating the optimality of a given clustering an appropriate distortion measure is required. In speech recognition such measures usually take advantage of the statistical nature of the components of the classifier from the perspective of their corresponding generating distributions. Frequently used are the *Bhattacharyya distance* (Kailath 1967; Rigazio et al. 2000) and the *Kullbak-Leibler divergence* (Myrvoll and Soong 2003; Li et al. 2005). For both measures optimal centroid algorithms exist (for Gaussian densities the previous references provide full details).

Considering two *pdfs*, p and q, these measures have the form

$$D_{Bhat}(p,q) = \int \sqrt{p(x)q(x)}\,dx$$

$$D_{KL}(p,q) = \int p(x)\log \frac{p(x)}{q(x)}\,dx$$

where a symmetrical version is often desirable for the divergence

$$D_{sKL}(p,q) = D_{KL}(p,q) + D_{KL}(q,p).$$

As example, for Gaussian densities $G_i(o,\mu_i,\sigma_i^2), i=1,2$, they have the closed form expressions below while the symmetrized KL can also be written as in the last equation.

$$D_{Bhat}(G_1,G_2) = \frac{1}{8}(\mu_1-\mu_2)^T\left[\frac{\Sigma_1+\Sigma_2}{2}\right]^{-1}(\mu_1-\mu_2) + \frac{1}{2}\ln\frac{\left|\frac{\Sigma_1+\Sigma_2}{2}\right|}{\sqrt{|\Sigma_1||\Sigma_2|}}$$

$$D_{KL}(G_1,G_2) = \frac{1}{2}(\log\frac{|\Sigma_2|}{|\Sigma_1|}+Tr(\Sigma_2^{-1}\Sigma_1-I)+(\mu_1-\mu_2)^T\Sigma_2^{-1}(\mu_1-\mu_2))$$

$$D_{sKL}(G_1,G_2) = \frac{1}{2}Tr\{(\Sigma_1^{-1}+\Sigma_2^{-1})(\mu_1-\mu_2)(\mu_1-\mu_2)^T + \Sigma_1\Sigma_2^{-1}+\Sigma_2\Sigma_1^{-1}-2I\}$$

11.5 Parameter Representations

Model parameters can be represented in three distinct forms: floating point, fixed point or quantized.

11.5.1 Floating Point Representation

Without special requirements on computation or storage, the floating point formats are, by far, the most frequently used. ANSI/IEEE Standard 754-1985 defines the two most commonly used floating point representations which require 32 bits of storage for single precision numbers and 64 bits for double precision. Both formats offer adequate range for storing any of the HMMs parameters provided that minor precautions are made (i.e., "well behaved" range for the input feature vectors, logarithm representation for transition likelihoods mixture weights and state emission likelihoods constants).

For most systems already the single precision format offers a more than adequate range for representing all types of model parameters. However, even if for density means and variances a more restricted floating point representation would have worked (for instance with only 16 bit size) such data type is not available in general, therefore fixed point representations are needed for a smaller memory footprint.

11.5.2 Fixed Point Representation

When using fixed point numbers a larger variety of such representations exist. The fixed point arithmetic makes use of integer numbers for which standard data types are available. For them the memory requirements are of 8, 16, 32 and 64 bits giving more storage options but also much stronger dynamic range constraints (please check Chap. 12 for more details).

The drawbacks to the increased storage flexibility are the additional processing and careful data normalization demanded by the fixed point computation mode. If a high performance floating point unit is also available then it is possible to avoid converting the whole recognizer into fixed point and only add a module for packing/unpacking the acoustic model data into the floating point format. However, these operations must be done efficiently and carefully scheduled within the data-flow of the classifier such as not to significantly increase its run-time complexity requirements.

11.5.3 Quantization

For parameter representations below 8-bits and/or for non-linear representations of the parameter values, quantization schemes are needed (Gersho and Gray 1992; Gray

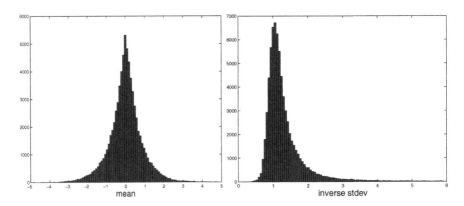

Fig. 11.3 Histogram for μ and σ^{-1} parameters

and Neuhoff 1998). In this case most of the memory expense is taken by indices in quantization codebooks. Considering typical model structures most of the parameters are represented by the mean and variance vectors from the mixture densities. As a result, all the quantization procedures focus on either a vector (subvector) or a scalar quantization of these values.

More complex model structures having Gaussians with full covariance matrices can also be addressed if first partitioning the densities, possibly at subspace level, into a reduced set of rotation classes (Gales 1999). In essence, the procedure aims at tying the covariance matrices among the densities within a class and forming the classes such that a minimal impact is seen on the model performance function Θ. For each such class an orthogonal transform is used which brings the associated densities into diagonal form, simplifying the quantization task as well as reducing the complexity for the likelihood computation.

In quantization the conventional procedure is to start with a set of CDHMMs optimized for the given recognition task, quantize the parameters by replacing them with index values into the newly constructed quantization codebooks and then append these codebooks to the model data. Retraining the models is seldom effective or even needed unless also a tying structure is introduced.

From a scalar quantization perspective, as example, typical distributions for μ and σ^{-1} parameters can be seen in Fig. 11.3.

11.6 Quantized Parameters HMMs

11.6.1 Scalar Quantization

The simplest form of parameter quantization is a scalar quantizer. For this case the main design decisions consist on how many quantizers to employ, at what rates and how to partition the data into these quantization classes.

For diagonal covariance densities, a natural approach is to separately consider the dimensions of the parameter space and create a mean and a variance quantizer for each. Although for low quantization rates the memory overhead of storing so many quantizers is manageable we can exploit the fact that classifiers based on Gaussian mixtures are invariant to invertible affine transformations of the observation vector space. If observation vectors are affine transformed

$$\underline{o} = Ao + b$$

and the matrix A is invertible, we obtain an equivalent classifier if mean values and covariance matrices are also transformed as

$$\underline{\mu} = A\mu + b$$
$$\underline{\Sigma} = A\Sigma A^{T}$$

We can use such a transformation to bring the model parameters into an optimized range, aiming at sharing the scalar quantizers among all the feature vector dimensions. With diagonal covariance matrices the matrix A should also be diagonal allowing then for the optimization of only two values for each feature dimension d. These are the scale a_{dd} and the shift b_d. An acceptable optimization criterion aims for a maximal overlap of the parameter distributions for each dimension. This, in practice, allows for sharing of a single mean and a single variance quantizer for all model dimensions. The sharing comes at a cost of storing two additional D dimensional vectors (the diagonal of A and the shift b) while saving the memory expense of $D-1$ mean and variance quantizers.

The quantizers themselves are of Lloyd-Max type and use an Euclidean distance measure for μ and σ^{-1} (the means and the inverse standard deviations are the quantization values). An example of how such scalar quantizers are maped over the 2 dimensional space of mean and variances can be seen on Fig. 11.4 while more details can be found in Vasilache (2000) and Vasilache and Viikki (2001).

Previously we have assumed that the same quantization rate is required for all feature vector components. In practice, not all the feature vector dimensions have similar impact on the classification performance therefore for the components with lower discriminating power we can assign lower rate quantizers as well. As example, if looking again at Fig. 11.4, half rate quantizers can be created by a selection of values from the full rate scalar quantizers (the full circles in the figure). A selection

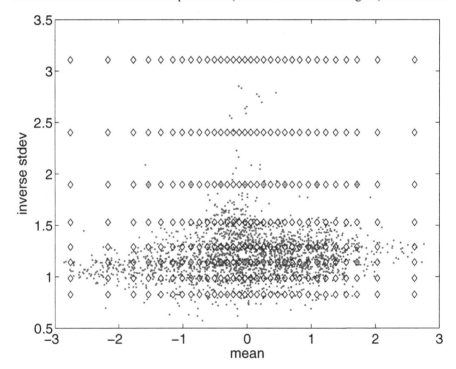

Fig. 11.4 Scalar quantizers over the space (μ, σ^{-1})

procedure which worked well in practice consisted in building an independent half rate scalar quantizer for the corresponding feature components and then replacing the quantizer values with the closest ones from the full rate quantizer (Vasilache in preparation).

11.6.2 Vector Quantization

When aiming for the highest possible compression, vector quantization becomes a necesity. From Fig. 11.4 it can be seen that the scalar quantizers are wasting a significant number of combination values. For this example a 2 dimensional vector quantizer might have done a better quantization job although at the expense of larger memory requirements for the quantizer codebook.

As before, the design decisions consists in how to split the parameter space into subspaces, what type of distortion measure to employ,[4] how many quantizers and at what rates. There are, therefore, many possibilities with some of them covered in the literature (Ravishankar et al. 1997; Pan et al. 2000; Lahti et al. 2003).

11.7 Subspace Distribution Clustering HMM

This type of models is created by partitioning the feature vector space into orthogonal subspaces. Under the assumption of statistical independence for these subspaces the likelihood for each density becomes a product of the subspace likelihoods. In this case, the state likelihood has the expression below

$$b_i(o) = \sum_{n=1}^{M_i} c_{in} \prod_{k=1}^{K} G(o_k, \mu_{kin}, \sigma_{kin}^2)$$

where we considered K subspaces each one with dimension d_k such that $\sum_{k=1}^{K} d_k = D$.

SDCHMM are formed by allowing density sharing at the subspace level. A tying structure is therefore created, the density components being formed by indexing within these subspace level codebooks.

When the tying structure is known it is possible to directly train SDCHMMs with the advantage of good performance even for smaller training set sizes (Mak and Bocchieri 2001a). However, most often SDCHMMs are obtained by converting a set of CDHMMs. The conversion process consists in two stages: subspace partitioning and density clustering.

---

[4] Especially if dealing with different types of parameters in the same quantizer (e.g., means and variances).

11.7.1 Subspace Partitioning

The problem of optimal partitioning the D dimensional feature space into *subspaces* does not have a direct solution. From combinatorial analysis it is known that the number of all possible partitions is the Dth Bell number. As example, when $D = 39$, which is a typical case in practice, there are about 7.4×10^{32} partitions.

In most cases the subspaces are empirically formed by grouping related dimensions of the feature stream. For instance, when the feature vector contains time derivatives the subspaces can each contain a static component together with its accociated time derivatives. Another option, as a limit case, is to create 1 dimensional subspaces. For such setup good results have been reported in practice (Leppänen and Kiss 2005).

When forming each subspace the objective is to achieve a very effective clustering with a minimal distortion induced to the original models. As we have seen, enumerating the subspaces and doing a clustering process for each case is not feasible therefore indirect approaches are required.

A possible approach consists in using as heuristic the measure of correlation between the feature space dimensions. Such a measure can be created based on the correlation for 2 dimensions

$$\rho_{ij} = \frac{\sigma_{ij}}{\sigma_i \sigma_j}, \quad R(i,j) = \rho_{ij}^2$$

which is then extended to k dimensions as

$$R(1,2,\ldots,k) = 1 - \begin{vmatrix} 1 & \rho_{12} & \rho_{13} & \cdots & \rho_{1k} \\ \rho_{21} & 1 & \rho_{23} & \cdots & \rho_{2k} \\ \rho_{31} & \rho_{32} & 1 & \cdots & \rho_{3k} \\ \vdots & \vdots & \vdots & \ddots & \vdots \\ \rho_{k1} & \rho_{k2} & \rho_{k3} & \cdots & 1 \end{vmatrix}$$

where σ_i, σ_j are variances and σ_{ij} is the covariance for the feature dimensions i and j.

Using this measure with a greedy algorithm it is possible to generate a subspace partitioning by repeatedly extracting the most correlated group of dimensions from the set of dimensions still available (Mak and Bocchieri 2001b).

Focusing now on the model parameters themselves, another measure used in subspace partitioning is their entropy. The target is now the formation of subspaces with minimal joint entropy and, as before, a set of greedy algorithms can be used (Filali et al. 2002, 2005).

11.7.2 Density Clustering

Once the subspaces are created the density clustering can be done following the same general principles as introduced in Sect. 11.4.5. A series of algorithms have been proposed in the literature (Mak and Bocchieri 2001b) or even been patented (Acero and Plumpe 2004).

Finally, we must emphasize again the fundamental difference between subspace vector quantized HMMs and SDCHMMs. When building these models the clustering and quantization procedures are, arguably, similar. However, although in both cases the densities are formed using a set of subspace densities taken from codebooks, in quantization the parameters are not tied while for SDCHMMs the tied structure is part of the model.

11.8 Computational Complexity Implications

A reduced set of parameters directly translates into significant computational gains as well. For SCHMMs the reduced number of densities, which is shared by all states, allows precomputing their likelihoods hence substantially reducing the costs of the state level computation from Eq. 11.3.

Subspace distribution clustering presents a similar advantage. In this case the density likelihoods are first computed for the subspace codebooks. For each of the full-space densities its score is then obtained by summing up the precomputed values using its associated subspace indexes. Even these summations can be significantly reduced by exploiting indexing similarities for groups of subspaces among the densities of the model. Savings of up to 50% in the number of additions have been reported (Aiyer et al. 2000).

The computational advantage of scalar quantization follows directly from the possiblity of tabulating the most expensive computational part, the evaluation of state emission likelihoods. For instance, for states with mixtures of Gaussian densities the state emission log-likelihood formula is

$$\log b(\mathbf{x}) = \log \sum_{k=1}^{K} \exp\{\log\left(c_k \frac{1}{\prod_{i=1}^{D}\sqrt{2\pi\sigma_{ki}^2}}\right) - \sum_{i=1}^{D}\frac{(x_i - \mu_{ki})^2}{2\sigma_{ki}^2}\}$$

where K represents the number of densities in the mixture and D is the dimension of the feature vector space.

The term containing the mixture weight and the Gaussian normalization factor is a constant with respect to the observed features therefore the most costly operation is the computation of the second term, the *Mahalanobis distance*.

With quantization, for any given feature vector, each of the terms $\dfrac{(x_i - \mu_{ki})^2}{2\sigma_{ki}^2}$ can take a limited range of values. For a typical rate of 5 bits for a mean component and 3 for a variance there are just $2^{5+3} = 256$ distinct values which, when computed in advance, will reduce the distance evaluation to an indexed summation from the precomputed tables.

With even lower rates the number of terms in the sum can be reduced by combining adjacent tables into a single one (e.g., with half the previous rate, combining two such tables results in the same number of distinct values but reduces the summation costs to half).

Computing the tables for each frame can be avoided if the feature vectors are also quantized (Vasilache et al. 2004). In this case the entire state likelihood evaluation is reduced to table lookup and summation with no other overhead costs per observation vector.

11.9 Practicalities and Conclusion

In practice the acoustic model compression methods are selected in close relation with the specifics of the problem at hand. As example, if the memory requirements are not very tight and/or the models need to support speaker or environment adaptation as well, a scalar quantization approach might work very well and it is also very simple. By accommodating larger missmatches between the trained model statistics versus the testing conditions it also allows more room for parameter updates. On the other hand, if we have a large model set to begin with and a higher compression ratio is required, then a vector quantization or subspace distribution clustering approach is needed. If also support for parameter adaptation is required then the tyied structure induced by SDCHMMs can help if it truly matches the intrinsic properties of the data since with tying it allows for faster, more effective model updates. However, with high compression it might happen that excessive tying severly reduces the degrees of freeedom for adaptation in which case vector quantization might be a better choice.

With respect to specific performance figures, scalar quantization at 5 bit for the mean parameters and 3 bits for the variances does not alter the original recognition performance and it also gives a good packing into one byte of the joint indices. At half this rate (3 bit means and 1 bit variances) a moderate recognition performance degradation must be tolerated for the substantial gain in memory (and computation). Even for extreme situations, globally tied variances and only 2 bit rates for the mean parameters more than 95% of the original recognition performance is preserved signaling a high degree of robustness and redundancy.

For subspace distribution clustering or vector quantization there are more alternatives to evaluate such as formation of subspaces and bitrate allocation for each codebook. Good results were reported with 6-bit rates for a mean-variance pair (Leppänen and Kiss 2005) or 4-bit rates for mean parameters in 2 dimensional codebooks with a global variance (Astrov 2002; Varga et al. 2002). When pushing the limits, as for the scalar case, it is surprising to observe that the performance is not dramatically decreased with rates as low as 1 bit per mean-variance pairs (Mak 2004, Sect. 3).

Of considerable practical importance during the state likelihood computation is the quick access into codebooks of parameters or of precomputed values. Due to this, byte sized indexes are desirable as these can avoid potentially costly bit unpacking operations. Here the costs are either the extra programming complexity in scheduling the unpacking in parallel with useful computation, or an unavoidable run-time complexity increase, or both.

With the massive market for portable devices and the growing interest for speech enabled user interfaces, embedded speech recognition has received considerable interest in recent years. A large body of work is targeting directly or indirectly the complexity reduction topic therefore we kindly ask the interested reader to explore the literature beyond the incomplete list of references included throughout this chapter. Although the methods introduced in this chapter were targeting acoustic model compression for speech recognition their range of application can be exten ded to other classification tasks, even if not HMM based. The model complexity criterions, the principles of parameter tying, formation of subspaces in multi-dimensional feature streams with data clustering and quantization are universal methods which find applicability in a wide context in pattern recognition.

References

Acero, A., Plumpe, M.D. (2004) Method for training of subspace coded Gaussian models. *United States Patent Application Publication* US2004/0181408A1.

Aiyer, A., Gales, M.J.F., Picheny, M. (2000) Rapid likelihood calculation of subspace clustered Gaussian components. In *Proceedings of the International Conference on Acoustics, Speech and Signal Processing.* vol. 3, Istanbul, Turkey, pp. 1519–1522.

Akaike, H. (1973) Information theory and an extension to the maximum likelihood principle. In *Proceedings of the 2nd International Symposium on Information Theory.* Budapest, Hungary, pp. 267–281.

Akaike, H. (1974) A new look at the statistical model identification. *IEEE Transaction on Automatic Control* vol. 19, nr. 6, pp. 716–723.

Astrov, S. (2002) Memory space reduction for hidden Markov models in low-resource speech recognition system. In *Proceedings of the International Conference on Spoken Language Processing.* Denver, USA, pp. 1585–1588.

Bahl, L., Brown, P., De Souza, P., Mercer, R. (1986) Maximum mutual information estimation of hidden Markov model parameters for speech recognition. In *Proceedings of the International Conference on Acoustics, Speech and Signal Processing.* vol. 11, Tokyo, Japan, pp. 49–52.

Barron, A., Rissanen, J., Yu, B. (1998) The minimum description length principle in coding and modeling. *IEEE Transaction on Information Theory* vol. 44, nr. 6, pp. 2743–2760.

Burnham, K.P., Anderson, D.R. (2002) *Model selection and multimodel inference: A practical-theoretic approach.* 2nd edition. Springer-Verlag.

Burnham, K.P., Anderson, D.R. (2004) Multimodel inference: Understanding AIC and BIC in model selection. *Sociological Methods and Research.* vol. 33, pp. 261–304.

Chen, S.S., Gopalakrishnan, P. (1998) Clustering via the Bayesian information criterion with applications in speech recognition. In *Proceedings of the International Conference on Acoustics, Speech and Signal Processing.* vol. 2, Seattle, USA, pp. 645–648.

Chien, J.-T., Furui, S. (2005) Predictive hidden Markov model selection for speech recognition. *IEEE Transaction on Speech and Audio Processing* vol. 13, nr. 3, pp. 377–387.

Chou, W., Reichl, W. (1999) Decision tree state tying based on penalized Bayesian information criterion. In *Proceedings of the International Conference on Acoustics, Speech and Signal Processing.* vol. 1, Phoenix, USA, pp. 345–348.

Digalakis, V., Monaco, P., Murveit, H. (1996) Genones: Generalized mixture tying in continuous hidden Markov model based speech recognizers. *IEEE Transaction on Speech and Audio Processing* vol. 4, nr. 4, pp. 281–289.

Duda, R.O., Hart, P.E., Stork, D.G. (2001) *Pattern classification and scene analysis.* 2nd edition. John Willey & Sons, New York.

Filali, K., Li, X., Bilmes, J. (2002) Data-driven vector clustering for low-memory footprint ASR. In *Proceedings of the International Conference on Spoken Language Processing.* Denver, USA.

Filali, K., Li, X., Bilmes, J. (2005) Algorithms for data-driven ASR parameter quantization. *Computer Speech and Language* vol. 20, nr. 4, pp. 625–643.

Gales, M.J.F. (1999) Semi-tied covariance matrices for hidden Markov models. *IEEE Transaction on Speech and Audio Processing*, pp. 272–281.

Ganapathiraju, A., Hamaker, J., Picone, J., Ordowski, M., Doddington, G.R. (2001) Syllable-based large vocabulary continuous speech recognition. *IEEE Transaction on Speech and Audio Processing* vol. 9, nr. 4, pp. 358–366.

Gersho, A., Gray, R.M. (1992) *Vector quantization and signal compression.* Kluwer Academic Press.

Gray, R.M., Neuhoff, D.L. (1998) Quantization. *IEEE Transaction on Information Theory* vol. 44, nr. 6, pp. 2325–2383.

Harju, M., Salmela, P., Leppänen, J., Viikki, O., Saarinen, J. (2001) Comparing parameter tying techniques for multilingual acoustic modelling. In *Proceedings of the Eurospeech.* Aalborg, Denmark.

Huang, X.D. (1992) Phoneme classification using semicontinuous hidden Markov models. *IEEE Transaction on Acoustics, Speech and Signal Processing* vol. 4, nr. 5, pp. 1062–1067.

Huang, X.D., Jack, M. (1989) Semi-continuous hidden Markov models for speech signals. *Computer Speech and Language* vol. 3, nr. 3, pp. 239–252.

Huang, X.D., Ariki, Y., Jack, M.A. (1990) *Hidden Markov models for speech recognition.* Edinburgh University Press, Edinburgh U.K.

Hui, J., Xinwei, L., Chaojun, L. (2006) Large margin hidden Markov models for speech recognition. *IEEE Transaction on Speech and Audio Processing* vol. 14, pp. 1584–1595.

Jelinek, F. (1998) Statistical methods for speech recognition. The MIT Press, Cambridge, Massachusetts.

Juang, B.-H., Chou, W., Lee, C.-H. (1997) Minimum classification error rate methods for speech recognition. *IEEE Transaction on Speech and Audio Processing* vol. 5, nr. 3, pp. 257–265.

Junqua, J.-C., Vassallo, L. (1996) Context modeling and clustering in continuous speech recognition. In *Proceedings of the International Conference on Spoken Language Processing.* Philadelphia, USA, pp. 2262–2265.

Kailath, T. (1967) The divergence and Bhattacharyya distance measures in signal selection. *IEEE Transaction on Communications* vol. 15, nr. 1, pp. 52–60.

Katagiri, S., Juang, B.-H., Lee, C.-H. (1998) Pattern recognition using a generalized probabilistic descent method. *Proceedings of the IEEE* vol. 86, nr. 11, pp. 2345–2373.

Lahti, T., Viikki, O., Vasilache, M. (2003) Low memory acoustic models for HMM based speech recognition. In *Proceedings of the Eurospeech*. Geneva, Switzerland, pp. 2489–2492.

Lanterman, A.D. (2001) Schwarz, Wallace, and Rissanen: Intertwining themes in theories of model order estimation *International Statistical Review* vol. 69, nr. 2, pp. 185–212.

Leppänen, J., Kiss, I. (2005) Comparison of low footprint acoustic modeling techniques for embedded ASR systems. In *Proceedings of the Interspeech*. Lisbon, Portugal.

Li, H.-B., Soong, F.K., Myrvoll, T.A., Wang, R.-H. (2005) Optimal clustering and non-uniform allocation of Gaussian kernels in scalar dimension for HMM compression. In *Proceedings of the International Conference on Acoustics, Speech and Signal Processing*. vol. 1. Philadelphia, USA, pp. 552–555.

Liu, X., Gales, M.J.F. (2007) Automatic model complexity control using marginalized discriminative growth functions. *IEEE Transaction on Speech and Audio Processing* vol. 15, pp. 1414–1424.

Mak, B.K.-W. (1998) *Towards a compact speech recognizer*. Ph.D. Thesis, Massachusetts Institute of Technology.

Mak, B.K.-W. (2004) An acoustic-phonetic and a model-theoretic analysis of subspace distribution clustering hidden Markov models. *International Journal of Speech Technology* vol. 7, nr. 1, pp. 55–68.

Mak, B.K.-W., Bocchieri, E. (2001a) Direct training of subspace distribution clustering hidden Markov model. *IEEE Transaction on Speech and Audio Processing* vol. 9, pp. 378–387.

Mak, B.K.-W., Bocchieri, E. (2001b) Subspace distribution clustering hidden Markov model. *IEEE Transaction on Speech and Audio Processing* vol. 9, pp. 264–275.

Myrvoll, T.A., Soong, F.K. (2003) Optimal clustering of multivariate normal distributions using divergence and its application to HMM adaptation. In *Proceedings of the International Conference on Acoustics, Speech and Signal Processing*. vol. 1, Hong Kong, China, pp. 552–555.

Nock, H.J., Gales, M.J.F., Young, S. (1997) A comparative study of methods for phonetic decision-tree state clustering. In *Proceedings of the Eurospeech*. Rhodes, Greece, pp. 111–114.

Normandin, Y., Cardin, R., De Mori, R. (1994) High-performance connected digit recognition using maximum mutual information estimation. *IEEE Transaction on Speech and Audio Processing* vol. 2, nr. 2, pp. 299–311.

Padmanabhan, M., Ban, L. (2000) Model complexity adaptation using a discriminant measure. *IEEE Transaction on Speech and Audio Processing* vol. 8, nr. 2, pp. 205–208.

Pan, J., Yuan, B., Yan, Y. (2000) Effective vector quantization for a highly compact acoustic model for LVCSR. In *Proceedings of the International Conference on Spoken Language Processing*. vol. 4. Beijing, China, pp. 318–321.

Rabiner, L., Juang, B.H. (1986) An introduction to hidden Markov models. *IEEE ASSP Magazine* vol. 3, 4–16.

Rabiner, L., Juang, B.H. (1993) *Fundamentals of speech recognition*. PTR Prentice-Hall, Inc., New Jersey.

Rabiner, L.R. (1989) A tutorial on hidden Markov models and selected applications in speech recognition. *Proceedings of the IEEE* vol. 77, nr. 2, pp. 257–286.

Ravishankar, M., Bisiani, R., Thayer, E. (1997) Sub-vector clustering to improve memory and speed performance of acoustic likelihood computation. In *Proceedings of the Eurospeech*. Rhodes, Greece, pp. 151–154.

Rigazio, L., Tsakam, B., Junqua, J. (2000) An optimal Bhattacharyya centroid algorithm for Gaussian clustering with applications in automatic speech recognition. In *Proceedings of the International Conference on Acoustics, Speech and Signal Processing*. vol. 3. Istanbul, Turkey, pp. 1599–1602.

Roberts, S.J., Husmeier, D., Rezek, I., Penny, W.D. (1998) Bayesian approaches to Gaussian mixture modeling. *IEEE Transaction on Pattern Analysis and Machine Intelligence* vol. 20, nr. 11, pp. 1133–1142.

Schwartz, G. (1978) Estimating the dimension of a model. *The Annals of Statistics* vol. 6, nr. 2, pp. 461–464.

Takahashi, S., Sagayama, S. (1995a) Effects of variance tying for four-level tied structure phone models. In *Proceedings of the ASI Conference* vol. 1-Q-23. Tokyo, Japan, pp. 141–142, (in Japanese).

Takahashi, S., Sagayama, S. (1995b) Four-level tied-structure for efficient representation of acoustic modeling. In *Proceedings of the International. Conference on Acoustics, Speech and Signal Processing*. vol. 1. Detroit, USA, pp. 520–523.

Varga, I., Aalburg, S., Andrassy, B., Astrov, S., Bauer, J., Baugeant, C., Hoge, H. (2002) ASR in mobile phones—an industrial approach. *IEEE Transaction on Speech and Audio Processing* vol. 10, nr. 8, pp. 562–569.

Vasilache, M. (2000) Speech recognition using HMMs with quantized parameters. In *Proceedings of the International Conference on Spoken Language Processing*. Beijing, China, pp. 871–874.

Vasilache, M. (2008) Multi-rate HMM quantization for speech recognition. In *Proceedings of the International Conference on Acoustics, Speech and Signal Processing*. Las Vegas, USA.

Vasilache, M., Iso-Sipilä, J., Viikki, O. (2004) On a practical design of a low complexity speech recognition engine. In *Proceedings of the International Conference on Acoustics, Speech and Signal Processing*. Montreal, Canada, pp. V.113–116.

Vasilache, M., Viikki, O. (2001) Speaker adaptation of quantized parameter HMMs. In *Proceedings of the Eurospeech-Scandinavia*. Aalborg, Denmark, pp. II. 1265–1268.

Vihola, M., Harju, M., Salmela, P., Suontausta, J., Savela, J. (2002) Two dissimilarity measures for HMMs and their application in phoneme model clustering. In *Proceedings of the International Conference on Acoustics, Speech and Signal Processing*. vol. 1. Orlando, USA, pp. 933–936.

Wallance, C., Boulton, D. (1968) An information measure for classification. *The Computer Journal* vol. 11, nr. 2, pp. 195–209.

Willett, D., Rigoll, G. (1997) A new approach to generalized mixture tying for continuous HMM-based speech recognition. In *Proceedings of the 5th European Conference on Speech Communication and Technology*. Rhodes, Greece, pp. 1175–1178.

Yang, Y., Barron, A. (1998) An asymptotic property of model selection criteria. *IEEE Transaction on Information Theory* vol. 44, nr. 1, pp. 95–116.

Young, S. (1992) The general use of tying in phoneme-based HMM speech recognisers. In *Proceedings of the International Conference on Acoustics, Speech and Signal Processing*. San Francisco, USA, pp. 569–572.

Young, S., Odell, J., Woodland, P. (1994) Tree-based state tying for high accuracy acoustic modelling. In *Proceedings of the ARPA Workshop on Human Language Technology*. pp. 307–312.

12

Fixed-Point Arithmetic

Enrico Bocchieri

Abstract. There are two main requirements for embedded/mobile systems: one is low power consumption for long battery life and miniaturization, the other is low unit cost for components produced in very large numbers (cell phones, set-top boxes). Both requirements are addressed by CPU's with integer-only arithmetic units which motivate the fixed-point arithmetic implementation of automatic speech recognition (ASR) algorithms. Large vocabulary continuous speech recognition (LVCSR) can greatly enhance the usability of devices, whose small size and typical on-the-go use hinder more traditional interfaces. The increasing computational power of embedded CPU's will soon allow real-time LVCSR on portable and low-cost devices. This chapter reviews problems concerning the fixed-point implementation of ASR algorithms and it presents fixed-point methods yielding the same recognition accuracy of the floating-point algorithms. In particular, the chapter illustrates a practical approach to the implementation of the frame-synchronous beam-search Viterbi decoder, N-grams language models, HMM likelihood computation and mel-cepstrum front-end. The fixed-point recognizer is shown to be as accurate as the floating-point recognizer in several LVCSR experiments, on the DARPA Switchboard task, and on an AT&T proprietary task, using different types of acoustic front-ends, HMM's and language models. Experiments on the DARPA Resource Management task, using the StrongARM-1100 206 MHz and the XScale PXA270 624 MHz CPU's show that the fixed-point implementation enables real-time performance: the floating point recognizer, with floating-point software emulation is several times slower for the same accuracy.

12.1 Introduction

There is an on-going world-wide powerful expansion of network technologies such as 3G cellular telephone networks, wireless data networks based on the *IEEE 802.11* (WLAN) and on the 802.16 (WIMAX) standards, and broadband networking to the home by fiber, DSL and wireless. There is a parallel growth of client devices such as cell and smart phones, PDA's, portable media players, set-top boxes, internet tablets, GPS systems, with applications in the areas of communication, entertainment and productivity. For example, the global volume of Short Message Services was about 1 trillion messages in 2005, and it is expected to grow to 3.7 trillion messages by 2012 yielding 67 billion USD of revenue.

Speech technologies can play a very significant role in these global developments by enhancing the user interface that is still limiting the device usability, in spite of continuous improvements over the years. Traditional interfaces based on screen, vision and keyboard arehindered by the physical size and by the typical on-the-go use of devices. The speech recognition algorithms can be implemented on the network server, distributed between server and device, or fully embedded on the device (Zaykovskiy 2006). The increasing computational power of the device will soon allow the real-time embedded implementation of high accuracy large-vocabulary continuous speech recognition (LVCSR), thus enabling access to data residing on the ever-growing memory of the device and on the network, through a speech-centric interface.

LVCSR on embedded platforms must overcome several and unique challenges (Novak 2004; Viikki 2001). To lower hardware cost and power consumption, as needed for longer *battery life* and miniaturization, the CPU's do not have floating-point arithmetic units. This motivates the study of the fixed-point implementation (for operation on the device) of high-accuracy, computationally intensive LVCSR algorithms that are traditionally implemented on the floating-point server. Relevant studies are Sagayama and Takahashi (1995) and Bocchieri and Mak (2001) concerning HMM parameter tying, (Kanthak et al. 2000; Vasilache 2000; Leppänen and Kiss 2005) for the state-likelihood computation in fixed-point. There are many other significant issues studied in the literature such as front-end implementation, noise robustness and memory reduction (Gong and Kao 2000; Kao and Rajasekaran 2000; Jeong et al. 2004; Rose et al. 2001; Vasilache et al. 2004), recognition of large lists (Novak et al. 2003) and rapid porting (Köhler et al. 2005). Custom hardware can also be designed to efficiently support speech recognition algorithms (Li et al. 2006).

Previous works on fixed-point decoding concern either small-vocabulary continuous-speech tasks or large-vocabulary tasks with deterministic grammars. This chapter also focuses on LVCSR tasks based on word N-gram language models. Section 12.2 presents the general principles of algorithm implementation in fixed-point arithmetic. Section 12.3 reviews the most popular LVCSR method, based on hidden Markov models (HMM), focusing on the system components needed for fixed-point recognition. Section 12.4 describes a systematic approach to the fixed-point representation of the parameters of the recognizer components, including frame-synchronous Viterbi beam-search, with stochastic and deterministic language models, HMM state and state-duration likelihood computations, and the acoustic front-end. Thefixed-point recognizer is shown to be as accurate as the floating-point recognizer in LVCSR experiments (Sect. 12.5) on the DARPA Switchboard task (http://www.nist.gov/speech) and on fluently spoken telephone speech from an AT&T customer care application. The design is quite general, and the same fixed-point parameterization is successfully used for different acoustic front-end features, feature transformations, and HMM's (ML and MMI trained), without the need of critical task-specific calibrations. The target hardware is 32-bit integer CPU's (e.g., StrongARM), but the approach may be suitable for 16-bit CPU's with 32-bit accumulators as well. Section 12.5 also reports about real-time recognition results on the DARPA Resource Management (RM) benchmark (1000 word vocabulary, speaker independent), as tested on two fixed-point devices, namely the 206 MHz Strong-ARM and the 624 MHz XScale PXA270 CPU's. Fixed-point implementation is necessary for real-time recognition because software emulation of floating-point operations is several times slower.

12.2 Fixed-Point Arithmetic

In computing, real numbers are commonly represented either in *floating-point* or in *fixed-point* notation. The latter is especially useful for CPU's capable of efficient computation on integer operands but lacking hardware support (thus inefficient) for floating-point types. Every-day life offers many examples of the fixed-point notation. The accepted tolerance for money amounts is half a cent and retail prices are rounded to the nearest cent. Money is therefore represented by an integer number of cents (e.g., ¢932) or, equivalently, as a number of dollars specified to two decimal places (e.g., $9.32). In this "dollars and cents" notation there are an integer part (dollars) and a fractional part (cents), and the position of the decimal point is fixed. This is perfectly suitable for most transactions, but it is inconvenient for large sums. For example the estimated 2005 U.S. GNP of $11,350,000,000,000.00 is more compactly expressed in the floating point notation, or $0.1135 \; 10^{14}$. In floating-point the position of the decimal point is variable as specified by the exponent, which allows for a wider range of values for a given number of digits in the number representation.

In computing, a real number is represented in fixed-point by storing its integer and fractional parts in a memory word. If the fractional part is stored in the p least significant bits, the number format is defined as $Q \; p$. Intuitively p denotes that an imaginary radix point (or decimal point in base 10) is between the pth and the $(p+1)^{th}$ least significant bits of the computer word. In practice, the choice of p is a compromise: larger values of p allow for smaller round-off errors $(\approx 2^{-p-1})$ and higher precision, but they give a smaller dynamic range. For example, in a 32-bit word the number of bits assigned to the integer part is $32 - p$, which limits the range of values to 2^{32-p}. In programming with the fixed-point notation, special care must be taken to avoid overflow problems while maintaining a suitable precision.

From this respect, floating-point types are much more convenient and flexible. In fact, the floating-point representation (by mantissa and exponent) of real numbers provides a much wider dynamic range because the radix point position, specified by the exponent, spans a wider range. In computations, the mantissa value is scaled up or down to use all available bits without overflowing, while the exponent keeps track of the radix point position.

12.2.1 Programming with Fixed-Point Numbers

Common programming languages, such as C and C++, do not have a native type for fixed-point numbers. From the above "dollars and cents" example it is evident that fixed-point numbers are essentially integers, and the programmer can store fixed-point variables as integer types in C. Arithmetic with fixed-point numbers uses integer operations, with additional rules concerning the position of the radix point. In operations on x, y and z, with radix-point positions p_x, p_y and p_z, these conditions apply:

$$z = x \pm y \quad : \quad p_z = p_x = p_y$$

$$z = xy \quad : \quad p_z = p_x + p_y$$

$$z = x/y \quad : \quad p_z = p_x - p_y$$

$$x > y \quad : \quad p_x = p_y$$

The operand radix point position can be changed to satisfy the above conditions. Arithmetic shifts to the left and to the right move the radix point to left and right, respectively:

$$x << i \quad \text{has fixed-point format} \quad Q(p_x + i)$$

$$x >> i \quad \text{has fixed-point format} \quad Q(p_x - i)$$

A right or left shift of a fixed-point number multiplies its integer representation by a negative or positive power of two, respectively. Therefore changes of the radix point position can avoid errors of overflow and underflow. In a typical example, the product of two integers representing two fixed-point numbers may overflow the range $[-2^{31}, 2^{31})$ of a 32-bit CPU: the problem can be avoided by suitable right-shifts of the factors before multiplication. For these several reasons arithmetic shifts are common in the implementation of fixed-point algorithms, and CPU's (e.g., StrongARM) may support integer arithmetic operations, shifts, and condition testing in a single instruction cycle.

A useful fixed-point programming technique makes use of the *block floating-point* format. The method consists in scaling a block of numbers so that the maximum absolute value uses all available bits (e.g., the full word length). The scale up (or down) operation is implemented by left (or right) shifts, with the number of shifts recorded as an "exponent" common to all the numbers of the block.

In the interest of computational speed, complex mathematical functions are implemented in fixed-point by *table look-up*. To reduce the table size, the programmer may exploit the properties of the desired function. For example, for $\log_{10}(.)$ one may store values over a limited input range, such as $[1, 2]$: to find $\log_{10}(x)$ the table look-up function performs the search (e.g., binary search) of the integer $k \ni 1 \le 2^k x \le 2$, and it computes $\log_{10}(2^k x) - k \log_{10}(2)$ from the stored logarithm values. The implementation of trigonometric functions may exploit their symmetry. To limit the table size and maintaining precision, the table look-up access method may use interpolation of stored values.

12.2.2 Fixed-Point Representation and Quantization

The $Q\,p$ *fixed-point* representation of a real number x can be determined as the nearest integer of $2^p x$: for example, in base 10, the $Q2$ fixed point representation of π is 314 (with an imaginary decimal point between the second and third significant digits) that is in fact the closest integer of $10^2 \pi$.

Later in the chapter we will use the following procedure to design a linear *quantizer* and to identify the $Q\,p$ fixed-point format of its output. Suppose that we want to quantize real values from the interval

$$[a,b] \tag{12.1}$$

using m bits, e.g., to the range of integers $[-2^{m-1}, 2^{m-1})$. We follow the procedure:

  i. Optional. Demean decimal values, by subtracting $\dfrac{a+b}{2}$:

$$[a,b] \rightarrow \left[-\frac{a+b}{2}, \frac{a+b}{2} \right]$$

  ii. Find the largest integer p, such that:

$$-2^{m-1} \leq 2^p x < 2^{m-1}, x \in [a,b]$$

  iii. Quantize $x \in [a,b]$ to the nearest integer of $2^p x$.

Step *iii* yields a fixed-point format of x with an average round-off error that is essentially scale invariant, because of the choice of p in step ii. We can then operate on the quantizer output with $Q\,p$ fixed-point arithmetic. Other quantizers, such as the Loyd-Max quantizer, are useful to minimize the average distortion for the desired number of bits. The output of these non-linear quantizers can be used in fixed-point arithmetic by mapping the quantizer code-words to a suitable fixed-point representation.

12.3 LVCSR MAP Recognizer

Speech recognition is the process of mapping the speech signal to a sequence of discrete symbols such as phonemes, words and sentences, the large variability of the signal being the obstacle to high accuracy. Causes of variability are the channel/ environment, the speaker, and various aspects of the language such as phonetics, phonology, syntax and prosody.

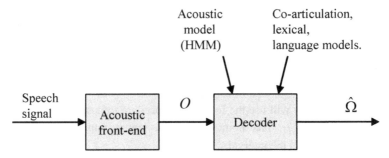

Fig. 12.1 Diagram of a speech recognizer

A typical recognition system based on the so called *noisy channel formulation* is shown in Fig. 12.1. The recognizer decodes the most likely word sequence given the acoustic signal, represented by a time sequence of feature or observation vectors $O = (\mathbf{o}_1, \dots \mathbf{o})_T$. An acoustic front-end typically outputs a vector every $10\, ms$, each vector providing a parametric representation of the short time spectrum over a time window of duration between 20 and $40\, ms$. If $\Omega = (\omega_1, \dots \omega_M)$ denotes a generic word sequence, the recognizer output according to the noisy channel formulation is $\hat{\Omega} = \arg\max_{\Omega} \Pr(\Omega \mid O)$, and after applying Bayes rule:

$$\hat{\Omega} = \arg\max_{\Omega} \Pr(O \mid \Omega) \Pr(\Omega) \qquad (12.2)$$

This probabilistic model relates the observed signal to the recognized sentence, and its implementation is based on several assumptions about speech and language. A sentence is thought as a word sequence with probability $\Pr(\Omega)$, as provided by the *language model* of the speech recognition application. Words are modeled according to a dictionary, or lexicon, as sequences of basic language units, the phonemes. In general, because of physical constraints of the human articulatory system, the acoustic realization of a certain phoneme is affected by neighboring sounds. This *co-articulation* phenomenon is represented as a mapping from the phonetic context-dependent acoustic realizations to the (context-independent) phonemic units of the language. Finally, the context-dependent phonetic units are related to the observation vectors through an *acoustic model*, typically based on techniques like *hidden Markov models* (HMM), neural-networks (NN), or NN-HMM hybrids. This chapter is concerned with the most popular HMM. HMM's are Markovian chains of observation probability density functions, known as *states*, that can be viewed as generative models of the observation vectors. In this interpretation the observation vectors are emitted from the state output densities thus modeling the signal variability caused by speaker and channel. Different states implicitly correspond to different parts of an acoustic unit, with state transitions controlled by the Markov chain topology and by the state duration model (Sect. 12.3.2). Thus, the HMM state sequence is related to the word sequence Ω by the HMM topology, context-dependency, lexical and language models; Eq. 12.2 can be solved by finding the sequence of HMM states that maximizes a suitable decoding function. After well known steps, Eq. 12.2 becomes:

$$\hat{\Omega} = \arg \max_{\Omega} F(\Omega) \qquad (12.3)$$

with the decoding function:

$$F(\Omega) = \ln(Pr(\Omega)) + \alpha \ln(A(S)) + \beta \ln(D(S)) \qquad (12.4)$$

and:

$Pr(\Omega)$: likelihood of language model,

$\quad S$: HMM state sequence corresponding to Ω,

$\quad A$: likelihood of O given S,

$\quad D$: likelihood of the durations of the states in S,

$\quad \alpha, \beta$: empirical state and state-duration multipliers.

12.3.1 HMM State Likelihoods

HMM's can be classified according to the type of output density functions. In *continuous* observation density HMM's, the state output observation densities are defined as a weighted mixture of base densities, typically Gaussians or Laplacians. Continuous HMM's are the most popular because they provide the highest accuracy in many tasks, and their fixed-point characterization is detailed in this chapter. Such a characterization can clearly be extended to the *semi-continuous* HMM's, where all mixtures are expressed in terms of a common set of base functions, with different mixtures characterized only by a different sets of weights. The semi-continuous approach, such as used for embedded ASR in Huggins-Daines et al. (2006), facilitates a smaller memory foot-print. In *discrete* HMM's the observations are vectors of symbols from a finite alphabet: for a given state, a discrete density is estimated for every observation component. The state observation density is obtained by multiplying the probabilities of the individual components under the assumption of independence. Typically the discrete models are the least accurate and normally used only in simple tasks. The computation of the discrete state likelihoods can be solely based on table look-ups of fixed-point numbers.

We define the generic HMM *state* s as a weighted mixture of N_s Gaussians with diagonal co-variances ($\boldsymbol{\sigma}$ denotes the vector of standard deviations):

$$Pr(\boldsymbol{o} \mid s) = \sum_{i=1}^{N_s} w_{s,i} \, N\left(\boldsymbol{o}, \boldsymbol{\mu}_{s,i}, \boldsymbol{\sigma}_{s,i}\right)$$

Given the sequence $S = (s_1, ..., s_T)$ of states corresponding to $O = (\boldsymbol{o}_1, ..., \boldsymbol{o})_T$, the total state log-likelihood contribution to Eq. 12.4 is:

$$\alpha \ln(A) = \alpha \sum_{t=1}^{T} \ln \left(\sum_{i=1}^{N_{s_t}} w_{s_t,i} \, \mathrm{N}\left(\boldsymbol{o}_t, \boldsymbol{\mu}_{s_t,i}, \boldsymbol{\sigma}_{s_t,i}\right) \right)$$

By approximation of the inner-most summation over the Gaussians probabilities (index i) with the maximum of its addenda, and after simple manipulations:

$$\alpha \ln(A) \approx \sum_{t=1}^{T} \frac{\alpha}{2} \max_{i=1,\dots,N_s} \left(2c_{s_t,i} + \sum_{j=1}^{d} \left(\left(o_t^j - \mu_{s_t,i}^j \right) \sigma_{s_t,i}^{j\ -1} \right)^2 \right)$$

d : feature vector dimension

$o_t^j, \mu_{s,i}^j, \sigma_{s,i}^j : \mathrm{j}^{\mathrm{th}}$ component of $\boldsymbol{o}_t, \boldsymbol{\mu}_{s,i}, \boldsymbol{\sigma}_{s,i}$

$c_{s,i}$ $: \ln(w_{s,i}) - \sum_{j=1}^{d} \ln\left(\sqrt{2\pi}\ \sigma_{s,i}^j\right)$

(12.5)

Standard deviation reciprocals are used in Eq. 12.5 because multiplications are computed more quickly than divisions. In Sects. 12.4.1 and 12.4.3 we address the fixed-point computation of Eq. 12.5, with the corresponding fixed-point representation of the HMM state parameters.

12.3.2 State Duration Model

When considering the state sequence $S = (s_1, \dots, s_T)$, corresponding to observations $O = (\boldsymbol{o}_1, \dots \boldsymbol{o}_T)$, let's suppose that starting at generic time t, exactly δ consecutive frames are generated by the state ψ, i.e.:

$$s_{t-1} \neq \psi, \quad s_t = s_{t+1} = \dots = s_{t+\delta-1} = \psi, \quad s_{t+\delta} \neq \psi$$

Then ψ is said to have duration δ, and we denote with $\Pr(\delta|\psi)$ such state duration probability. The *duration models* are estimated from data, typically as gamma probability density functions. For run-time access during recognition, $\Pr(\delta|\psi)$ are stored in look-up tables for a suitable range of durations, such as $1 \leq \delta \leq 32$.

Given the states $\Psi = (\psi_1, \dots, \psi_\Theta)$, with durations $\Delta = (\delta_1, \dots, \delta_\Theta)$, the contribution of the duration model to Eq. 12.4 is:

$$\beta \ln(D) = \beta \ln\left(\Pr(\Delta|\Psi)\right) = \sum_{\theta=1}^{\Theta} \beta \ln\left(\Pr(\delta_\theta|\psi_\theta)\right)$$

(12.6)

The fixed-point representation of the duration probabilities for the computation of Eq. 12.6 is discussed in Sect. 12.4.1.

12.3.3 Language Model

A simple yet successful stochastic language modeling basis for LVCSR is the *N-gram* model. In general, because of the chain rule of probability:

$$\Pr(\Omega) = \Pr(\omega_1, \ldots \omega_M) = \prod_{m=1}^{M} \Pr\left(\omega_m \mid \omega_1, \ldots, \omega_{m-1}\right)$$

The *N*-gram model assumes that the conditional probability of ω_m depends only on the *N* preceding words. The language model probability becomes:

$$\Pr(\Omega) \approx \prod_{m=1}^{M} \Pr\left(\omega_m \mid \omega_{m-N}, \ldots, \omega_{m-1}\right)$$

Therefore, the language model log-probability is:

$$\ln\left(\Pr(\Omega)\right) = \sum_{m=1}^{M} \ln\left(\Pr\left(\omega_m \mid \omega_{m-N}, \ldots, \omega_{m-1}\right)\right) \qquad (12.7)$$

There is a vast body of literature concerning both the *N*-gram model estimation from large text corpora and various *N*-gram model extensions. This chapter is concerned with the fixed-point representation of the N-gram log-probability contribution to the ASR decoding function (Eq. 12.4), as detailed in Sect. 12.4.1.

12.3.4 Viterbi Decoder

The acoustic HMM's are related to the word sequence by the context-dependency, dictionary and language models, as briefly discussed in Sect. 12.3. The recognized word sequence $\hat{\Omega}$ in Eq. 12.3 is determined by searching for the HMM state sequence that maximizes $F(\Omega)$ as in Eq. 12.4. Today, the most adopted *decoder* is based on the time-synchronous Viterbi search where *all* partial state paths are extended (using the Markovian assumption of the model) in parallel from generic time t to $t+1$, until all the T observation vectors are processed. In the Viterbi decoder implementation we adopt the formulation based on weighted *finite-state transducers* (Mohri et al. 2002). A finite-state transducer is a finite automaton whose between-state transitions are labeled with input and output symbols. Therefore a path through the transducer maps an input symbol sequence to an output symbol sequence. In a weighted transducer, quantities (such as probabilities) are encoded into transition weights. These are accumulated along the transducer path to provide the total weight for mapping the input sequence to the output sequence. The speech recognition transducer uses the state observation distributions as input and the words as output symbols, respectively. The transducer encodes all aspects of the recognition model, such as the HMM topology, context dependency, lexicon and language model. The arc weights encode the HMM state likelihoods, and the pronunciation and language model probabilities. The decoding process, or searching for the best HMM state path, becomes therefore equivalent to searching for the transducer path with maximum total weight.

12.3.5 Acoustic Front-End

There are many different parametric representations of the speech signal for the purpose of speech recognition and a respective vast literature. As a working example, we consider one of the most popular parameterizations, namely the vector of *Mel-frequency cepstrum coefficients* (*MFCC*) first proposed for ASR in Davis and Mermelstein (1980). MFCC's are derived from a cepstral analysis of the speech signal. The principal difference from the standard cepstrum is that the frequencies are equally spaced on the mel auditory scale to approximate the response of the human auditory system. The MFCC computation is depicted in Fig. 12.2.

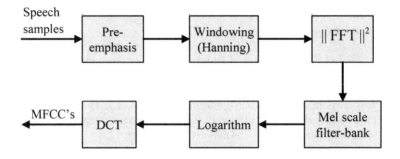

Fig. 12.2 Computation of MFCC's

The filters are simulated by weighted sums of the square magnitudes of the FFT components. The weighting function of the generic i^{th} filter is triangularly shaped, with maximum at the center-frequency

$$f_i = 100i \text{ Hz, } 1 \le i \le 10; \qquad f_i = 1.1 f_{i-1}, \ i > 10$$

and linearly tapered to zero at frequencies f_{i-1} and f_{i+1}. The key-points of the fixed-point computation of the MFCC's are discussed in Sect. 12.4.4. Dynamic aspects of the MFCC's are also parameterized in the observation vectors, either explicitly as first and second time differentials (delta-delta coefficients) or through discriminatively trained transforms (Saon et al. 2000).

12.4 Fixed-Point Implementation of the Recognizer

This section describes the application of *fixed-point* arithmetic (Sect. 12.2) to the speech recognition problem (Sect. 12.3). In general, a practical approach to fixed-point implementations is to examine histograms of the algorithm variables to choose the fixed-point formats giving the required numerical precision without overflow problems. However, the statistical properties of the decoding function (Eq. 12.4) suggest a more systematic approach that generalizes to different observation vector types, HMM's and language models. Intuitively, a crucial role in the evaluation of the search hypotheses during decoding is played in Eq. 12.5 by the normalized difference

$$\left(o_t^j - \mu_{s_t,i}^j\right)\sigma_{s_t,i}^{j\ -1} \tag{12.8}$$

and by the distance

$$\left(\left(o_t^j - \mu_{s_t,i}^j\right)\sigma_{s_t,i}^{j\ -1}\right)^2 \tag{12.9}$$

Distributions of Eq. 12.8 and Eq. 12.9 are respectively Gaussian (zero mean and unit variance) and χ^2, regardless of the type of observation vector and vector component: thus the same fixed-point representation may be appropriate for different feature vector types and components, and for different HMM's. Our fixed-point design of Eq. 12.4 is parametrized by e, m and v, as summarized in Table 12.1.

12.4.1 Log-Likelihoods

For the weighted difference of Eq. 12.8 and its square (Eq. 12.9) we adopt Qe and $Q2e$ fixed-point representations, respectively, as in Table 12.1. Since Eq. 12.5 accumulates Eq. 12.9 into the *Mahalanobis distance*

$$\sum_{j=1}^{d}\left(\left(o_t^j - \mu_{s_t,i}^j\right)\sigma_{s_t,i}^{j\ -1}\right)^2 \tag{12.10}$$

and then into the state log-likelihoods, we also represent Eq. 12.10 and the state log-likelihoods in $Q2e$ format. The HMM log-terms $2c_{s,i}$ (Eq. 12.5) are also $Q2e$ fixed-point numbers, because they are added into Eq. 12.10.

In the implementation of the multiplication by $\alpha/_2$, an appropriate arithmetic shift yields a $Q2e$ fixed-point product. The constant $\alpha/_2$ is typically optimized to the speech recognition task, type of observation vectors and HMM's. In our systems this parameter varies between *0.025* and *0.05*. To help regressing the 32-bit fixed-point implementation against the floating-point decoder, it is useful to represent $\alpha/_2$ with a significant precision, choosing its fixed-point format to yield an integer representation between *1024* and *2048*. Before multiplication by $\alpha/_2$ in Eq. 12.5, the radix-point position of the log-likelihood factor is adjusted to prevent overflow. Similarly, the duration log-probabilities (Eq. 12.6) are $Q2e$ fixed-point numbers, and when multiplying by β, the $Q2e$ format is maintained by arithmetic shift.

Table 12.1 Fixed-point parameters m, v and e

---

e : fixed point Qe format for:

- normalized error $\left(o_t^j - \mu_{s_t,i}^j \right) \sigma_{s_t,i}^{j \; -1}$

- and $Q2e$ format for:
- HMM state log-likelihoods,
- duration model log-likelihoods,
- language model probabilities, and
- cumulative log-probabilities of partial state paths
- during decoding, and related parameters such as
- beam threshold.

m : bits for the d quantizers of $\mu_{s,i}^j, j = 1,...,d$.

v : bits for the d quantizers of $\sigma_{s,i}^{j \; -1}, j = 1,...,d$.

The language model log-probability (Eq. 12.7) is added to the state and state-duration likelihoods into the cumulative log-probability (Eq. 12.4). Therefore Eq. 12.7, its N-gram log-probabilities addenda, and the decoding function (Eq. 12.4) are represented in the $Q2e$ format. The $Q2e$ format is also adopted for cumulative log-probabilities of the partial state-path hypotheses that are evaluated during decoding. In the weighted finite state formulation of (Mohri et al. 2002) the language model (e.g., N-gram) log-probabilities are encoded in the arc weights. The transducer semi-ring can be implemented in fixed-point: for example, if the transducer arc weights share the same fixed-point representation, the product operator of the tropical semi-ring is the integer addition, and the sum operator is the *max* function. Delayed composition (Mohri et al. 2002) can be supported in the fixed-point implementation, which is useful to reduce run-time memory in many applications such as (Novak et al. 2003).

12.4.2 Viterbi Frame-Synchronous Search

As motivated above, the cumulative log-likelihoods of partial state paths are represented in the $Q2e$ format. To save computation while extending the paths from an observation vector to the next, it is common practice to *prune* (ignore) the paths whose log-likelihoods fall below a certain threshold (or *beam-width*). The fixed-point representation of the beam-width is therefore $Q2e$.

A *normalization* procedure is important to avoid that the cumulative log-likelihoods of the state paths grow too large and overflow the fixed-point representation. Before extending the state paths ending at a certain time t, one may simply subtract the maximum log-likelihood score from all path scores. The search for the most likely path is not affected, because the same value is subtracted from all the hypotheses. The frequency of this normalization step depends on the word size and on the adopted $Q\,2e$ representation of the log-likelihoods. With 16-bit words it may be necessary to normalize the scores after processing every observation vector, or at least every few vectors. With 32-bit words several minutes of speech may be processed without normalization.

12.4.3 Gaussian Parameters

We need to address the *fixed-point representation* of the Gaussian mean and variance parameters that are required in the computation of the *Mahalanobis distance* (Eq. 12.10). To account for the different dynamic ranges of the Gaussian mean components, we build a quantizer for every j^{th} $(j=1,...,d)$ component, as in *i*, *ii* and *iii* of Sect. 12.2.2, with the range of Eq. 12.1:

$$\left[\min_{\text{State } s,\text{ Gaussian } i} \mu_{s,i}^{j}, \max_{\text{State } s,\text{ Gaussian } i} \mu_{s,i}^{j}\right]$$

Parameter m in Table 12.1 specifies the number of bits of the mean quantizers. We denote by $Q\,p^{j}$ the fixed-point format of $\mu_{s,i}^{j}$ induced by the jth quantizer. The same format is adopted for o_{t}^{j} because it is subtracted from the mean component in Eq. 12.10. Similarly, we build another set of d quantizers, one for every $\sigma_{s,\square}^{j\ -1}$ (the jth inverse standard deviation component) using steps ii and iii of Sect. 12.2.2, with the range as in Eq. 12.1:

$$\left[\min_{\text{State } s,\text{ Gaussian } i} \sigma_{s,i}^{j\ -1}, \max_{\text{State } s,\text{ Gaussian } i} \sigma_{s,i}^{j\ -1}\right] \tag{12.11}$$

Parameter v specifies the number of bits, with output range $[0,2^{v})$, of these quantizers. We denote by $Q\,r^{j}$ the fixed point format of $\sigma_{s,\square}^{j\ -1}$ induced by its quantizer. Because of artifacts in the training data, there may be a small number of variance estimates that are exceedingly small. These incorrect estimates are problematic because they may cause exceedingly large (and erroneous) likelihood values during recognition. In fact, the estimated variance values should be suitably floored: this is good practice in floating-point recognizers and even more so in fixed-point systems because of the limited dynamic range. In particular the artifact of small variance values may determine an exceedingly large range of the inverse standard deviations in Eq. 12.11, and very large average distortions of the corresponding linear *quantizer*. To avoid these problems we simply floor the estimated variance of the generic j^{th} feature component to one thousandth of its average value across the Gaussians of the HMM states.

The pseudo-code of the fixed-point computation of Eq. 12.10 is shown in Table 12.2.

Table 12.2 Pseudo-code for the fixed-point implementation of Eq. 12.10

$$
\begin{aligned}
&sum = 0;\ j = 0;\\
&while\ (j < d)\ \{\\
&\quad j = j + 1\\
&\quad temp = \left(o_t^j - \mu_{s,i}^j\right)\sigma_{s,i}^{j\ -1} \qquad // \mathrm{Q}\left(p^j + r^j\right)\\
&\quad temp = temp >> shift_j \qquad\qquad // \text{ change to } \mathrm{Q}e\\
&\quad sum = sum + temp * temp \quad // \text{ sum is } \mathrm{Q}2e\\
&\}
\end{aligned}
$$

The integer product $\left(o_t^j - \mu_{s,i}^j\right)\sigma_{s,i}^{j\ -1}$ is $\mathrm{Q}\left(p^j + r^j\right)$ fixed-point, then it is changed to $\mathrm{Q}e$ with a right arithmetic shift of $shift_j = p^j + r^j - e$ bits, and finally it is squared and accumulated into the sum with the desired $\mathrm{Q}2e$ format.

In Table 12.2 a negative $shift_j$ designates a left shift of $-shift_j$ bits. In practice we can choose e, m and v so that $shift_j$ is positive for every component. Typically Eq. 12.10 is the most computational intensive operation in the speech recognition process. Depending on the CPU architecture, other implementations, for example based on multiply-add operations, may give a higher throughput.

12.4.4 MFCC Front-End

This section reviews the most critical details of the fixed-point implementation of the *MFCC* computation with 32-bit word arithmetic. A description of the MFCC computation with 16-bit words and 32-bit accumulators can also be found in Gong and Kao (2000). Empirically, as tested on speech data, our fixed-point implementation typically approximates the floating-point computation to the third decimal digit. This approximation is sufficient. In fact the Gaussian means, subtracted from the MFCC's in Eq. 12.9, are more coarsely quantized with no loss of recognition accuracy, as observed in the experiments of the next section. We experimented with speech sampled at 8 and 16 KHz: before FFT input the analysis window ($20\,ms$) is zero-padded to 512 and 1024 samples, respectively. Scaling operations are essential in the fixed-point computation of the FFT (Oppenheim and Schafer 1975). We store the sine values of the FFT butterflies as $\mathrm{Q}15$ numbers, and we scale the outputs of the butterfly banks so that the maximum absolute value lies between 2^{13} and 2^{14}. Scaling is by *block floating point*, see Sect. 12.2.1. We also store the weights of every mel filter in block floating-point format, with maximum filter weight not to exceed

128. Then, knowing the maximum number of weights of the filters, we can apply the block floating-point representation to the input of the mel-filter to prevent overflow with 32-bits. The logarithm function for the MFCC computation is implemented by *table look-up*, with values stored as $Q12$ fixed-point numbers, as discussed in Sect. 12.2.1. The cosine values of the discrete cosine transform are stored in $Q10$ format.

12.5 Experiments

In this section we perform several recognition experiments on different tasks and we verify that the described fixed-point approach is as accurate as the floating point recognizer. The design is largely based on the fixed-point representation of Eq. 12.8, whose statistics are independent of the ASR task, type of observation vector and HMM: in fact we also verify that the fixed-point parameters do not require critical task-specific calibrations. The experiment setup is shown in Fig. 12.3. The floating-point HMM and the language model are converted to the fixed-point representation by quantizers, as described in Sect. 12.4. Thus we compare the accuracies of the fixed-point and floating-point recognizers with the same models. To experiment with different types of acoustic front-ends (besides the MFCC front-end of Sect. 12.3.5 that is implemented in fixed-point), we convert the front-end output to fixed-point by quantization. The Gaussian mean quantizers (Sect. 12.4.3) are applied to the front-end feature components, as these are subtracted from the Gaussian means in Eq. 12.8.

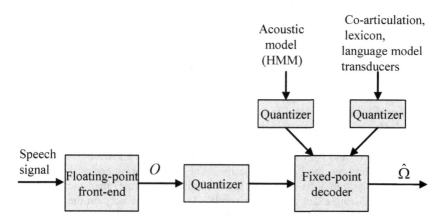

Fig. 12.3 Testing the fixed-point decoder

The recognition experiments are performed on the following tasks:

SWBD : *DARPA Switchboard* task, tested on the 2003 real-time test set (recognition from first-pass only),
CCAPP: fluent telephone speech from a customer-care application, with word tri-gram language model (perplexity of 60), vocabulary of 7,000 words, 5,000 test sentences, and up to 50 words/sentence,
RM   : *DARPA Naval Resource Management*, word-pair grammar, 40 speakers, speaker independent task, 1,200 test sentences, speech sampled at 16 KHz (Lee 1989).

In the experiments we used these types of feature vectors:

MFCC : *mel-frequency cepstrum coefficients,*
PLP   : *perceptual linear prediction* cepstra (Hermansky and Morgan 1994),

with feature transformations:

DD   : cepstra and energy with 1st and 2nd differentials, 39 components,
HDA   : discriminative linear transformation of the cepstrum and energy features (Saon et al. 2000), 60 components,
VTLN : vocal tract length normalization (Lee and Rose 1996).

The HMM's are context-dependent triphonic models, estimated either by maximum likelihood (ML) or maximum mutual information (MMI) methods. The CCAPP and SWBD HMM's were trained on 170 and 300 hours of audio, respectively.

Table 12.3 compares the LVCSR accuracies of the fixed-point and of the floating-point recognizers on a Pentium 4 PC, for different tasks. For example, the CCAPP system, with MFCC features, discriminative transformation, vocal tract length normalization, and MMI-trained HMM, is denoted by CCAPP_MFCC-HDA-VTLN_MMI. On the PC, fixed-point implementations may be faster than floating-point, as shown in Kanthak et al. (2000) for the state likelihoods. Our target is the StrongARM CPU, and we have not optimized the fixed-point software for speed on the Pentium. However, the Pentium is convenient for measuring accuracies, because it runs the recognition software much faster (higher clock rate) than the embedded CPU while producing exactly the same results. The large memory on the PC allows for testing the recognition accuracy of the fixed-point recognizer for very large tasks, such as the DARPA Switchboard.

The accuracies (Table 12.3) of the fixed-point and of the floating-point recognizers (equal beam-width), are the same, within 0.1%, for all tasks. All systems use the same configuration of the fixed-point parameters, without task-specific tuning.

Table 12.3 Accuracies of floating-point and fixed-point decoders (fixed-point parameters: $m = v = 8, e = 5$)

ASR system	Word accuracy (%)	
	Floating	Fixed
RM_MFCC-HDA_MMI	96.4	96.4
RM_MFCC-DD_ML	95.7	95.6
RM_PLP-DD_ML	95.6	95.5
CCAPP_MFCC-HDA-VTLN_MMI	80.5	80.6
CCAPP_MFCC-HDA_MMI	78.4	78.4
SWBD_MFCC-HDA_MMI	59.2	59.1
SWBD_MFCC-HDA_ML	56.7	56.6
SWBD_PLP-HDA_ML	55.7	55.6

The values of the fixed point parameters m, v and e can be changed over a wide range without affecting the recognition accuracy, (Table 12.4 and 12.5). Even though the tables show results only for the system CCAPP_MFCC-HDA-VTLN_MMI, the accuracies of the other tasks are equally affected by m, v and e.

Means and variances are linearly quantized to 5 bits, without significant loss of accuracy (Table 12.4).

Table 12.4 Word accuracy (%) as function of m and v ($m = 5$) of fixed-point system CCAPP_MFCC-HDA-VTLN_MMI

	$v = 3$	$v = 4$	$v = 5$	$v = 6$	$v = 7$	$v = 8$
$m = 3$	43.5	53.3	56.9	57.0	56.6	56.4
$m = 4$	44.1	74.4	77.8	77.6	78.0	77.8
$m = 5$	45.0	76.3	80.0	80.1	80.2	80.3
$m = 6$	45.3	76.4	80.1	80.2	80.5	80.4
$m = 7$	45.7	76.5	80.2	80.4	80.6	80.5
$m = 8$	45.6	76.6	80.2	80.5	80.5	80.6

Nonlinear quantization would provide additional compression (Leppänen and Kiss 2005; Vasilache 2000), at the cost of additional indirections in the computation. Our goal was to quantize HMM means and variances to no more than eight bits to reduce the HMM size to a relatively small fraction of total run-time memory. The accuracy doesn't change (within 0.1%) for $2 \le e \le 6$ (Table 12.5). Accuracy is affected by truncation errors for $e < 1$, and by overflows for $e > 7$. Larger e's, would require normalization of the cumulative log-likelihoods in the Viterbi search, as discussed in Sect. 12.3.4. In any case the decoder operates correctly over a wide range of e, on the various tasks. We use 32-bit fixed-point arithmetic, but the good performance for *e* as small as 2 or 1, suggests that the implementation is suitable for 16-bit CPU's with 32-bit accumulators.

Table 12.5 Word accuracy (%) as function of *e* ($m = v = 8$) of fixed-point system CCAPP_MFCC-HDA-VTLN_MMI

$e = 0$	$e = 1$	$e = 2$	$e = 3$	$e = 4$	$e = 5$	$e = 6$	$e = 7$	$e = 8$
75.6	80.2	80.5	80.5	80.5	80.6	80.5	79.9	69.2

12.5.1 Real-Time on the Device

The fixed-point recognizer has been benchmarked on the RM task (RM_MFCC-DD_ML) using these devices:

- A desk-top telephone with a *StrongARM-1100* CPU, running at 206 206 MHz, with 30 MBytes of RAM, and Linux 2.6.6.

- A Pocket Pc iPAQ hx4700 with an *XScale PXA270* CPU, running at 624 MHz and 64 Mbytes of RAM. We installed the Linux OS (Familiar distribution) following the instructions at *www.handhelds.org*.

The device executables were cross-compiled on the PC (see Fig. 12.4), with the GNU "tool-chain" and gcc version 3.4.2. For testing ASR on the device, access to executables and to fixed-point speech feature files is through the network link. The Unix command "top" shows a run-time memory use for the RM_MFCC-DD_ML of 7.5 Mbytes (virtual and resident). The HMM (with mean and variance parameters stored in one byte) and the pre-compiled transducer require one mega-byte each.

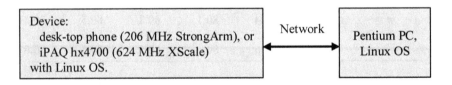

Fig. 12.4 Device development system

Real-time recognition performance for the RM_MFCC-DD_ML task is shown in Fig. 12.5.

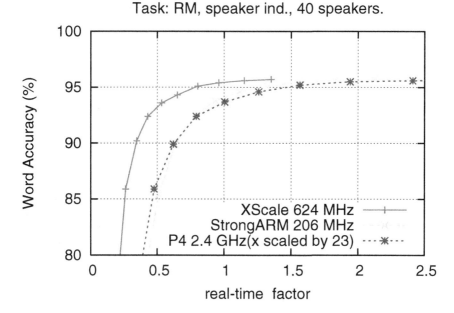

Fig. 12.5 Word accuracy of task RM_MFCC-DD_ML as a function of recognition time normalized by input audio time

Here the recognition accuracy is plotted as a function of time (normalized by the duration of the input audio). The recognition times and the corresponding accuracies were measured in various experiments using different beam-width values. Two plots are shown for the fixed-point recognizer running either on the 206 MHz StrongARM or on the 624 MHz XScale, respectively, both showing recognition in real-time. As reference, the third plot shows the accuracy of the floating-point recognizer running on the PC. The time axis of this plot was scaled by a factor of 23, to account for the speed difference between the 2.4 GHz Pentium and the embedded CPU. The tick marks on the plots correspond to specific values of the beam-width, showing that the fixed-point implementation provides the same accuracy (within a small difference) as the floating-point recognizer over the entire range of beam-widths.

It should be noted that a program containing floating-point operations can be executed on a fixed-point CPU by means of *software emulation* of the floating-point instructions. However, the program will be slower (depending on the number of floating-point instructions contained in the code) than the corresponding fixed-point implementation. We measured that the floating-point recognizer running on the 206 MHz StrongARM-1100 CPU is about 40 times slower than the fixed-point implementation (task RM_MFCC-DD_ML). Floating-point emulation can be obtained in two ways. In one method (used in our experiments) the compiler generates floating-point instructions that trigger run-time instruction faults: these are caught and properly handled by the Linux kernel. In the other method the programmer provides a library of floating-point emulation functions that are invoked at compile-time (compiler flag –msoft-float). This second emulation method is generally more efficient than the former (there is no kernel overhead, the actual improvement depends on the user-supplied library), but it still is several times slower than the fixed-point implementation.

12.6 Conclusion

The on-going expansion of network technologies and the need of enhanced user interfaces for the client-devices motivate the fixed-point implementation of speech recognition algorithms, for operation on CPU's without floating-point arithmetic units. This chapter has reviewed problems and it has proposed methods concerning the fixed-point implementation of ASR algorithms. In particular it has described a practical approach to the implementation of the frame-synchronous beam-search Viterbi decoder, N-grams language models, HMM likelihood computation and mel-cepstrum front-end, typical of large vocabulary continuous speech recognition (LVCSR) systems. The described methods are also useful to prototype ASR applications in embedded systems. In fact, the decoder fixed-point parameters do not need critical task-dependent calibrations, and the language and acoustic models, trained with the standard floating-point algorithms, can be automatically ported to the required fixed-point representation.

The presented fixed-point implementation of the LVCSR algorithms is as accurate as the floating-point recognizer, in medium and large vocabulary continuous speech recognition tasks. The chapter results demonstrate real-time recognition of the standard DARPA Resource Management on two embedded CPU's namely the 206 MHz StrongARM-1100 and the 624 MHz XScale PXA270. The fixed-point implementation enables real-time operation: the floating point recognizer, with floating-point software emulation, is several times slower for the same accuracy.

Acknowledgments

Grateful thanks to S. Kanthak and J. Schroeter for many insightful discussions and support.

References

Bocchieri, E. and Mak, B. (2001) Subspace distribution clustering hidden Markov model. *IEEE Transactions on ASSP*, vol. 9, pp. 264–275.
Davis, S.B. and Mermelstein, P. (1980) Comparison of parametric representations for monosyllabic word recognition in continuously spoken sentences. *IEEE Transactions on ASSP*, vol. ASSP-28, no. 4, pp. 357–366.
Gong, Y. and Kao, Y. (2000) Implementing a high accuracy speaker-independent Continuous speech recognizer on a fixed-point DSP. In *Proceedings of ICASSP*, pp. 3686–3689.

Hermansky, H. and Morgan, N. (1994) Rasta processing of speech. *IEEE Transaction on ASSP*, vol. 6, pp. 578–589.

Huggins-Daines, D., Kumar, M., Chan, A., Black, A.W., Ravishankar, M. and Rudnicky, A.I. (2006) Pocketsphinx: A free, real-time continuous speech recognition system for hand-held devices. In *Proceedings of ICASSP*, vol. 1, pp. 185–188.

Jeong, J., Han, I., Jon, E. and Kim, J. (2004) Memory and computation reduction for embedded ASR systems. In *Proceedings of ICSLP*.

Kanthak, S., Schütz, K. and Ney, H. (2000) Using SIMD instructions for fast likelihood calculation in LVCSR. In *Proceedings of ICASSP*, pp. 1531–1534.

Kao, Y.H. and Rajasekaran, P.K. (2000) A low cost dynamic vocabulary speechrecognizer on a GPP-DSP system. In *Proceedings of ICASSP*, pp. 3215–3218.

Köhler, T., Fügen, C., Stüker, S. and Waibel, A. (2005) Rapid porting of ASR systems to mobile devices. In *Proceedings of INTERSPEECH*, pp. 233–236.

Lee, K.F. (1989). Automatic Speech Recognition Recognition. The Development of the SPHINX System, Kluwer Academic.

Lee, L. and Rose, R.C. (1996) Speaker normalization using efficient frequency warping procedures. In *Proceedings of ICASSP*, vol. 1, pp. 353–356.

Leppänen, J. and Kiss, I. (2005) Comparison of low foot-print acoustic modeling techniques for embedded ASR studies. In *Proceedings of INTERSPEECH*, pp. 2965–2968.

Li, X., Malkin, J. and Bilmes, J. (2006) A high-speed, low-resource ASR back-end based on custom arithmetic. *IEEE Transaction on Speech and Audio Processing*, vol. 14, issue 5, pp. 1683–1693.

Mohri, M., Pereira, F. and Riley, M. (2002) Weighted finite-state transducers in speech recognition. *Computer, Speech and Language*, vol. 16 issue 1, pp. 69–99.

Novak, M. (2004) Towards large vocabulary ASR on embedded platforms. In *Proceedings of ICSLP*.

Novak, M., Hampl, R., Krbec, P. and Sedivy, J. (2003) Two-pass search strategy for large list recognition on embedded speech recognition platforms. In *Proceedings of ICASSP*, vol. 1, pp. 200–203.

Oppenheim, A.V. and Schafer, R.W. (1975) *Digital signal processing*, Prentice-Hall.

Rose, R., Parthasarathy, S., Gajic, B., Rosenberg, A. and Narayanan S. (2001) On the implementation of ASR algorithms for hand-held wireless mobile devices. In *Proceedings of ICASSP*, vol. 1, pp. 17–20.

Sagayama, S. and Takahashi, S. (1995) On the use of scalar quantization for fast HMM computation. In *Proceedings of ICASSP*, Vol. 1, pp. 213–216.

Saon, G., Padmanabhan, M., Gopinath, R., and Chen, S. (2000) Maximum likelihood discriminant feature spaces. In *Proceedings of ICASSP*, vol. 2, pp. 1129–1131.

Vasilache, M. (2000) Speech recognition using HMM's with quantized parameters. In *Proceedings of ICSLP*, vol. 1, pp. 441–444.

Vasilache, M., Iso-Sipilä, J. and Viikki, O. (2004) On a practical design of a ow complexity speech recognition engine. In *Proceedings of ICASSP*, vol. 5, pp. V-113–16.

Viikki, O. (2001) ASR in portable wireless devices. In *Proceedings of ASRU*, pp. 96–99.

Zaykovskiy, D. (2006) Survey of the speech recognition techniques for mobile devices. In *Proceedings of 11th International Conference Speech and Computer, SPECOM'2006*, pp. 88–92.

Part IV

Systems and Applications

13

Software Architectures for Networked Mobile Speech Applications

James C. Ferrans and Jonathan Engelsma

Abstract. We examine architectures for mobile speech applications. These use speech engines for synthesizing audio output and for recognizing audio input; a key architectural decision is whether to embed these speech engines on the mobile device or to locate them in the network. While both approaches have advantages, our focus here is on networked speech application architectures. Because user experience with speech is greatly improved when the speech modality is coupled with a visual modality, mobile speech applications will increasingly tend to be multimodal, so speech architectures therefore must support multimodal user interaction. Good architectures must reflect commercial reality and be economical, efficient, robust, reliable, and scalable. They must leverage existing commercial ecosystems if possible, and we contend that speech and multimodal applications must build on both the web model of application development and deployment, and the large ecosystem that has grown up around the W3C's web speech standards.

13.1 Introduction

In this chapter we explore architectures that support multimodal user interaction on *mobile devices*. Our particular emphasis is on those architectures that rely on speech engines located in the network instead of on the device. We will briefly survey the current state of speech recognition, then describe how voice-only applications have rapidly shifted to a standards-based web model of development and deployment. Because mobile devices already have very capable visual modalities, and because combining a voice and a visual modality greatly improves the user experience on mobile devices, mobile *voice applications* will increasingly be *multimodal*. After providing this background we present a conceptual model for categorizing multimodal architectures, describe several commercial multimodal systems, and discuss the *standards* needed before wide adoption of multimodal systems can occur. Our discussion is informed by a commercial-grade multimodal system developed by Motorola and partner companies.

13.1.1 Embedded and Distributed Speech Engines

Mobile devices first supported "speech recognition" via simple template matching: to enable voice dialing, the user first trained an embedded template matching algorithm by providing it with an audio input sample to associate with each phone book entry. To voice dial, the user would repeat the name or phrase associated with that contact, and the algorithm would compare the new audio sample against the stored waveforms to determine the contact to call. This approach is speaker-dependent and suffices for up to a few hundred contacts.

True speech recognition became practical commercially about a decade ago, and took two forms. *Transcription systems* on desktop PCs were speaker-dependent and required high-quality microphones and a quiet environment. The user would spend perhaps 10 or 15 min reading sample sentences to train the system, which would then do a fairly credible job of transcribing what the user said.

The second form of commercial speech recognition to arrive in the mid to late 1990s was the *network-based speech recognizer*. These were reached over circuit-switched voice calls and were speaker-independent. Instead of being able to transcribe all of a single user's speech based on a relatively large dictionary, network-based speech recognizers could understand many users but had to be given strict constraints on what to expect them to say. Constraints were specified by context-free grammars much like those used to specify programming languages. So while a grammar-based speech recognizer achieved speaker-independence by limiting what users can say, a transcription base speech recognizer achieved grammar independence by limiting the users who can speak with it.

As mobile devices have become more powerful, it became possible to embed grammar-based speech recognition systems on them. They remain less capable than their larger cousins running on network-based computers. They typically support vocabularies in the ten to twenty thousand word range, and take up roughly ten megabytes of storage. As a rule of thumb, network-based and desktop speech recognizers have vocabularies ten times larger than embedded recognizers, and have proportionately greater hardware requirements.

Transcription systems are now just starting to appear on mobile devices, where voice entry of SMS messages and email is a very valuable use case. So far these have had mixed results. Transcription systems are also moving into network-based server farms in configurations that support speaker-independent recognition, which is potentially a very significant development.

Speech recognition has steadily improved over the years, both in the network and on devices. We will see more speaker-independence, less restriction on what people can say, and other advances, although challenges remain (Deng and Huang 2004).

Speech synthesis has made parallel gains. Older formant-based systems synthesized speech from acoustic models, and tended to sound rather unnatural. But these are giving way to concatenative systems that string together segments of prerecorded speech samples to sound far more human. As with speech recognition, speech synthesis systems based in the network are more advanced than those embedded on mobile devices.

13.1.2 The Voice Web

By the mid-1990s, speech technologies had matured to a point where voice applications could begin to displace existing touch-tone (DTMF) applications. Voice applications were initially deployed on proprietary *interactive voice response* (IVR) systems, which were connected to the public switched telephony network (PSTN) with specialized hardware that supported banks of incoming analog lines or digital T1 or E1 lines.

In addition to integration with the PSTN, an IVR system contained speech engines, one or more voice applications, and also the back-end business logic, database interfaces, and legacy application interfaces needed to integrate the voice applications with the existing infrastructure. The proprietary nature of IVR systems meant that voice applications were costly to deploy, and difficult to port to other platforms.

In the mid 1990s, researchers at AT&T exploring ways to best implement web services realized that the web model for application development and delivery was as well suited for voice applications as it was for visual ones: it made no difference at all if the user was interfacing with microphone and speaker instead of a keyboard and display (Atkins et al. 1997). The web model enables and encourages a clean division between each application's interface and its back-end business logic. All the application's legacy system integration, database access, and business logic could be factored out of the IVR platform and onto standard application web servers, using the rich variety of tools developed for visual web applications, and leveraging the simplicity of web application deployment.

This factoring required standards that would enable any IVR platform to render the same backend web application to callers in the same way. Standard web protocols such as HTTP and TCP/IP would be used of course, and resources such as audio files would be delivered to the IVR platform the same way as they would to a visual web browser. But how would the web application convey voice dialogs to the IVR platform? Some researchers proposed augmenting HTML with voice dialog constructs, but most concluded that the unique aspects of voice dialogs—the need to manage temporal flow, handle input errors, resolve ambiguous inputs, specify timings, and so on—required a new markup language.

Some early voice markup languages were AT&T's PML (Atkins et al. 1997), HP's TalkML (Raggett 1999), IBM's SpeechML, and Motorola's VoxML (Ladd et al. 1999). Commercial realities dictated there be only one, so in early 1999 AT&T, IBM, Lucent, and Motorola created the *VoiceXML Forum*, whose purpose was to develop a standard language. The Forum published VoiceXML 1.0 (Boyer et al. 2000) and then gave it to the *World-Wide Web Consortium* (W3C) which published the VoiceXML 2.0 Recommendation (McGlashan et al. 2004).

The industry eagerly adopted the VoiceXML standards, because they were a first major step in the disaggregation of proprietary IVR platforms into interchangeable components based on open standards. The IVR platform was transformed into a generic VoiceXML *voice server*, and it now rendered standard voice web applications hosted on standard application web servers (see Fig. 13.1). Web development and deployment technologies, coupled with these new standards, dramatically drove down the cost of voice applications, so that today literally billions of calls are processed each year by VoiceXML-based voice servers, and applications as large as the North American Directory Assistance service are based on VoiceXML.

These new voice server platforms have become further commoditized by standards closely related to VoiceXML. The first is the Internet Engineering Task Force (IETF) *Media Resource Control Protocol* (MRCP), whose goal is to provide a standard control interface to speech engines, to make them easily interchangeable (Shanmugham et al. 2006). Developed with VoiceXML servers in mind, MRCP can be adapted to other contexts. It is comparable to HTTP, with each textual request to the speech engines specifying the prompts to play and the speech grammars to listen for, and each corresponding textual responses from the speech engines giving the recognition results. MRCP makes it far easier to integrate new speech engines into a voice server, to give it better speech technologies, customize it for new locales, or simply switch to a lower cost supplier.

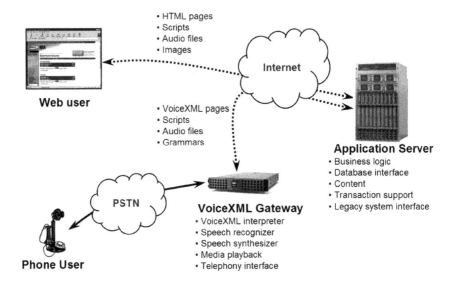

Fig. 13.1 Voice web architecture

While MRCP is the protocol for the control of speech engines and other media resources, *voice over Internet Protocol* (VoIP)   standards like the Session Initiation protocol (SIP) (Rosenberg et al. and Schooler 2002) and RTP are protocols for directing audio streams to and from the speech engines (Sutherland and Danielsen 2006).

The first benefit of VoIP to the voice platform architect is that specialized hardware terminating incoming PSTN lines no longer needs to be located inside the voice server platform itself. A *media gateway* can now terminate the lines and convert their time-division multiplexed (TDM) audio streams into VoIP. This significantly drops the hardware costs and makes the overall system more flexible. Without VoIP, the incoming audio channels need to be terminated at a telephony hardware card attached to some machine in the voice server, either one running speech engines or one doing media gateway-like conversion of the TDM audio into IP packets for processing in speech engines on another machine. This means that machines must handle some multiple of the incoming PSTN line size. In North America this typically means the speech engine box has to support one or more 23-channel T1 lines, while in Europe it has to handle one or more 31-channel E1 lines. If the box could comfortably handle, say 80 incoming calls instead of 69 (three T1s) or 62 (two E1s), that extra capacity is wasted.

With *VIP*, the media gateway ca n deliver exactly the right amount to each box so that fewer are needed. And VoIP-based voice platforms can serve pure IP traffic such as Skype calls directly, with no need for a media gateway.

The combination of MRCP and VoIP also allows the architect of a *voice server* platform to cleanly separate the VoiceXML dialog interpreter from the speech engines and place them in various convenient and efficient topologies. For example, platforms are usually composed of self-contained "pods" of machines, each of which operates independently and handles several hundred callers. A pod supporting two hundred callers had to dedicate a speech recognizer channel to each possible incoming call, but now with MRCP and VoIP they can easily get by with, say, fifty speech recognizers, each of which is shared among many calls. For each prompt and collect cycle, a VoiceXML interpreter will use SIP to establish a session to an available speech recognizer and a media player and to set up the RTP audio pathways to each. Then the interpreter will use MRCP to tell the engines what to play to the user and what to listen for. When MRCP returns the recognition results are returned to the VoiceXML interpreter, the interpreter closes the SIP session to release the speech engines for another caller to use. This greatly increases the scalability and flexibility of the voice server platform architecture. This efficiency is possible because people interacting with voice applications spend much more time listening and thinking than they do speaking.

The W3C's *Call Control XML* (CCXML) is a final standard used to open up voice server platform architectures. VoiceXML cleanly separates the application and business logic from the voice platform, MRCP provides a generic "plug-and-play" control interface to the speech engines, and VoIP standards enable very flexible internal audio pathways in the voice server platform. But the VoiceXML interpreter is still coupled to platform-dependent call control operations for accepting incoming calls, placing outgoing calls, disconnecting calls, transferring calls, etc. Call control needs to be factored out in a standard way, and this is what CCXML enables (Auburn 2007). A CCXML interpreter now becomes part of the platform, and is driven by web pages in the CCXML markup language. These tell the interpreter how to establish and tear down call legs between two or more human and computer endpoints. The platform then uses CCXML to start up sessions and to bring in new participants as needed (as in teleconferences). A VoiceXML interpreter participating in such a session implements its call control operations by sending markup to the CCXML interpreter. Platform dependent call control interfaces are now encapsulated inside the CCXML interpreter.

We covered these topics at some length to convey something of the scale and commercial importance of the voice web, and to lay groundwork to return to later in our discussion. It turns out that this sophisticated network infrastructure, with only minor change, can support multimodal applications as well as voice-only applications. Multimodal architectures that leverage the VoiceXML-based voice web ecosystem will therefore have significant commercial advantages.

13.1.3 Multimodal User Interfaces

Mobile devices are physically small, making interaction with the keypad, stylus, and display relatively difficult. These difficulties are compounded when we have *accessibility* problems like arthritis or poor eyesight. "Situational impairments" are also problematic: we may be wearing gloves, walking on an uneven sidewalk, or trying to read the screen in bright sunlight. Various user studies quantify these difficulties: a joint study at Columbia and Google analyzed one million Google Mobile Search queries and found that the average time to enter even short one to four character search terms on a mobile keypad was over 40 s, with 30–34 character searches taking over 90 s. Stylus input was faster, at 25 s and 50 s respectively (Kamvar and Baluja 2005).

These times are very problematic from a *usability* standpoint. But while mobile devices are poor at keypad entry, they are highly optimized for audio interaction, which makes voice input especially attractive in mobile search: assuming the speed and accuracy of the system is high enough, speech entry of search terms can take just a few seconds.

But pure speech applications have their own issues. We do not want to blurt out personal information, and complex spoken output is much harder to remember than visual output. Speech interfaces have their own accessibility issues, e.g., for people with accents and hearing problems, and they have associated situational impairments such as background noise and laryngitis.

Conveniently, the weaknesses of mobile visual user interfaces are offset by the strengths of speech user interfaces: while it is slow and difficult to type (or even spell) *Albuquerque* in a mobile airline application, it is quite fast and easy to say it. And likewise, the strengths of visual interfaces offset the weaknesses of speech interfaces: visual information often is faster to process and remember than spoken information, while disambiguation of speech input can be done quickly with a visual drop-down menu of the alternatives (Oviatt 2000). The weaknesses of one modality are offset by the strengths of the other, which makes mobile multimodal applications very attractive (Suhm et al. 2001).

13.1.4 Distributed Speech Recognition

Speech recognizers should be given the highest quality audio input to reduce misrecognition, but telephony channel quality is generally not of the best quality. Landlines deliver only about 4 kHz of bandwidth, though they are circuit-switched and tend not to drop segments of audio. Mobile audio channels use codecs that favor low bandwidth over audio fidelity, and they also drop packets. IP telephony channels can also drop packets, but their codecs can use more bandwidth.

To deliver high-quality audio to speech recognizers over mobile channels, the *ETSI Aurora* group developed the Distributed Speech Recognition (DSR) standards (Pearce 2000). They achieve this by moving the earliest stage of audio processing from the speech recognizer to the mobile device. This stage converts the raw audio into a digital stream of audio samples, called feature vectors. These are encoded in an RTP stream transmitted by a UDP/IP data channel to the speech recognizer. In this way channel loss is reduced, the fidelity of the audio signal is kept high, and bandwidth is reduced. Moreover, DSR front-ends on mobile devices can do special processing to eliminate background noise, approximately halving the error rate due to background noise (Pearce 2004).

These benefits are substantial, but compete against alternative approaches such as using existing audio channels and accepting higher recognition error rates, or shipping the full raw audio over reliable broadband connections to the speech server. The best chance for widespread adoption of DSR will be to pair it with distributed multimodal systems, since its benefits are synergistic with those of multimodal systems.

13.1.5 Multimodal Architectures

The *Open Mobile Alliance* (OMA) is a standards group formed in 2002 to develop open standards for the mobile phone industry. It consists of mobile operators, device manufacturers, software vendors and others. One of their working groups is Browser Technologies, and a subgroup called Mobile Application Environment recently published a conceptual multimodal architecture (Open Mobile Alliance 2006).

This architectural view is at a high enough level to cover cases where the speech engines are in the network and cases where they are embedded on the device. Figure 13.2 illustrates the key entities in their architecture. Each user interface modality is controlled by a *user agent* (UA), which has zero or more *processing engines* (PEs) supporting it. A web browser is a canonical example of a user agent for the visual modality; one of its processing engines might be its input processing subsystem, another processing engine could be the subsystem that renders output using HTML, CSS, and so on. Similarly, a VoiceXML-based voice browser is a user agent: its speech recognizer is a processing engine for speech input, and its audio output subsystem, which includes speech synthesis, forms a second processing engine.

Multimodal systems are those that have at least two user agents (modalities). Typically, they are comprised of a visual modality and a voice modality, but many other combinations are possible. For example, we can add in a third modality for handwriting recognition and, perhaps, cursive handwritten output. A haptic modality can be driven by motion input detected by a three-axis accelerometer and can generate motion output by causing a transducer to vibrate at various frequencies: using it you could turn pages by flicking the phone left and right, and get feedback when you try to go beyond the first or last page through feeling a particular vibrational pattern. A pulse sensor could be part of an input-only modality used in wellness applications: during physical activity, the pulse modality can be linked to an audio output modality that coaches the user on how intensely to exercise.

Some fairly unusual multimodal systems have been developed using modalities other than speech. A head-mounted sensor can track one's gaze to determine what is being looked at, and therefore forms a sort of ocular input modality. An "emotional" input modality is even within reach of current technology: several startups are working with low-cost electroencephalogram (EEG) sensors that measure "focus" and "tranquility." At the Consumer Electronics Show in 2006, a startup called NeuroSky demonstrated a multimodal computer game with three modalities: (1) a standard computer display showing a 3D world of objects, (2) a head-mounted gaze sensor to pick out what object the player is looking at, and (3) a head-mounted "emotion" sensor that measured focus and tranquility. The system caused objects looked at with a high degree of focus to be moved closer to the player, and caused objects looked at with tranquility to float off the floor (NeuroSky 2007).

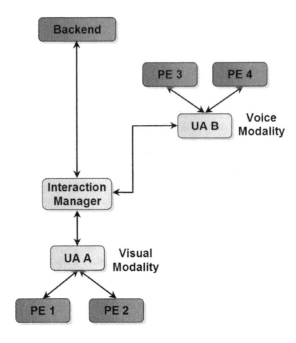

Fig. 13.2 Conceptual multimodal architecture (OMA)

When using a speech modality, the visual modality need not be a standard form-based interface or web browser. It could be a game engine (Zyda et al. 2007) or an avatar interface.

Modalities may or may not support both input and output. A voice modality can have speech recognition but not generate audio prompts. Or the visual modality can be used for output but not for input. An example illustrating this point is a Bluetooth service discovery application that features speech input and visual output to connect rapidly to location-based services (Engelsma and Ferrans 2007).

Returning to the OMA conceptual architecture in Fig. 13.2, there is a need to coordinate the user agents (modalities). For instance, the results of a speech recognition may affect the visual display (i.e., field values are updated) or a typed input value affects the active speech recognition grammar. The *interaction manager* (IM) is the OMA architectural element that effects this coordination: it synchronizes the data and execution flow between the user agents.

Multimodal applications generally do not operate in a vacuum and therefore must obtain external information and update external state. A weather application needs to look up weather conditions, download spoken weather reports, download radar images and other visuals, and download advertisements. It also might upload user preferences. The OMA architectural element representing the external world is called the *backend*, and the IM communicates with the backend. The backend is generally the web and all its applications and services, but in a self-contained system it might be a local web server, a set of local files, etc. The minimal backend is probably a static specification file defining a multimodal dialog.

In practice of course a real multimodal system will differ from this ideal view. A visual web browser's processing engines are not necessarily distinctly separable, since input and output have close cross linkages. And it is very common for each user agent to fetch needed resources directly from the backend rather than use the IM as a client-side proxy. But overall, the OMA model is a very helpful tool for understanding and comparing variant multimodal architectures.

Looking again at Fig. 13.2, one can draw a horizontal line across it at various heights to effect divisions between client and server components. Each possible division defines a class of multimodal architectures. Draw the line at the top, with only the backend above it, and it describes the family of multimodal architectures with everything resident on the device. Draw the line above the visual modality's user agent ("UA A"), and you describe a family where everything but the visual user agent is in the network. We will explore these families in more depth later.

13.1.6 Simultaneous and Sequential Multimodality

Multimodal systems are divided into two broad categories depending on whether the user interacts with the modes simultaneously or not. In a *simultaneous multimodal* system, more than one mode is active at the same time. In a *sequential multimodal* system only one mode is active at any time. A simultaneous multimodal map application could both display a map and play a voice prompt at the same time, and allow input by keypad, touch screen, or voice at any time. For instance the user could select a "zoom in" menu item or say "zoom in" (Maes and Saraswat 2003).

A sequential multimodal map application would only have one mode active at a time. For example, the user could place a voice call to establish the current location and the destination for a trip, hang up, and then start a visual application that downloads this information and the turn-by-turn directions for that route. Or in a sequential stock trading application the user might again interact first by voice, then later get an SMS or multimedia message containing a trade confirmation. Sequential multimodal systems offer some of the same advantages as simultaneous multimodal systems, but are less complex to architect and implement.

Simultaneous multimodal systems are further subdivided into *composite* and *non-composite* multimodal systems. In a non-composite multimodal system the inputs from the various modalities are independent and are presented to the application in the order that they occur, even if they occur at nearly the same time. In a composite multimodal system, inputs from two or more modalities that occur at or close to the same time are considered to be a single coordinated input, so they must be composed or "fused" into a single input before being given to the application. Take an application for finding out movie theater shows and show times. In a non-composite approach the user might first select a theater from a list or a map, and then a moment later say "show times, please." In a composite approach, the user might draw a circle around a theater push-pin on the map while saying "show times" (Maes and Saraswat 2003). Composite multimodal systems are potentially faster and easier to use, but have not yet been introduced commercially.

13.1.7 Mode Composition

The OMA architectural model supports hierarchical decomposition: a user agent can itself be decomposed into an interaction manager and two or more lower level user agents. For example, consider adding a new voice modality to an existing user interface that supports visual output, input from the keypad, input from a virtual keyboard with touch screen and stylus, and input by handwriting recognition with the stylus. The existing user interface is already multimodal, and so must consist of an interaction manager, a couple of lower level user agents, and some internal processing engines (e.g., the handwriting recognizer). To add the new voice modality then, one has to add the voice modality's user agent and processing engines, and couple the voice user agent to the existing system with a new higher level interaction manager.

13.2 Classes of Multimodal Architectures

We now turn to how best to architect a multimodal system. We consider only simultaneous multimodality: sequential systems are a kind of "degenerate" case of simultaneous multimodality where a relatively lengthy context switch has to take place to shut down one mode and activate another. Simultaneous multimodal systems require much tighter coordination, and hence are more difficult to architect than sequential multimodal systems.

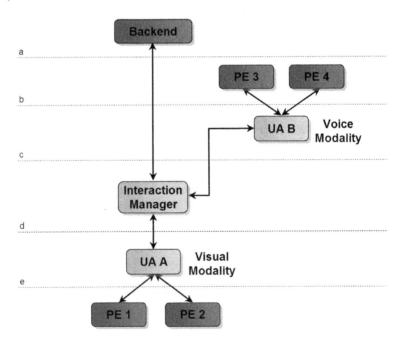

Fig. 13.3 Five families of simultaneous multimodal architectures

Architecting a multimodal system is a complex process with no one right solution: each family of multimodal architectures has its own comparative advantages.

Figure 13.3 shows the OMA conceptual multimodal architecture with five alternative horizontal dividing lines between client and server. Each division identifies a family of simultaneous multimodal architectures. We consider only the very common case of a visual modality plus a voice modality: other architectural families are possible when combining other modalities.

13.2.1 Fully Embedded or "Fat Client" (a)

Let's consider each class in turn. First we consider the case where every component is placed on the mobile device.

Dividing line (a) places only the backend on the server.[1] All other components are on the mobile client: the visual modality, the voice modality, and the interaction manager linking them together. This approach is necessary if multimodal applications must operate when the device is not connected to a network, but it requires a fairly powerful device. On the surface it would seem to be the class of multimodal architecture that makes the least use of network bandwidth, but that depends in large part how self-contained the speech applications are. An embedded driving direction application with voice entry of addresses would need to download huge speech grammar files for each town or postal code, but a networked driving direction application would only have to send a relatively short audio stream up to the voice server.

One instance of this architecture is a prototype created by IBM and Opera on a Windows Mobile handset (Kennedy 2005). In this prototype, the visual user agent is the Opera XHTML browser, and the voice user agent is an IBM embedded *VoiceXML* 2.0 interpreter. The processing engines for the voice user agent are from IBM (embedded ViaVoice). The interaction manager is IBM client middleware.

This prototype's demonstration application was voice-activated local search. Search terms were entered by voice, and after each term was recognized on the device, it was sent to the Yahoo local search web service to obtain the results. Mobile local search is a very compelling multimodal application: it is valuable to people, requires rich visual output, and works far better with voice input than with keypad input. The IBM application is authored in the *XHTML+Voice Profile* (X+V) markup language, a clean unification of XHTML for the visual modality and VoiceXML for the voice modality (Axelsson et al. 2004).

13.2.2 Distributed Processing Engines (b)

Dividing line (b) defines the class of multimodal architectures where the speech engines are distributed to a network-based voice server, but nothing else is. (A variant on

---

[1] Here and subsequently we gloss over the special, and relatively rare case where the application backend is entirely local to the client device.

this would be to distribute the speech recognizer, but leave the speech synthesizer on the device.) The natural protocol to communicate with distributed speech engines is MRCP, which as described above is the IETF's textual protocol, patterned after HTTP, that sends prompt-and-collect requests to the speech engines and gets recognition results in the corresponding responses (Shanmugham et al. 2006). Before we talk about this family of architectures in particular, we will take a lengthy discursion into the benefits of placing speech engines in the network.

There are some very highly significant advantages to distributing speech engines. If they are on the device they take up substantial memory, even though only a minority of device owners may be using them. They are also compute-intensive, which can make battery drain an issue (Delaney et al. 2005). Administration is far easier if speech engines are on the network: it is much more efficient to patch the speech recognizer on a thousand voice servers than ten million mobile devices. The speech application itself is much easier to tune and update in a distributed architecture: usability experts can listen to recorded sessions to find places where users run into difficulties, and use that data to revise prompts and tune speech grammars. Testing itself is much easier, since only one set of speech engines must be tested, not a multiplicity of speech engines and versions on scores and hundreds of different types of mobile device.

A final advantage of distributing speech engines to the network is that it can greatly minimize network traffic and delay in many common scenarios. The speech recognizer needs to have both the audio to recognize, and compiled speech recognition grammars to tell it what to look for. The audio originates on the handset, while the grammars originate in the backend application. There are two pathways into the speech recognizer: the inexpensive high-speed wired network, or the expensive, slower-speed wireless network. The relative size of the audio and the speech grammars, the frequency of change in the speech grammars, and the speeds of the two networks all must be taken into account by the architect in deciding where to place the speech recognizers.

One anti-pattern to avoid is using an embedded speech recognizer driven by huge frequently changing speech grammars generated in the network. For example, if the user is browsing an online music store with five million songs divided into a hundred categories of 50,000 titles per category, with new titles added each day, then each day and for each category the user triggers a speech recognition grammar download of perhaps five megabytes, and an embedded grammar compilation step that together might take 5–10 min and substantial battery power. In this case, it is far better to send up a couple of kilobytes of DSR-compressed audio to the voice server: the results will be back in a couple of seconds, and the battery will barely be affected. This tradeoff turns out to be fairly common: think of mobile search, map applications with points of interest being added and removed each day, corporate directory access, access to back end enterprise data, looking for auctions on eBay, ordering books from Amazon, and so on.[2]

---

[2] One optimization would be to do the grammar compilation in the network instead of the device, but then each application needs to have the grammar compiler for each

On the other hand, if the application backend is on the mobile device, it is better to do the speech recognition on the device, otherwise the device would have to generate a potentially large speech grammar for the network-based speech recognizer.

This tradeoff is captured in the *Pearce Principle* (Pearce 2002), which states that speech recognition should be done at the point closest to the location of the speech grammar being listened for. This provides a rational for *hybrid speech recognition* systems, which leverage local recognition for local applications, and remote recognition for network-based applications.[3] This is similar to the data-intensive supercomputing principle of locating computation where the data resides, rather than moving the data to the point of computation (Bryant 2007). This insight has also long been known in the area of query processing in distributed databases.

Returning from our discursion into the virtues of distributed speech engines, the Distributed Processing Engines family of distributed multimodal architectures has a significant disadvantage in that it requires MRCP or a protocol at the same level to go over the wireless network to the server hosting the speech engines. This is relatively expensive in terms of bandwidth and round trips: MRCP was designed to be a lower-level protocol used within a voice server platform and hence it has many more, and much larger messages than a higher level protocol would have.

13.2.3 Thin Client (d)

We will return to architectural family (c) after we discuss (d) and (e). Family (d) is the "thin client" multimodal architecture. This places the full voice modality in the network, along with the interaction manager. This approach is fairly balanced for contemporary mobile devices and networks. It turns out to be second best in terms of network bandwidth, but there can be some awkwardness in writing applications where some logic has to be broken out into an explicit interaction manager off in the network. But overall it shares many virtues with family (c), which we believe edges it out in desirability.

13.2.4 Remote Visual Interface (e)

With the dividing line drawn beneath the visual user agent, as in (e), we have an architecture class where everything but the visual user interface rendering and the input subsystem is distributed to the server. A protocol for driving a remote user interface

---

possible mobile device configuration, and know each device's configuration, a complex task. Even if this reduced data transmission, battery drain, and elapsed time by an order of magnitude, the resulting delay would still make the experience very painful for the user.

[3] When high-quality embedded transcription engines become practical, and application developers take advantage of them, the dynamics change: transcription systems do not use speech grammars.

needs to be developed, and the mobile device just contains a module that does the lower levels of the visual user interface. Most of the logic driving the visual user interface is in the network.

This is the same approach that the X Window System takes. One instantiation of the Remote Visual Interface architecture would be to put an X Server on the mobile device and drive it from an X Client in the network. They would communicate with the X11 protocol. In this approach, the X Server corresponds to the OMA visual processing engines and the X client corresponds to the OMA visual user agent.

Auvo, an early multimodal startup (ca. 2000–2002), used this architecture, but unfortunately they were years ahead of the market and ran out of funding.

The main drawback of remoting the visual user interface over a mobile data network is of course bandwidth and latency. Bandwidth is becoming less and less important, but a protocol that introduces many round trip delays will be less useful than one that has few delays. Another drawback is that this architecture makes it very hard to expose the full power of the native visual user interfaces on each device: it almost invariably presupposes that each device runs a client that understands a "least common denominator" protocol and API.

13.2.5 "Pudgy" Client (c)

The final major family of multimodal architectures is described by dividing line (c), the so-called "Pudgy" Client. This is a slight variation on Thin Client, moving the interaction manager from the server over to the mobile client. This makes it a bit fatter than Thin Client, hence the name.

This approach is more optimal in terms of network usage (the interaction manager has somewhat more work to do to drive the visual interface than the voice interface, and hence should be located with it). It is more intuitive for developers, who tend to view the mobile client as the proper locus of control, just as it is for purely visual applications. The notion that voice is a sort of supplemental input method under control of the client software has proven to be especially appealing. We cover an implementation of Pudgy Client at length in Sect. 13.3.

13.2.6 Discussion

We have just described five main families of distributed architectures that support simultaneous multimodal interaction.

The Fully Embedded architecture is well-suited for more powerful devices and applications that reside on the device itself. It has trouble running applications that require significant fetching of speech grammars from a network-based source, since these can take very significant amounts of bandwidth and time to download and compile. Embedded speech engines lead to various administration difficulties, and also make voice application testing more complex and problematic. Nevertheless, devices and networks are both gaining in power and speed, diminishing some of these difficulties. As device-based speech recognition becomes more transcription-based (open vocabulary), the need for speech grammars will diminish. We therefore believe that this will be an effective architecture going forward.

The four remaining approaches leverage network speech engines and do not share many of the above limitations. On the other hand, they cannot be used in a disconnected mode. Of the four, Distributed Processing Engines requires the client to exercise detailed low-level control and therefore requires more network message round trips, introducing delay. The Remote User Interface also requires a lot of network traffic, and pushes off on the multimodal server a lot of user interface control logic, which means that the server has to support a single generic abstract visual user interface with least common denominator functionality, and that therefore the native user interface capabilities of each device cannot be fully leveraged.

The Thin Client and Pudgy client architectures are both nice balances that minimize network traffic and are easy to develop applications for. Of the two we have a moderate preference for Pudgy: it is more natural to have the interaction management done close to the visual modality than the voice modality, as the client application is the natural locus of control.

13.3 The "Plus V" Distributed Multimodal Architecture

Motorola began working on a distributed multimodal system connected to a standard *VoiceXML* server in the network in 2001 (Pearce et al. 2005). Our initial architecture was primarily Thin Client (d), with the interaction manager consisting of a few extensions to the VoiceXML Form Interpretation Algorithm, so the voice dialog actually drove the visual dialog as a side effect. We quickly found this to be awkward and unnatural, as developers believed interaction management belonged in the client device.

In early 2002 we tried another approach, where the interaction management was explicitly made a module in the client software. This was an implementation of Pudgy Client. A major motivating factor was a series of unpublished simulation studies we did to evaluate the architectural families. The goal was to determine bandwidth and latency costs of each approach on GPRS networks. We found that Pudgy Client was much better overall than the others we tested. Our subsequent implementations confirmed this: on the 2.5G GPRS and the 2G iDEN networks, our system takes between 0.8 s and 2.0 s between the end of speech and the visual display of the recognition result, substantially faster than even today's multimodal systems running on 3G data networks. This speed is due to Pudgy's low messaging requirements, its terse binary message format, and the use of the DSR codec, which takes only 5.6 kbps of bandwidth, on the audio channel.

In this approach, the client is fully in charge of the interaction. The networked VoiceXML server is under its control and merely adds the voice modality to the interaction, hence the architecture's more formal name "Plus V." A key advantage of Plus V is that it supports any visual user interface. We have created three instances: one that connects a *Java J2ME* MIDlet on the handset to the networked voice server (J+V), another that connects a C++ application using Qt user interface on Linux handsets to the voice server (Qt+V), and a third that connects a version of the Konqueror XHTML browser using the *X+V* multimodal markup language (Axelsson et al. 2004). Any visual interface can be supported: for instance the Torque 3D game engine could be used in a "Torque+V." This agnosticism to the graphical user interface is a strong advantage.

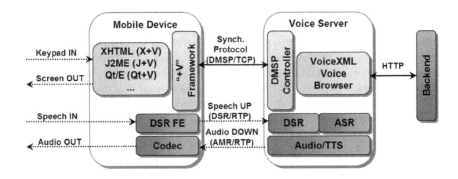

Fig. 13.4 Plus V multimodal architecture

Figure 13.4 shows the Plus V multimodal architecture at the next level of detail. The voice server is a very slightly modified VoiceXML server. We started with the commercial SandCherry Voice Platform (see www.sandcherry.com) and dropped in the commercial Motorola VoxGateway VoiceXML 2.0/2.1 interpreter (Ferrans 2003) and the Nuance OSR 3.0 speech recognizer, which supports the DSR codec.

On the client we have the standard codec for audio output, a DSR front end to do the encoding of the audio, a native user interface, and one of the Plus V implementations as described above.

In the OMA terminology, the visual user interface and the VoiceXML voice browser are the user agents, the speech engines are the voice modality's processing engines, and the visual modality's processing engines are elements of the graphical user interface software. The interaction manager is represented by the Plus V device-side framework (the client application can do some interaction management).

The client drives the voice server using the Distributed Multimodal Synchronization Protocol (Engelsma and Cross 2007) over a reliable TCP/IP channel. The client tells the voice server which VoiceXML page to load and which VoiceXML dialog on that page to run. Once the dialog is running, if the user speaks to the system, the voice server uses DMSP to convey the recognition result back to the client. If the user types, the Plus V Framework sends the new field value to the voice server via DMSP, where it causes the VoiceXML dialog to advance. If the user scrolls through the visual form's fields, the client also tells the voice server the new focus field. This level of coordination is necessary because each visual field may have a distinct audio prompt introducing it, and each field typically also has a speech grammar associated with it. Mixed initiative dialogs are also supported by this approach. DMSP is currently an IETF Draft, and for performance it seeks to minimize messages, message size, and round trips. The message format is a very condensed binary format, with an optional XML format for use when message size is not an issue.

The efficiency of DMSP and of the DSR speech recognition codec makes Plus V the fastest distributed multimodal architecture we are aware of in terms of recognition response latency. As mentioned above the time between the raising of the push-to-talk key and visual confirmation of the user's speech runs between 0.8 s and 2.0 s on the 2G iDEN network, and 1.0 s and 3.0 s on the 2.5G GPRS network. These times are also at least as fast as the embedded speech approaches we are familiar with.

The DMSP protocol has its client endpoint inside the Plus V Framework; its voice server endpoint is the DMSP Controller. The Controller in turn has some hooks inside the VoiceXML interpreter's main loop: the Form Interpretation Algorithm (FIA), which determines what field to prompt and collect at each iteration. The FIA just needs to stop and check for commands coming from the client, and if it is in its listen phase when control commands come in, it needs to break out of that speech recognition to see what to do next. It was not at all hard to make this modification: we estimate that the effort needed to multimodal-enable a VoiceXML interpreter is at most 2% or 3% of the effort needed to write that interpreter.

13.4 Other Distributed Multimodal Architectures

Plus V is by no means the only way to architect a multimodal system. In this section we briefly sketch several other commercial distributed architectures. The goal is to show how varied the solutions are, not to exhaustively enumerate them.

13.4.1 Video Interactive Services with VoiceXML

In 2005 several people realized that VoiceXML could be adapted to video telephony quite easily. It already supported the playback of recorded audio, identified by URL and media type. It also already supported the recording of audio, of a given media type, and the posting of that audio to a web server. Why not plumb the voice platform to carry mixed audio and video streams via SIP and RTP, link those streams to the mobile handset, and support the idea of video prompts and video recordings?

This turned out to be relatively straightforward, and the only impact on VoiceXML itself was a desire to generalize the name of the "audio" prompt element.

The resulting platforms support multimodal applications that combine voice and video modalities. A video answering machine application can play different video prompts based on the caller, and take video messages from callers. Support applications can now show videos of procedures and accept videos showing problems to support representatives. Many other interesting multimodal applications are enabled by this approach (Burke and McGlashan 2006).

Because the modes used are voice and video, these systems do not fit neatly into our architectural families, but it is somewhat analogous to the Remote User Interface (e). The drawbacks of the Remote UI approach do not apply when using a video user interface instead of a graphical user interface: video playback is very standard and not highly interactive.

13.4.2 Multimodal for Set-Top Boxes

PromptU (www.promptu.com) began a few years ago as a company specializing in using voice to interact with the electronic program guide (EPG) displayed on televisions via the cable operator's set-top boxes. The EPG application runs on the head-end equipment in the operator's infrastructure, and is controlled by keys on the television remote.

In the PromptU system, the remote is augmented by a microphone and a push-to-talk button. When the user speaks ("Find actress Penelope Cruz"), the audio from the remote goes to the set-top box, where it is encoded by an Aurora DSR Front End (Pearce 2000) and sent up to a voice server located in the head-end. The voice server runs an application that maps the voice commands into actions on the EPG, and the output is sent back to the set-top box for display on the television.

More recently, PromptU has been moving into the general mobile multimodal application space, supporting music download, ring tones, games, and so on. The PromptU architecture is in the Thin Client family.

13.4.3 Bare Minimum Mobile Voice Search

Plus V was developed at a mobile handset company, where we had luxuries to do things that others cannot. We wrote DSR front end encoders for DSP chips, ensured that audio packets could be streamed using RTP, and even influenced the future MIDP 3.0 J2ME implementation.

A company that wanted to get a *multimodal application* out to its customers on as many handsets as possible would have to start from a different point, deploying a system that made the least possible assumptions about those handsets, and then influencing the industry to add the sort of enablers that we put into Plus V. Let's assume this company is doing a mobile multimodal search application.

By necessity, this company would choose a distributed architecture, since that offloads a huge amount of complexity and variability from the mobile devices. On the client they would probably select *Java JME* for its ubiquit y. Their Java client application would present the visual interface, use the JSR 135 Mobile Media API to gather voice input, and use HTTP to post that audio up to the server. Along with the audio, the HTTP request would contain the location, from GPS or the carrier's cell tower ID information. The request might contain a cookie identifying the user, and perhaps other contextual information.

On the server receiving this request runs the server side of their application. This first would send the audio over to a speech server for recognition, a process that probably would take into account the user's desired search location and radius. The speech server sends back the results, and the server-side search application feeds them to the existing web services API for the search service. At the same time the server-side application could interact with an ad server to get contextually relevant advertising to show the user. The server-side application then sends back the HTTP response with the search results, advertisements, and other response information.

The architecture described is not highly optimized or general, but it can be improved on handsets that support streamed audio, and if the application is successful, the industry will quickly try to add enablers to improve the user experience.

13.4.4 A Transcription-Based Architecture

Our last example architecture is from Mobeus, a startup just coming out of stealth mode in May 2007 (http://www.podtech.net/scobleshow/search/Mobeus). They have server side technology for doing speaker-independent transcription, which is ideal for mobile multimodal search applications, voice to SMS and email applications and so on. Their view of how this should be integrated with a visual user interface on the client is radically simple: provide a text entry widget connected to this transcription server, plus controls for speaking into it and editing the result to correct any errors or select from the "n-best" alternatives for each word. The results are very impressive, and while again it may not be the fastest or most general system (audio prompting is not addressed, e.g.), at this stage these sorts of approaches can unlock a lot of value.

13.5 Toward a Commercial Ecosystem

The *World-Wide Web Consortium*   (W3C) has been working in the area of multimodal standards since early 2002. Progress has been slow mainly because of a lack of early proprietary implementations, but as we have seen above this is soon going to change. As the value of multimodal systems becomes apparent, there will be a renewed push to create interoperability standards to grow the industry. Where do things stand today?

The W3C Voice Browser and Multimodal Interaction working groups (www.w3.org) are working on a future markup language. This will be philosophically similar to *X+V* (Axelsson 2004) in that a combination of *XHTML* and VoiceXML is called for. The framework that integrates the two markup languages will be a markup language called State Chart XML (SCXML) which is closely patterned on David Harel's State Chart formalism (Harel 1987). The challenge will be to create a language accessible enough to attract developers from ad hoc approaches.

The W3C is also working to "modularize" VoiceXML into a subset appropriate for use in a multimodal system (for instance it makes no sense for the executed VoiceXML to do call control operations like disconnect in a multimodal configuration). They are also revising VoiceXML's stand-alone event model to allow control events to come in from external sources, a task necessary if VoiceXML interpreters need to be controlled by interaction managers.

The *IETF* is to protocols what the W3C is to web markup languages. We have described at a very high level one such control protocol between the interaction manager and user agents: DMSP (Engelsma and Cross 2007) which has been submitted to the IETF as an Internet Draft. The outcome of this submittal is not yet clear, but it will probably take the upcoming impetus of successful proprietary multimodal systems to push this forward.

The 3GPP, an industry standards body focused on GSM standards, has approved the use of DSR for multimodal applications, and 3GPP2, the parallel organization for CDMA standards, is also considering it. DSR should offer continued incremental benefits even in a world of huge bandwidth.

Other standards would be needed to mature this ecosystem. There needs to be a standard for how a control protocol like DMSP drives a VoiceXML interpreter, perhaps a standard for authoring languages other than the W3C's StateChart-based one (e.g., for Java or C++ application authoring), standard APIs for integrating XHTML browsers on the mobile device, and so on.

13.6 Conclusion

Multimodal user interaction is very natural and is about to become a common part of our lives. Systems like our Plus V platform demonstrate conclusively that multimodal technology is practical, fast, and efficient even on older mobile data networks. Speech recognition has advanced to the point where complex and commercially important applications like mobile voice search, voice media search, and voice to SMS and email transcription can be implemented.

Commercial interest from companies like Google, Microsoft, and Nuance is very high and focused in the area of *multimodal local search*. It seems inevitable that Google will merge their new 1.800.GOOG411 voice directory assistance application in with their visual Google Local Mobile. Microsoft paid $800 million in early 2007 to acquire TellMe for their deep experience in voice directory assistance and driving directions. Nuance has acquired at least two companies with multimodal capabilities, Lobby7 and Mobile Voice Control, and acquired BeVocal for their application hosting capability. Japanese mobile operator KDDI deployed the EZ Navi Walk pedestrian navigation multimodal application (with DSR) in late 2006. Other players like Yahoo, PromptU, V-Enable, Kirusa, and VoiceBox are entering this arena. All of these are deploying distributed multimodal architectures.

This wide range of proprietary architectures will inform standards efforts at the W3C and elsewhere. Multimodal interaction will remain a fruitful area of research, especially as other innovative modalities are developed.

References

Atkins, D., Ball, T., Baran, T., Benedikt, M., Cox, K., Ladd, D., Mataga, P., Puchol, C., Ramming, J.C., Rehor, K., and Tuckey, C. (1997) Mawl: Integrated web and telephone service creation. *Bell Labs Technical Journal*, 2(1), pp. 19–35.

Auburn, R. (2007) Voice browser call control: CCXML version 1.0, W3C Working Draft, http://www.w3.org/TR/ccxml/

Axelsson, J., Cross, C., Ferrans, J., McCobb, G., Raman, T., and Wilson, L. (2004) XHTML+Voice Profile 1.2, VoiceXML Forum, March 2004, http://www.voicexml.org/specs/multimodal/x+v/12/spec.html

Boyer, L., Danielsen, P., Ferrans, J., Karam, G., Ladd, D., Lucas, B., and Rehor, K. (2000) Voice Extensible Markup Language (VoiceXML) version 1.0, VoiceXML Forum.

Bryant, R. (2007) Data-intensive supercomputing: The case for DISC, CMU Technical Report CMU-CS-07-128. May 10, 2007.

Burke, D. and McGlashan, S. (2006) Video interactive services with VoiceXML. *VoiceXML Review*, 6(2), March/April 2006, http://www.voicexml.org/Review/Mar2006/features/video_interactive_services.html

Delaney, B., Simunic, T., and Jayant, N. (2005) Energy-aware distributed speech recognition for wireless mobile devices. *IEEE Design and Test of Computers*, 22(1), pp. 39–49.

Deng, L. and Huang, X. (2004) Challenges in adopting speech recognition. *CACM*, 47(1), pp. 69–75.

Engelsma, J. and Cross, C. (2007) Distributed multimodal synchronization protocol, IETF Internet Draft, (Work in Progress), January 2007.

Engelsma, J. and Ferrans, J. (2007) Bypassing bluetooth device discovery using a multimodal user interface, In *Proceedings of the 4th Annual International Conference on Mobile and Ubiquitous Systems: Computing, Networking and Services* (Mobiquitous 2007), Philadelphia, PA.

Ferrans, J. (2003) The Motorola VoxGateway, lessons learned. *VoiceXML Review*, 3(4), July/August 2003, http://www.voicexmlreview.org/Jul2003.

Harel, D. (1987) Statecharts: A visual formalism for complex systems. *Science Computer Programming*, 8, pp. 231–274.

Kamvar, M. and Baluja, S. (2005) A large scale study of wireless search behavior: Google Mobile Search. In *Proceedings of ACM SIGCHI Conference on Human Factors in Computing Systems* (CHI 2005), pp. 701–709.

Kennedy, N. (2005) Igor Jablokov interview on multimodal search, October 16, 2005, http://www.niallkennedy.com/blog/archives/2005/10/igor_jablokov_interview_on_mul.html

Ladd, D., Hay, M., McClaughrey, P., and Ferrans, J. (1999) VoxML 1.1 Language Reference, http://www.w3.org/Voice/1999/VoxML.pdf

Maes, S. and Saraswat, V. (2003) Multimodal interaction requirements, W3C Note, http://www.w3.org /TR/mmi-reqs

McGlashan, S., Burnett, D., Carter, J., Danielsen, P., Ferrans, J., Hunt, A., Lucas, B., Porter, B., Rehor, K., and Tryphonas, S. (2004) Voice Extensible Markup Language (VoiceXML) version 2.0, W3C Recommendation, http://www.w3.org/TR/voicexml20

Neurosky (2007) http://www.neurosky.com

Open Mobile Alliance (2006) OMA multimodal and multi-device enabler architecture, OMA-AD-MMMD-V1_0-20061011-D, October 2006, http://member.openmobilealliance.org/ftp/Public_documents/BT/MAE/Permanent_documents/OMA-AD-MMMD-V1_0-20061011- D.zip

Oviatt, S., (2000) Taming recognition errors with a multimodal interface. *CACM*, 43(9), pp. 45–51.

Pearce, D. (2000) Enabling new speech driven services for mobile devices: An overview of the ETSI standards activities for distributed speech recognition front-ends. In *Proceedings of Applied Voice Input/Output Society Conference* (AVIOS 2000), San Jose, CA.

Pearce, D. (2004) Robustness to transmission channel—The DSR approach. In Proceedings COST278 & ISCA Research Workshop on Robustness Issues in Conversational Interaction.

Pearce, D., Engelsma, J., Ferrans, J., and Johnson, J. (2005) An architecture for seamless access to distributed multimodal services. In *Proceedings of 9th European Conference on Speech Communication and Technology* (Interspeech 2005), pp. 2845–2848.

Pearce, M. (2002) Pearce principle, private communication, January 2002.

Raggett, D. (1999) Introduction to TalkML, http://www.w3.org/Voice/TalkML/

Rosenberg, J., Schulzrinne, H., Camarillo, G., Johnston, A., Peterson, J., Sparks, R., Handley, M., and Schooler, E. (2002) SIP: Session Initiation Protocol. IETF RFC 3261, June 2002, http://www.ietf.org/ rfc/rfc3261.txt

Shanmugham, P., Monaco, P., and Eberman, B. (2006) A media resource control protocol (MRCP). IETF RFC 4463, April 2006, http://www.rfc-editor.org/rfc/rfc4463.txt

Suhm, B., Myers, B., and Waibel, A. (2001) Multimodal error correction for speech interfaces. *ACM Transactions on Computer-Human Interaction*, 8(1), pp. 60–98, March 2001.

Sutherland, I. and Danielsen, P. (2006) VoiceXML and voice-over-IP. *VoiceXML Review*, 6(3), September/October 2006. http://www.voicexml.org/Review/Oct2006/features/voip.html

Zyda, M., Thukral, D., Jakatdar, S., Engelsma, J., Ferrans, J., Hans, M., Shi, L., Kitson, F., and Vasudevan, V. (2007) Educating the next generation of mobile game developers. *IEEE Computer Graphics and Applications*, 27(2), pp. 95–96.

14

Speech Recognition in Mobile Phones

Imre Varga and Imre Kiss

Abstract. Speech input implemented in voice user interface (voice UI) plays an important role in enhancing the usability of small portable devices, such as mobile phones. In these devices more traditional ways of interaction (e.g. keyboard and display) are limited by small size, battery life and cost. Speech is considered as a natural way of interaction for man-machine interfaces. After decades of research and development, voice UIs are becoming widely deployed and accepted in commercial applications. It is expected that the global proliferation of embedded devices will further strengthen this trend in the coming years. A core technology enabler of voice UIs is automatic speech recognition (ASR). Example applications in mobile phones relying on embedded ASR are name dialling, phone book search, command-and-control and more recently large vocabulary dictation. In the mobile context several technological challenges have to be overcome concerning ambient noise in the environment, constraints of available hardware platforms and cost limitations, and necessity for wide language coverage. In addition, mobile ASR systems need to achieve a virtually perfect performance level for user acceptance. This chapter reviews the application of embedded ASR in mobile phones, and describes specific issues related to language development, noise robustness and embedded implementation and platforms. Several practical solutions are presented throughout the chapter with supporting experimental results.

14.1 Introduction

As in virtually every area, manufacturers of mobile phones are interested to enrich their product portfolio for offering added value to end users. This includes additional features as well as improving existing ones. For this, clear user benefit is balanced with additional costs on the manufacturing side. In certain market segments end users may refuse to accept an increase in price even if the improvement of the feature set makes the product much more attractive.

ASR is considered as a comfortable input modality of man-machine-interfaces. Meeting the expectation of end users fully is the target that we mean by the term natural man-machine-interface.

Some typical applications covered by ASR are supported by other means already. Indeed, speech input is an alternative method of user interface in mobile phones which compares and measures against existing methods like keypad or joystick. That is the main reason why it is not obvious how to implement ASR technology in a generally accepted and successful way in consumer products like mobile phones.

Attracting end users also includes meeting the user expectation of virtually perfect accuracy of ASR—they expect a similar level of perfection to a human. ASR demonstrated significant advances during the past decade and achieved excel-ent performance in certain application areas. However, it is still has not yet reached the level of human performance.

Speech input in mobile phones seems especially attractive in combination with other features, e.g. hands-free operation. Various factors increase the importance of hands free. One of them is the introduction of new, mobile multimedia applications: for example, video telephony requires hands-free mode. A further aspect is the use of mobile phones in cars, which has been recognized as dangerous due to conventional methods of user interface. Violation of basic traffic safety requirements motivated many countries to prohibit the use of mobile phones by law while driving.

On the other hand, the typical acoustic environment when using mobile phones, especially in hands-free operation, makes it much more difficult to achieve high ASR accuracy. Indeed, noise robustness in adverse conditions is one of the key issues of designing ASR for mobile phones.

Some further specialities of ASR in mobile phones are important to mention as well. Miniaturization resulted in keypads shrunk in size making the role of speech input more important. Terminals without any keypad may stand at the end of an evolution path where voice control is the only method of user control. Support of multiple languages is needed in mobile phones. Cost sensitivity represents a further important aspect in consumer product implementation. This includes hardware cost, such as fixed-point DSP and memory and implementation cost components as part of the unit end price.

Based on the elaboration above, we can state the challenge we are facing is to achieve a high quality (virtually error-free) ASR under adverse conditions at virtu-ally no extra costs for the user, competing with existing and already accepted user interface techniques—all at the same time. Even though this is an extremely tall order, in certain practical applications this challenge can be coped with successfully (Varga et al. 2002).

14.2 Applications of Speech Recognition for Mobile Phones

The various applications of speech recognition in mobile phones make the handling of the devices more user-friendly. First we address the basic functionalities.

The *name dialling* feature seems very useful since it supports the basic function-ality of a mobile phone, i.e., to place a call. After activation of the function, the user says the name of the person to be called. This implies immediate action. For name dialling, a pre-defined register of names with associated phone numbers (contact database) is needed. A less user-friendly (although simpler) method is digit dialling where the user must have the phone number of the person to be called in mind and speaks the digits one after the other. A more comfortable variant of digit dialling is natural number dialling (22). Name and digit dialling can be implemented directly in the phone or in a car kit associated with the phone.

Command-and-control improves the user experience by flattening complex menu structures. This makes the multi-step approach of navigation by keypad or joystick super-fluous. The user just inputs the desired action which is passed to an interpreter causing the action performed. Voice control is especially attractive in combination with hands-free operation.

Speech-to-text or dictation is fundamentally different in scope since this functionality is not basic to mobile phones originally. Dictation systems exist for a long time for desktop computers and their performance is continuously improving. However, these systems have been mostly successful in applications where dictation has already been an established practice, such as legal or medical domains (Fenn 2005). For general-purpose text entry ASR systems are more likely to succeed in the mobile environment, where there is a stronger motivation for users to adopt the new technology due to cumbersome traditional input mechanisms. ASR for dictation may be fully implemented in the network with speech transmitted over the wireless network by usual transmission techniques, or be partly located in the mobile phone and partly in the network. The latter approach is usually referred to as distributed speech recognition (DSR). For more details see Chap. 5 of this book. Alternatively, large vocabulary dictation systems can also be implemented in mobile phones as mobile computing platforms become more and more powerful.

Next, we review the basic technologies relevant for the above mentioned applications. *Isolated word recognition* means the capability of recognizing a single word (the command). Typically, isolated word recognition is useful for name dialling and command-and-control applications; however, it results in a rather artificial (machine-like) speaking style. The complexity is rather low, both in terms of algorithmic processing power and vocabulary size (below 100 in most cases).

Keyword spotting allows for a much more user-friendly operation because the user is not required to speak isolated words anymore. The speaker may speak natural phrases which contain dedicated keyword(s), the actual command(s). The speech recognizer separates the useful information (keyword) from the non-useful information (classified as garbage). The vocabulary size can be kept still restricted as with isolated word recognition, up to 100 words.

Keyword spotting is hence a kind of connected word recognition. The term *connected word recognition* stands in contrast to isolated word recognition. It refers to a technique which allows the speaker to speak several keywords in a connected manner. Connected word recognition greatly improves the value of the user interface feature. The difficulty is that words are pronounced differently when connected than when separated which causes an increase of the algorithmic complexity.

Continuous speech recognition allows natural speaking style by requiring no pauses between words. The difference to keyword spotting from technical point of view is the vocabulary size. Large vocabulary makes dictation possible. On the other hand, the computing complexity and memory requirements increase significantly.

A further dimension to consider is the distinction between *speaker-dependent*, *speaker-independent* and *speaker-adapted* systems. In a speaker-dependent system, the user may include any new word in the vocabulary at the expense of training. In this way, the language and pronunciation behaviour of the speaker are automatically taken into account. Speaker-dependent recognition is independent of languages, dialects, and pronunciations. Examples of speaker-dependent ASR are name dialling applications. Dynamic Time Warping (DTW) (Ning et al. 2002) and Hidden Markov Model (HMM) based (Laurila 1997) speaker dependent name diallers have been widely used in mobiles.

The need for training in speaker-dependent systems impacts their usability greatly. Users may not be willing to train the system, or may forget the voice tags trained. Speaker-independent systems overcome this difficulty by pre-training speech models on large amounts of training data and they are predominantly HMM-based. To cope with various languages, dialects, speaker behaviours, high efforts are spent in the algorithmic design and pre-training of HMM-based speaker-independent systems. In order to achieve a good performance over a wide range of speaker variations, databases in various languages containing a large set of speech samples taken from different speakers in different conditions are needed for pre-training (Höge 2000).

The combination of speaker-independent and speaker-dependent recognizers leverages the benefits of both systems: user-friendliness due to pre-trained vocabulary and high performance due to user-trained additions to the vocabulary. Furthermore, advanced speaker independent systems support on-the-fly adaptation of the acoustic models. These speaker-adaptive systems maximize the accuracy of the system for user and environment variations, while maintaining a low level of user interaction. Adaptation is typically carried out during the normal course of use, in a transparent manner to the user. Table 14.1 illustrates some typical mobile ASR applications in function of speech recognizer capabilities.

In the context of ASR application in mobile phones we emphasize that voice UIs provide convenience and ease of use as an alternative to small sized keypad and display. In addition, in developing regions, with low rate of literacy among the population, voice (and graphical) UIs lower the usage entry barrier for people. Many

Table 14.1 Typical applications in terms of speech recognition capabilities

	Isolated word	Keyword spotting	Continuous
Speaker independent (SI)	Basic digit or natural number dialling, basic command-and-control	Flexible command-and-control	SMS and/or Email dictation
Speaker dependent (SD)	Basic name dialling	High accuracy voice activation	n.a.
Mix of SI and SD and/or speaker adaptive	Advanced digit and name (or natural number and name) dialling	Flexible digit dialling, name dialling, and command-and-control	High accuracy SMS and/or email dictation

semi-literate or illiterate users' first experience with voice communication and/or the Internet may be a portable mobile device. Since small portable devices, such as mobile phones and multimedia computers are produced in large volumes for a global market, it is essential to offer them with a wide set of languages.

There are two important factors that make this challenging. First, as already mentioned, speech recognition systems rely on statistical techniques that usually require large training corpora for providing sufficient performance. This includes both textual and acoustic databases. For some languages the necessary language resources are readily available. For some others, they may be difficult to find or collect. Second, as in practical configurations a set of languages needs to be supported by a mobile device, several languages have to coexist in the limited memory space. Therefore, a suitably compact representation has to be developed—we address these issues next.

14.3 Multilinguality and Language Support

In the context of this paragraph, we refer to multilinguality as simultaneous support for several languages for ASR and TTS. Multilingual systems typically also possess the capability of easy adaptation to unseen (or scarcely resourced) languages (Schultz and Waibel 2000, 2001).

14.3.1 Multilingual Speaker Independent Name Dialing

In this section we discuss multilinguality in the context of a typical embedded application: speaker independent name dialling. A typical architecture for a multilingual isolated word speech recognition engine is shown in Fig. 14.1. The system consists of the following modules: text-based language identification (LID), pronunciation modelling or text-to-phoneme conversion (T2P), acoustic modelling (AM) and isolated word decoder (DEC).

Written entries (e.g., name tags from a contact database) are first fed into language identification, which assigns the most likely languages to the word in question. Language identification may be based on e.g., statistical models using neural networks or N-gram probabilities. Normally the character set used in the entries also limits the possible language choices.

Next, the words and language tags are inputted to the T2P module that produces the respective pronunciations for the word. To account for possible LID errors, as well as the possible ambiguity of some names (e.g., Peter may be an English or Swedish name) several pronunciation variants are provided for a given word. The methods applicable for T2P depend heavily on the language in question. Simple pronunciation rules can be used for regular languages, while more sophisticated models, i.e., decision trees (Quinlan 1993), neural networks (Sejnowski and Rosenberg 1987; McCulloch et al. 1987; Häkkinen et al. 2000) or finite-state transducers (Caseiro et al. 2002) can be applied for less regular ones.

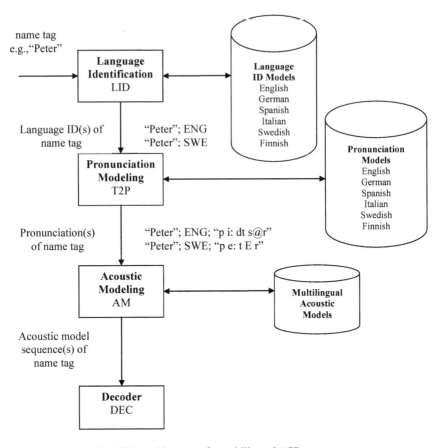

Fig. 14.1 Architecture of a multilingual ASR system

Finally, the complete set of pronunciations (all words, all pronunciation variants) and the sub-word acoustic models are used to build a recognition network. To save space for acoustic models, as well as to provide robustness for languages with less or no training data, acoustic models are trained in the following manner. First, overlaps between language specific phoneme sets are identified. This can be done in a knowledge-based manner (e.g., based on the IPA phoneme definitions), a data driven manner by clustering phonemes based on the statistical properties of their realizations in the acoustic model and database, or as a third option, a combination of these methods can be used. It is also a good practice to verify the resulting phoneme set by recognition experiments on a test database and compare the performance to a mono-lingual setup. In many cases, some problematic phoneme combinations can be identified this way.

The task of the decoder is to match the incoming sequence of features (representing the words uttered by a speaker) to the recognition network. Most likely word hypotheses can then be displayed to the user for selection or confirmation.

Usually the number of languages a system has to support depends on the market a product variant is produced for. In the case of name dialling, certain pronunciation variants of names can be eventually eliminated from the recognition network based on usage statistics. If the owner of the mobile device pronounces an ambiguous word consistently, the system may be able to identify which pronunciation variant is the most likely and discard the rest.

As the number of languages grows, the benefits of multilingual phoneme set become more and more dominant (Fischer et al. 2000). In addition, when no acoustic data is available for some language, but a knowledge-based mapping can be created between the phonemes of the language and the phoneme-set in the multilingual system the language coverage can be easily extended for the unseen language.

Figure 14.2 illustrates the performance of a practical multilingual ASR system (Kiss and Vasilache 2002). The name dialling system supports 25 languages. The speech data for multilingual acoustic model training (using a common multilingual phoneme set) consisted of Danish, Dutch, English, Finnish, French, German, Portuguese and Spanish material. Altogether 11 databases were used for training. Depending on the language, the material contained natural sentences, phonetically rich words and command words. For the rest of the languages no acoustic training data was available. Only clean speech was used in the training phase. The quantized (Vasilache 2000) acoustic models were 8-mixture monophone HMMs with 76 phonemes as defined in our in-house multilingual phoneme set.

A small vocabulary, isolated word recognition task was chosen for the evaluation. For each of the languages, a 120-word lexicon was defined containing both native and non-native name entries. The recognition tests were carried out using an in-house isolated-word database comprising of 1,000–8,000 test utterances from several speakers for each of the languages. The actual number of utterances depended on the language. The performance evaluation was carried out both under clean and noisy operating conditions. The noisy test data was obtained by artificially mixing noise to the clean test utterances. Four kinds of noise (car, café, car noise with background speech and/or music, airport hall) were used at randomly chosen SNRs between +5 dB and +20 dB. The SNR distribution was set to be uniform. To reduce the effects of speaker, language, pronunciation and environmental mismatches, on-line, supervised, maximum a posteriori (MAP) adaptation was applied to the acoustic models as described in (Vasilache and Viikki 2001).

The average recognition accuracies are 93.61% (SI) and 97.01% (SA) in clean, 85.73% (SI) and 93.01% (SA) in noise. The figure also shows that the lack of native training data did not necessarily imply worse recognition accuracy. Some languages, e.g., Romanian and Slovak even outperformed e.g., Danish and Dutch for which native training data was available. For this small vocabulary recognition application, the 25-lingual phoneme set proved to be well performing and robust solution.

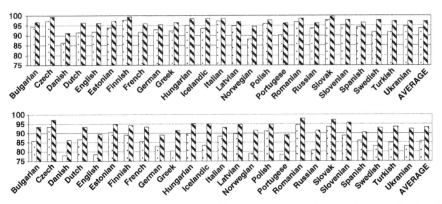

Fig. 14.2 Multilingual name dialling results with 25 languages in clean environment (top) and noise (bottom). The bars reflect the accuracy for each language for speaker independent (SI) and speaker adapted (SA) models

14.3.2 Multilinguality in Other ASR Applications

In this section we address more complex ASR tasks that are attractive for small portable devices. More sophisticated command-and-control type of applications include voice control for music player or radio ('play song X', 'tune to station Y'); or entering calendar entries or natural numbers by voice ('meeting on Monday, 22nd of May at 8 AM', 'two hundred and thirty five'). Most of these applications can be efficiently realized by an ASR engine using a recognition network defined by compact context-free grammars. Multilinguality in these cases may require language-specific variants of the grammars. Extending a system to new languages is relatively straightforward.

One of the most demanding ASR applications for embedded systems is large vocabulary dictation. These systems can significantly improve the ease and speed of text input on devices with limited (small sized qwerty, or only numerical) keypads. On contrary to simple name dialling and command and control type of applications, dictation systems require statistical language models (in many cases in the form of statistical N-grams). There are two difficulties that arise from language modelling. First, the size of these language models in most cases is quite significant (the other significant factor usually being the context-dependent acoustic model set). Second, languages are different and they may call for different types of language models. For example, highly inflecting languages, such as Hungarian, Finnish, Turkish are best represented by morpheme-based language models with longer context size, while analytic languages, such as English can be well described by word models with relatively short contexts.

14.3.3 Language Resources

As we discussed above, language resources play an important role in state-of-the-art ASR systems. Automatic text-to-phoneme mapping requires large pronunciation lexica, acoustic modelling and language modelling require acoustic and text data. In various Frame Programs of the European Community there have been several projects targeting at language resource creation for ASR applications. The *SpeechDat-Car* (http://www.speechdat.org/SP-CAR/) project partners collected large multi-channel speech databases in automotive environments for several languages. The *Speecon* (http://www.speechdat.org/speecon/index.html) project focused on collecting linguistic data for speech recogniser training in the consumer devices. The *LC-STAR* and *LC-STAR II* projects (www.lc-star.com) created lexica and text corpora. Recently, it has been demonstrated in an increasing number of application that language models can significantly be improved by using freely available text resources from the World Wide Web (Bulyko et al. 2003; Sarikaya et al. 2005; Sethy et al. 2006; Sethy et al. 2007).

14.4 Noise Robustness

Robustness of speech recognizers in mobile phones is a key requirement to achieve a high recognition rate needed for user satisfaction. The term 'robustness' in general reflects the desired high-quality system behaviour in adverse conditions which include the presence of environmental and background noise, transmission channel characteristics, speaker specific variations (Lombard reflex, male/female/child, spontaneous speech, dialects etc.).

In mobile phones, use of single channel techniques seems feasible. In car environment, microphone array has found to be an effective means to perform noise reduction by directional characteristics. The combination of microphone array with beamforming signal processing proves very effective (Balan et al. 2004). In case of severe disturbances in car, speech recognition rate is so low that the application of noise reduction is mandatory. The improvement of speech recognition rate is substantial and varies as a function of input SNR.

A proven method to cope with adverse conditions is to follow a multi-step approach including the use of robust HMM models, feature extraction, and noise reduction (Varga et al. 2002).

14.4.1 Robust HMM Models

Robust HMM models are an effective means to capture the variability. Robustness is achieved when the emission probabilities observed from the real speech data come close to the emission probabilities incorporated in the used HMM models. In order to reduce the probability differences and hence increasing recognition rate, training by appropriate databases (Höge 2000; SpeechDat 2000) is essential in order to produce robust HMM models.

14.4.2 Feature Extraction

Feature extraction algorithms in the front-end are implemented to adapt to varying channel characteristics, various background noises, and to extract tonal features as well. A Maximum Likelihood channel adaptation algorithm proved to be efficient

(Varga et al. 2002). In the enhanced MFCC (Mel Frequency Cepstral Coefficients) analysis the parameters are aug-mented by first and second order derivatives. Optionally two tonal features (voicing parameter and pitch value) can be added to the feature set. Two parameter sets resulting from the analysis of two adjacent frames are transformed via a Linear Discriminant Analysis (LDA) leading to a 24-dimensional feature vector. The main purpose of the LDA is to reduce the dimensionality of the vector to achieve a memory efficient solution although it is effective in improving noise robustness as well (Westphal 1997).

14.4.3 Noise Reduction

Among different kinds of noises, non-stationary noises are the most difficult to compensate for. Examples of non-stationary noises are background speech in a cafeteria, music and street noise. Their spectral and temporal properties overlap with speech and hence it is difficult to separate the speech from noise.

Spectral attenuation and subtraction algorithms have proven as effective means in reducing acoustic noise. These schemes regard noise as an additive uncorrelated component over clean speech in the captured signal. Noise reduction forms a time-varying filter whose parameters are calculated from estimated short-term signal and noise spectrum. Various versions of the noise reduction algorithms were proposed.

In the method of cascading of two stages in combination with a *frame dropping* scheme (Andrassy et al. 2001), in the first stage, a Wiener filter is calculated for every spectral bin as the attenuation function. For the second stage of spectral attenuation, the noise power spectrum is estimated by the minima of the smoothed power spectrum within a moving interval having the advantage that no explicit detection of non-speech segments is needed. For every frequency bin the noise estimate is subtracted of the noisy speech signal. In both stages the noise estimate is weighted by an oversubtraction factor pending on the frequency and on the signal to noise ratio in order to reflect the uncertainty of the noise estimate. To prevent the thus processed signal from being negative flooring is employed. The channel compensation reduces signal changes due to the different characteristics of the transmission channels. The signal distortion caused by the transmission channel is assumed to lead to an offset in the cepstral domain. This offset is estimated using a Maximum Likelihood Estimator. Finally a frame drop algorithm is contained in the front-end in order to reduce insertion errors by dropping non-speech frames. The speech/non-speech decision is based on an energy criterion.

Next, we describe a noise reduction algorithm called *spectral subtraction* in the modulated spectral domain in more detail. Modulation spectral enhancement approaches like RASTA and high-pass filtering have been effective in reducing channel distortions. As shown in (Sivadas 2006), spectral subtraction can also be successfully applied in the modulation spectral domain.

Linguistic information in speech signals is concentrated between 0.4 and 20 Hz in modulation frequency domain (Houtgast 1989), the region around 4 Hz being the most significant. Many of the popular temporal filtering algorithms such as

RASTA (Hermansky, Morgan 1994), dynamic features (Furui 1986) suppress the modulation frequencies outside the required modulation frequency domain. These techniques assume that the channel distor-tion is predominantly linear time invariant and convolutive and that additive noise is minimal. Channel distortion becoming additive in the log spectral domain, is removed by linear filtering. The most common approach to minimize the effect of both convolutional and additive noises is to cas-cade the algorithms to remove each of them. First, spectral subtraction is applied to reduce the additive noise followed by RASTA filtering or some other bandpass filtering to remove the convolutional noise.

For effective suppression of non-stationary noise, spectral subtraction needs special improvement like noise spectrum update during non-speech intervals. An alternative noise compensation approach tackles the effect of spectrally overlap-ping non-stationary noises. Modulation spectrum gives the temporal spectral charac-teristics of the signal within each (mel) frequency band. By applying the Wiener filter to the time trajectories of each mel frequency filter output it is possible to alleviate the effect of non-stationary noises. A possible use case for this is voice activated name dialling in mobile phones. The user may be trying to dial a number in a crowded cafeteria or in a subway. Due to the non-stationary nature of the back-ground noise, the effectiveness of conventional noise robustness approaches is limited.

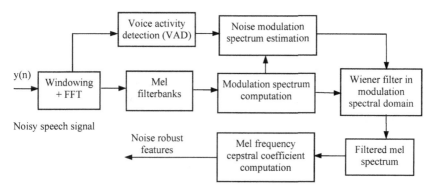

Fig. 14.3 Additive noise reduction by filtering in modulation spectral domain

Figure 14.3 shows the block diagram of the noise reduction scheme. First, Short Time Fourier Transform (STFT) of the noisy speech is computed. Triangular mel filter weights are applied to the magnitude spectrum to obtain a mel-spectrogram. A voice activity detector (VAD) is used to keep track of non-speech regions. Let $x(k,n)$, $y(k,n)$ and $w(k,n)$ represent instantaneous energy of clean speech, noisy speech and noise respectively at frame index n for mel-frequency bin k.

Assuming that speech and noise are uncorrelated, the spectral energy of speech corrupted by additive noise is given by

$$y(k,n) = x(k,n) + w(k,n) .$$ (14.1)

Modulation spectrum for each mel filter bank output is computed using N point FFT window for every sample.

$$X(k,\Theta) = Y(k,\Theta) + W(k,\Theta),$$ (14.2)

where Θ is the modulation frequency. Subtractive noise reduction algorithms can be expressed as (Virag 1999)

$$|X(k,\Theta)| = G(k,\Theta) \cdot |Y(k,\Theta)| \quad with \ 0 \le G(k,\Theta) \le 1.$$ (14.3)

Dropping the mel-frequency bin index k, the gain function can be written as

$$G(\Theta) = G\left[SNR_{post}(\Theta)\right]$$

$$= \begin{cases} \left(1 - \alpha \cdot \left[\frac{|\hat{W}(\Theta)|}{|Y(\Theta)|}\right]^{\gamma 1}\right)^{\gamma 2} \\[3mm] if \left[\frac{|\hat{W}(\Theta)|}{|Y(\Theta)|}\right]^{\gamma 1} < \frac{1}{\alpha + \beta} \\[3mm] \left(\beta \cdot \left[\frac{|\hat{W}(\Theta)|}{|Y(\Theta)|}\right]^{\gamma 1}\right)^{\gamma 2} \end{cases}$$ (14.4)

where, $|\hat{w}(\Theta)|$ is the modulation spectral magnitude of noise estimated during non-speech segments, α is the oversubtraction factor, β is the noise floor and exponent $\gamma = \gamma 1 = \frac{1}{\gamma 2}$ determines the rate of change of gain $G(\Theta)$ from 0 to 1. The *a posteriori* Signal to Noise Ratio (SNR) is given by

$$SNR_{post}(\Theta) = \frac{|Y(\Theta)|^2}{|\hat{W}(\Theta)|^2}.$$ (14.5)

Modulation spectral component of clean speech is estimated as

$$\hat{X}(k,\Theta) = |\hat{X}(k,\Theta)| e^{j \arg(Y(k,\Theta))}.$$ (14.6)

The mel-spectral energy trajectory of clean speech is given by

$$\hat{x}(k,n) = FFT^{-1}(\hat{X}(k,\Theta)).$$ (14.7)

The algorithms were tested on a multilingual small vocabulary isolated word recognition task. Test set comprises of ~40,000 words from seven European languages: Finnish, Swedish, German, English, Danish, Icelandic and Norwegian. The size of vocabulary per language was ~120.

The baseline front-end used in the experiments was based on 13 FFT-derived Mel-frequency cepstral coefficients (MFCC) and their first and second order derivatives (39 coefficients in total). Recursive mean removal was applied on all components of the resulting feature vectors, and the variance of all the components was normalized to unity (Viikki et al. 1998). A generalized Wiener filter is applied to the magnitude of the modulation spectrum. The mel filter bank output is reconstructed from the filtered magnitude spectrum and phase of the noisy speech modulation spectrum using overlap-add method. Modulation spectrum of the noise is computed during the non-speech segments using a VAD.

Table 14.2 Performance of noise reduction front-ends on an isolated word recognition task. The numbers represent Word Error Rate (WER) in percentage

		Car	Cafe	Street	Clean
Baseline		8.79	13.56	12.27	3.67
Spectral subtraction		6.77	9.90	9.12	3.69
Modulation Spectral Wiener filter	FFT = 16	6.79	9.34	8.39	3.72
	FFT = 32	6.05	8.44	7.88	3.69
	FFT = 64	6.02	8.29	7.81	3.68

The acoustic model set consists of three state monophone models with eight Gaussian densities per state. The model sets were trained on an in-house training set containing clean speech data from various European languages. Both sets contained a total of 75 multilingual phone models that were used to model the basic acoustic units of the seven European languages mentioned above.

The Word Error Rates (WER) of Wiener filter based front-end with noise compensation in spectral domain and in modulation spectral domain are tabulated in Table 14.2. Noise is artificially added to the clean utterances at Signal to Noise Ratios (SNR) ranging from 5 dB to 20 dB in steps of 5 dB. The non-stationary noises are cafeteria and street noise. Car noise is the stationary one. Results in Table 14.2 are the average WER for the 5 dB to 20 dB SNR conditions.

The highest detectable modulation frequency is half the analysis frame rate. In our experiments, the frame rate was kept at 100 Hz (= 10 ms frame shift), resulting in maximum modulation frequency of 50 Hz. The effectiveness of the modulation frequency noise suppression approach depends on the resolution of the FFT to obtain the modulation spectrum. With 50 Hz Nyquist frequency, FFT length of 16 provides resolution of ~6 Hz. Since we want to resolve modulation frequency components less than 4 Hz, a longer analysis window is required. From the table it can be seen that longer FFT windows give better noise robustness.

Comparing the relative improvements obtained using spectral subtraction and modulation spectral Wiener filter for different types of noises, it can be seen that the improvements in the case of non-stationary noises is higher in the case of modulation spectral filtering.

14.5 Footprint and Complexity Reduction

Small memory footprint and low computational complexity are essential for any embedded ASR system. The reasons are several-fold: cost reduction, competing applications for limited resources and preserving *battery life* in portable devices.

Concerning memory footprint, we differentiate between static (Flash) and dynamic (RAM) memory. In most embedded ASR systems, the Flash memory footprint is largely determined by three factors: the size of acoustic models (AM), the size of language model (LM) and the size of pronunciation lexicon (Lex) and/or automatic text-to-phoneme model.

14.5.1 Footprint Reduction of Acoustic Models

In speaker independent, sub-word based systems acoustic models are usually context-independent or context-dependent phonemes. These models aim at capturing the statistical properties of phoneme realizations (phones) in real-life speech. There are two major options for reduction of AM size: to reduce the number of parameters in the model set, and to reduce the memory necessary to represent model parameters. Fortunately, the first approach coincides with the goals of robust model training, i.e. the amount of training data available usually limits the number of parameters in the model set that can be estimated in a reliable manner.

There are two widely used parameter tying methods to reduce the size of acoustic models while retaining as high modelling accuracy as possible. The first method is called decision tree-based state-tying (Young et al. 2002), where states sharing some common properties in the model set are pooled together, so a common pool of training data can be used to estimate state parameters more reliably. In most cases, state-tying is performed in a knowledge and data-driven manner, whereby pre-defined set of questions (phonetic questions) are used to train a binary decision tree. At each branching point, the tree is grown (phonetic question is selected) in a manner to maximize the likelihood of the training data given the final set of state tyings.

The second method is called density tying. It is usually done in a completely data-drive manner, and aims at reducing the number of acoustic densities in the model space by combining densities that are closer to each other (as defined by a suitably chosen distance metric) than a certain threshold. Density tying does not take into account the state structure in the model space, and can be used in combination with state-tying. In many practical cases, more compact AM models can be trained by using subsequently state and density tying, than either of these methods alone. As we shall see in Sect. 14.6, these techniques can be successfully applied to build a practical embedded dictation system.

The next large category of footprint reduction techniques consist of various quantization schemes aiming at representing model parameters at a reduced resolution to save memory.

In HMM-based systems the model parameters are density means, variances (assuming diagonal covariance matrix), and mixture weights. Usually the first two are considered more important, because in practice mixture weights contribute only little to the overall state emission probabilities. These density parameters can be quantized in several ways. Popular techniques include scalar quantization (Vasilache 2000), joint vector quantization of mean and variance, and subspace vector quantization (Bocchieri and Mak 1997). A comparison of these techniques is presented in (Leppänen and Kiss 2005). As we shall see later in this section, the quantization of model parameters can be combined with quantization of feature vectors to result in fast state emission probability computation.

14.5.2 Footprint Reduction of Language Models

For language modeling, we focus our attention to statistical N-gram language models as they are widely used in many practical systems (for the sake of simplicity we assume that the basic modeling unit in the LM is a word and the model is a back-off bi-gram):

$$p(i,j) = \begin{cases} (N(i,j)-D)/N(i) & \text{if } N(i,j) > t \\ b(i)p(j) & \text{otherwise} \end{cases} \tag{14.8}$$

where N(i; j) is the number of times word j follows word i and N(i) is the number of times that word i appears. Essentially, a small part of the available probability mass is deducted from the higher bi-gram counts and distributed amongst the infrequent bi-grams. This process is called discounting. When a bi-gram count falls below the threshold t, the bi-gram is backed-off to the unigram probability suitably scaled by a back-off weight in order to ensure that all bi-gram probabilities for a given history sum to one (Young et al. 2002). Bi-gram N-gram models need the following parameters to be stored: word pairs, bi-gram probabilities and back-off scaling factors. An LM can be considered as a sparse graph where vertices are words from the vocabulary and edges represent bi-grams. As such, it can be efficiently represented using adjacency lists.

For language model compression, the same two principles can be applied as we showed for acoustic models. First, the number of parameters in a language model can be effectively reduced (and the model be made more robust) by using e.g., entropy-based pruning schemes (Stolcke 1998).

LM model parameters can be represented in an efficient manner by using profile-based compression (Olsen and Oria 2006) and quantization. The idea behind profile-based com-pression is simple. When N-grams with the same history are ordered according to decreasing probability, the resulting probability profiles are remarkably similar. Some profiles may be identical (especially when the probability values are quantized), or they may be prefixes of longer profiles. Therefore, the same profile can be re-used to represent the probability distribution of N-grams for several different word histories, thereby saving memory.

Profiles in effect act as a codebook. Profiles themselves can also be compressed. As they tend to follow an exponential decay, non-uniform sampling can effectively be applied. Values in between samples can be interpolated. Table 14.3 shows the LM footprint reduction achieved for representing N-gram probabilities in a practical large vocabulary embedded dictation system. In the table Q represents the quantization levels for probability values (e.g., Q16 corresponds to 4-bit quantization), while S represents the parameter for non-uniform (logarithmic) sampling of profiles. So for i_k denoting the location of the kth sample in the profile:

$$i_0 = 1$$
$$i_k = i_{k-1} + \max\left(1, round\left(\log_{10}\left(i_{k-1}\right)/S\right)\right). \tag{14.9}$$

It can be seen that the memory footprint can be reduced 12-fold from 436 KB (S = 1,000, NoQ) to 36 KB (S = 0.5, Q16) with a minor loss in word accuracy (85.1% vs. 85.0%).

Another useful practical technique to reduce the footprint and improve the performance of language models in embedded systems is clustering. The use of semantic classes has been proposed in (Oria and Olsen 2006). In a large vocabulary (33Kwords) embedded dictation task for US English, the use of semantic classes reduced the model size by 16%, while at the same time also reduced the word error rate by 12% relatively.

Table 14.3 Number of profiles, average profile length, size of profile codebook and word accuracy for different quantization and profile compression settings (From Olsen et al. 2006, © 2006 IEEE)

	S = 1,000	S = 0.5	S = 0.1	S = 0.05	S = 0.01
No Q	4033/52	4017/10	3849/4	3814/3	3810/2
	436 kb	103 kb	51 kb	45 kb	39 kb
	85.1%	85.0%	84.6%	84.0%	79.5%
Q32	3108/64	2797/13	1279/6	905/5	456/3
	213 kb	48 kb	13 kb	8 kb	4 kb
	85.0%	85.0%	84.6%	84.1%	81.2%
Q16	2454/78	1773/16	566/8	344/6	151/4
	202 kb	36 kb	7 kb	4 kb	1 kb
	84.9%	85.0%	84.5%	84.1%	81.1%
Q8	937/164	479/30	154/12	90/9	42/5
	157 kb	17 kb	2.6 kb	1 kb	0.4 kb
	73.7%	84.7%	84.2%	83.7%	81.9%
Q4	297/356	119/61	44/22	26/13	14/5
	107 kb	8 kb	1 kb	0.5 kb	0.1 kb
	84.0%	83.7%	82.4%	81.8%	78.3%

14.5.3 Footprint Reduction of Pronunciation Lexicon

In addition to acoustic and language models, the large size of pronunciation lexica can also affect the footprint of embedded ASR system. In most cases, however, word labels (written form) and corresponding phoneme sequences (pronunciations) can be effectively compressed by relatively simple means. In a large vocabulary system, it is likely that several words in the lexicon have the same starting characters, or starting phonemes in their pronunciation. This property can be used to store written word labels (or pronunciations) in a tree structure. Words can share the common starting letters, and these prefixes need to be stored only once. The benefit of the tree representation is that it lends itself to efficient search.

In addition to storing the pronunciation lexicon explicitly, automatic T2P methods, based on decision trees (Quinlan 1993), neural networks (Sejnowski and Rosenberg 1987; McCulloch et al. 1987; Häkkinen et al. 2000), or finite-state transducers (Caseiro et al. 2002) can also be applied.

14.5.4 Reduction of Computational Complexity in Embedded ASR Systems

Next we look into methods to reduce the computational complexity of embedded ASR algorithms. We divide these methods into two categories. The first category contains algorithms for efficiently computing state emission probabilities, while the second category focuses on efficient search in the recognition network.

In state emission probability computation, first we focus on continuous density HMMs (CDHMMs) with state densities consisting of a mixture of diagonal Gaussian densities. The logarithmic Gaussian density likelihoods are computed as follows

$$L_k = C_k - \frac{1}{2}\sum_{i=1}^{N} \frac{(x_i - \mu_{ki})^2}{\sigma_{ki}^2}, \qquad (14.10)$$

where L_k is the log-likelihood of the density k, x_i denotes the ith component of the feature vector, μ_{ki} and σ_{ki} stand for the ith mean and standard deviation component of density k. N denotes the total number of components in the feature vector. The additive constant C_k is given by

$$C_k = \log\left(\frac{1}{\prod_{i=1}^{N}\sqrt{2\pi\sigma_{ki}^2}}\right). \qquad (14.11)$$

Finally, the emission probability for one state is expressed as

$$B_s = \log\left(\sum_k \exp(W_{sk} + L_k)\right) \approx \max_k(W_{sk} + L_k), \qquad (14.12)$$

where W_{sk} is the mixture weight for density k in state s and the summation is performed for all mixture densities corresponding to s. In practice the log sum operation can be avoided by taking into account only the best scoring density in every state. This significantly reduces the state score computation without significant effect on recognition accuracy.

In (Kiss and Vasilache 2002) three methods for simplifying the computation of state emission probabilities of continuous density-based HMMs are proposed. Feature component masking, variable-rate partial likelihood update and density pruning all resulted in significant savings in the decoding complexity with marginal impact on the recognition performance. A combination of feature component masking and density pruning was evaluated in a small vocabulary, 25-lingual, speaker-independent, isolated word recognition system. With a computational complexity reduction of 62% compared to the baseline system, a marginal, 1.6%/6.5% relative error rate increase was obtained without/with on-line Maximum A-Posteriori (MAP) adaptation on the average in clean and noisy operating environments. The presented framework can also be extended to larger vocabulary systems.

An often used technique to speed up density score computation is Gaussian selection (Bocchieri 1993; Gales et al. 1999). The idea behind GS is to reduce the search space for density score computation by clustering the densities in the model space. During decoding, in a two-level GS setup, first the cluster centroids are matched to the incoming feature vector, and in the second step only members of the best matching clusters are used for score computation. The number of densities in the best clusters can be significantly less than in the entire model space. Usually densities are clustered into overlapping clusters (one density may belong to more than one cluster). This improves accuracy, but also increases the memory needed for storing the model. The reason is that storing the identities of densities belonging to a given cluster can be excessive when the overall density count in the model set is large. In (Leppänen and Kiss 2006) a novel Gaussian selection algorithm is proposed. It uses non-overlapping clusters, therefore cluster members can be identified by a starting and ending index in a linear array. The overhead to store the identity of cluster members is thus minimal. The scheme achieves 66% computational savings with a relative increase in word error rate (WER) of 4%. The GS scheme is also combined with frame rate reduction and feature masking provides further savings in computation. 75% (4% increase in WER) and 68% (3.5% increase in WER) savings were obtained by adding frame rate reduction and feature masking, respectively.

All of the above schemes can be applied to HMM models with continuous or discrete probability distributions. However, in case of discrete HMMs state emission probability computation can also speeded up significantly, if in addition to model parameters, feature vectors are also quantized. In the simple case of scalar model quantization (Vasilache 2000) with 5-bit allocated to mean and 3-bit allocated to variance (more precisely inverse standard deviation), any given pair of mean and inverse standard deviation component can take 256 different values. Feature vectors can usually be quantized at a rate of 3–4 bits per component, without any loss in recognition accuracy. In a system utilizing $5 + 3$ bit quantization for model parameters and 4-bit quantization for feature vector components state emission probabilities can be calculated as a simple lookup operation in a 256×16 table. The model parameters and the feature vector can be used to address the table, which stores pre-computed state emission probabilities.

14.5.5 Low Memory, Fast Decoding

A large network recognition network will have many nodes and one way to make a significant reduction in the computation needed is to only propagate tokens (Young et al. 1989) which have some chance of being amongst the eventual winners. This process is called token pruning (Young et al. 2002) and is widely applied in embedded ASR systems. It can be implemented at each time step by keeping a record of the best token overall and de-activating all tokens whose log probabilities fall more than a beam-width below the best. In certain cases, it may be necessary to limit the worst-case RAM memory use for the decoding network (not to risk running out of memory when several applications are used concurrently). In this case, a maximum limit can be set for the overall number of active tokens. In the beginning of decoding this token buffer is filled up, but when progressing in the utterance, the log-probability-based thresholding may limit the number of active tokens below the maximum level. At first it may seem risky to set a hard limit in the number of tokens for a decoding network. In practice, however, tree-structured networks prove quite robust to this, because in the initial phase of decoding tokens are active close to the root of the tree (shared prefixes of words), and when progressing further, only the most likely word candidates must be covered by active tokens.

Finally, a practical and important approach to reduce active RAM footprint of embedded ASR systems is to dynamically load large components of the system (e.g., lexicon or LM). The idea is similar to caching, whereby only the most actively used parts are kept in RAM memory, and less frequently used parts are stored in slower access Flash memory.

14.6 Platforms and an Example Application

In Table 14.4 we summarize the basic technical properties of some recent embedded platforms. Nokia's smartphones and N800 Internet tablet are ARM-based devices running around 220–330 MHz featuring 10–128 MB of SDRAM and 10–256 MB of NAND Flash memory that can be extended up to 1–4 GB using removable memory cards. The smartphones use Symbian OS, while the Internet tablet is Linux-based. Depending on the model and OS used, the free RAM memory available for user applications varies between ~4 and 22 MB for smartphones, and is ~112 MB for the Internet tablet. Internal Flash memory has 8.5–160 MB free space for applications on the smartphones and ~176 MB on the Internet tablet. ARM 9 (OMAP1710) supports only fixed-point arithmetics, while ARM11 (OMAP2420) has a built-in floatingpoint co-processor, making porting easier.

The ARM11 core also includes hardware acceleration for 3D graphics, but this feature is not easily used for ASR purposes. Both the OMAP 1710 and 2420 processors include a powerful DSP in addition to the ARM MCU. For easy application development (and portability) reasons, however, the DSP is not used by add-on ASR applications.

Table 14.4 Basic properties of example embedded portable platforms

Device	Nokia 6630	Nokia E60	Nokia N93	Nokia N95	Nokia N800	HP iPAQ HX2495	Phrase-lator P2	HP iPAQ H2750	HP iPAQ H3800
OS	Symbian OS	Symbian OS	Symbian OS	Symbian OS	Linux	Windows	Windows	Windows	Linux
SW platform	S60 2.6	S60 3.0	S60 3.2	S60 4.1	Internet Tablet OS 2007	Windows Mobile 5.0	Windows CE 3.0	Windows Mobile 2003 SE	Familiar v0.6.1
Processor	ARM926 (OMAP 1710)	ARM926 (OMAP 1710)	ARM11 (OMAP 2420)	ARM11 (OMAP 2420)	ARM11 (OMAP 2420)	Intel Xscale PXA270	Intel XScale PXA 255	Intel XScale PXA 270	Intel StrongARM
Clock rate	220MHz	220MHz	332MHz	332MHz	332MHz	520MHz	400MHz	624MHz	206MHz
SDRAM	10MB	64MB	64MB	64MB	128MB	64MB	256MB	128MB	64MB
SDRAM available for apps.	~4-5MB	~21MB	~22MB	~18MB	~112MB	N/A	N/A	N/A	~35MB
Flash / Memory card (max)	10MB / 1GB	128MB / 2GB	128MB / 2GB	256MB / 2GB	256MB / 4GB	192MB / N/A	N/A / 1GB	82MB / N/A	32MB / N/A
Fash available for apps.	~8.5MB	~64MB	~50MB	~160MB	~176MB	N/A	N/A	N/A	N/A
Virtual memory	no	no	no	no	yes	yes	yes	yes	yes

HP's PDAs are built around Inter's Xscale processors (PXA255, PXA270) running around 400–620 MHz. An older model released in 2002 used Intel Strong-ARM processor running at 206 MHz. They include 64–128 MB of RAM (with the exception of custom made Phraselator P2 device having 256 MB of RAM). The amount of available Flash memory ranges from 32 MB up to 1 GB using external memory card. All Xscale processors are fixed-point. Most systems use various versions of Windows CE/Mobile, however some devices have been used with Linux and Pocket Mac operation systems. Linux and Windows operating systems support virtual memory, while the presented versions of Symbian do not.

14.6.1 Example Application: Large Vocabulary Isolated Word Dictation

In this section we describe the work done in Nokia Research Center on low footprint embedded dictation (Karpov et al. 2006). The system supports five languages: US English, UK English, French, Spanish and Mandarin Chinese. It runs in 1.5–2 MB of RAM and requires 1.7–2.4 MB of Flash storage depending on the language. The system works reliably in speaker independent mode, but for the best accuracy a few minutes of speaker enrollment is advised. The word accuracy of the system (after enrollment) varies between 85% and 92% on the average for western languages and ~90% character accuracy for Chinese. The size of the vocabulary is between 23,000 and 43,000 words depending on the language (Fig. 4.4).

Fig. 14.4 Screen shots of the embedded dictation application running on a Nokia 6630 mobile phone. The left and right panels respectively show the dictation process and the enrollment

The system is designed for isolated word dictation; users are required to leave short pauses between words. This dictation style allows word segments to be identified reliably, and feedback can be given to the user between word segments. Isolated word decoding also made it possible to keep the RAM footprint very low.

Isolated word dictation is computationally simpler than continuous word dictation because the word segmentation can be decoupled from the recognition process. In our engine we do this by using a VAD module for identifying word segments based on the pauses between the words. Word segments are decoded left to right in real time as they become available by an isolated word decoder. The words to be scored in a segment are selected by using a langu-age model for predicting likely word continuations of the words in the previous segment. Depending on the UI mode, a sentence decoder can optionally be used for computing the overall most likely sentence hypothesis given the scored word lists in each word segment.

Voice activity detection: The VAD algorithm measures the long-term spectral divergence (LTSD) between speech and noise (Ramírez et al. 2004). It formulates the speech/non-speech decision rule by comparing the long-term spectral envelope to the average noise spectrum. The decision threshold is adapted to the measured noise energy while a controlled hangover is activated only when the observed SNR is low. It uses a long-term speech window to track the spectral envelope and is based on the estimation of so-called long-term spectral envelope (LTSE). The decision rule is then formulated in terms of the long-term spectral divergence (LTSD) between speech and noise.

Word decoder: For every word in the dictated message, the system predicts a number of possible follower words using a language model. These words form the system vocabulary for the next word. The recognition network is composed as follows: first phonetic pronunciations are fetched from the pronunciation lexicon for all words in the vocabulary. Next, a tree-structured phoneme-decoder network is created in such a way that common prefixes in the phonetic pronunciations are shared (i.e., Dave d-eI-v and David d-eI-v-I-d will share the three initial phonemes d-eI-v). This representation reduces memory footprint and increases decoding speed. The decoding is carried out in the conventional token-passing way with pruning.

Language modelling: The language model (LM) used in the demonstrator is based on a second order n-gram model: bi-grams and unigrams. The LM has two roles: vocabulary selection and sentence modelling. When the beginning of a new speech segment has been identified by the VAD module, the LM is used for selecting the words that are to receive acoustic scoring by the decoder in that segment. In the demonstrator, selection is based on bi-gram 'prediction'. Assuming the correct word in the previous segment is known, the most likely word continuations will be the bi-grams that start with that particular word. In case the list of predicted words is short, the word list is padded by backing off to unigram prediction. This minimizes the probability of not having the correct word in the list that is selected for acoustic scoring. Vocabulary prediction selects the most likely words, but at that stage the LM probabilities are not needed. However, when the selected words have received acoustic scoring, the word probabilities are required so that a correct ranking can be made which takes both acoustic probability and LM probability into account. Hence, both the word and the n-gram probabilities need to be stored in the language model. The probabilities do not have to be represented very accurately, and therefore they can be heavily compressed (Olsen and Oria 2006).

Acoustic modelling: The acoustic model set used in the demonstrator consists of bi-phone HMMs. Each model is made up of three states with 16 Gaussian densities in each state. All bi-phones are left-context. To make the models more compact and to enable proper training of all parameters the models have been tied using decision-tree state-tying and density-tying. For fast observation probability computation, both model parameters and feature vectors are quantized.

Language resources: A large amount of domain-specific data is required for training reliable acoustic and language models. This is especially true for LMs that show a very strong dependency on the genre, style and topic of the data they are trained on. This means that an LM that performs well in one specific domain will most probably perform quite poorly on test data from a different domain. On the other hand, a relatively small amount of domain-specific training data is considerably more efficient than a large amount of generic training data. A database of Personal Communication (PCOM) data was collected for training both acoustic and language models. The text database for LM training consists of 2 million words of simulated SMS messages that were submitted by native speakers of the language. The messages cover 12 topics representative of typical messaging communication ('vacation report', 'change of plans', 'family communication', 'invitation', 'congratulations', 'travel plans', 'business', 'feedback', 'teenagers', 'school', 'notes/reminders', and 'open domain').

The acoustic model training data consisted of off-the-shelf databases (Speech-Dat-Car, Speecon, Wall Street Journal, etc.) and the in-house PCOM databases for each language. The PCOM acoustic databases were recorded from 100 native speakers per language (evenly distributed by gender, age and dialectal region) in the office environment. Each speaker has 30 enrollment utterances for acoustic adaptation and 240 test utterances (SMS messages). Enrollment and SMS messages were read by the speakers both in continuous and isolated word manner. We used on the average 70 speakers for training and 30 speaker for testing. The total amount of training material was ~200 h e.g., for US English. The addition of the PCOM data base increased performance quite significantly. This is mainly because part of this data is spoken in an isolated manner, which matched well the recognition task.

14.7 Conclusion and Outlook

Speech recognition is an important technology that can greatly improve the *usability* of mobile phones. In order to make this possible, several problems ranging from adverse noise conditions and implementation constraints to wide language support have to be overcome.

In this chapter, we reviewed the typical mobile application scenarios and presented some advanced solutions to address the above problems. We described methods for robustness impro-vement by robust HMM modelling, feature extraction, and noise reduction techniques.

Simultaneous support of multiple languages is important for mobile phones dis-tributed on the global market. We described the necessary technology components in the framework of a practical multilingual speaker-independent name dialling system. We also briefly reviewed the implications of multilinguality to more complex ASR applications, such as embedded dictation.

Our presentation also focused on implementation aspects including effective te-chniques for small memory footprint and low computational complexity. We ad-dressed the charac-teristics of typical mobile phone platforms from the perspective of speech recognition and presented the details of an example application on large vocabulary isolated word dictation system.

Voice UIs clearly compete with already accepted UI methods of mobile phones. Due to the significant advances in embedded ASR technology over the past years, the technology is becoming more and more widespread. However, in practical de-ployments ASR technology often faces extremely high level user expectation. Human-like performance still remains a challenge for speech recognizers. As ASR algorithms improve and mobile phone platforms become more and more powerful, we can expect embedded ASR to become a viable and widely used solution for more and more complex applications in mobile phones.

References

Andrassy, B., Vlaj, D., and Beaugeant, Ch. (2001). Recognition performance of the siemens front-end with and without frame dropping on the Aurora 2 database. In *Proc. Eur. Conf. Speech Comm. Technol.* (Eurospeech), vol. 1, pp. 193–196.

Balan, R., Rosca, J., Beaugeant, Ch., Gilg, V., and Fingscheidt, T. (2004). Generalized stochastic principle for microphone array speech enhancement and applications to car environments. In *Proc. Eur. Signal Proc. Conf.* (Eusipco), September 6–10, 2004.

Bocchieri, E. (1993). Vector quantization for efficient computation of continuous density likelihoods. In *Proc. of ICASSP*, Minneapolis, MN, vol. 2, pp. II-692–II-695.

Bocchieri, E., and Mak, B. (1997). Subspace distribution clustering for continuous observation density hidden Markov models. In *Proc. 5th Eur. Conf. Speech Comm. Technol.*, vol. 1, pp. 107–110.

Bulyko, I., Ostendorf, M., and Stolcke, A. (2003). Getting more mileage from web text sources for conversational speech language modeling using class-dependent mixtures. In *Proc. 2003 Conf. North Amer. Chapter Assoc. Comput. Linguistics Human Language Technol.: Companion Volume Proc. HLT-NAACL 2003*—short papers—vol. 2.

Caseiro, D., Trancoso, L., Oliveira, L., and Viana, C. (2002). Grapheme-to-phone using finite-state transducers. In *Proc. 2002 IEEE Workshop Speech Synthesis*.

Fenn, J. (2005). Speech recognition on the desktop: still niche after all these years. Gartner Research Report, G00132456.

Fischer, V., Gonzalez, J., Janke, E., Villani, M., and Waast-Richard, C. (2000). Towards multilingual acoustic modeling for large vocabulary continuous speech recognition. In *Proc. IEEE Workshop Multilingual Speech Comm.*, pp. 31–35.

Furui, S. (1986). Speaker independent isolated word recognition using dynamic features of speech spectrum. *IEEE Trans. Acoust. Speech Signal Process.*, vol. 34, pp. 52–59.

Gales, M.J.F., Knill, K.M., and Young, S.J. (1999). State-based Gaussian selection in large vocabulary continuous speech recognition using HMM's. *IEEE Trans. Speech Audio Process.*, vol. 7, no. 2.

Häkkinen, J., Suontausta, J., Jensen, K., and Riis, S. (2000). Methods for text-to-phoneme mapping in speaker independent isolated word recognition. Technical Report, Nokia Research Center.

Hermansky, H., and Morgan, N. (1994). RASTA processing of speech. *IEEE Trans. Speech Audio Proc.*, vol. 2, no. 4, pp. 578–589.

Höge, H. (2000). Speech database technology for commercially used recognizers-status and future issues. In *Proc. Workshop XLDB LREC2000*, Athens.

Houtgast, T. (1989). Frequency selectivity in amplitude-modulation detection. *J. Acoust. Soc. Amer.*, vol. 85, pp. 1676–1680.

Karpov, E., Kiss, I., Leppänen, J., Olsen, J., Oria, D., Sivadas, S., and Tian, J. (2006). Short message dictation on symbian series 60 mobile phones. In *Proc. Workshop Speech Mobile Pervasive Environments (SiMPE) Conjunction MobileHCI 2006*.

Kiss, I., and Vasilache, M. (2002). Low complexity techniques for embedded ASR systems. In *Proc. ICSLP*, Denver, Colorado, pp. 1593–1596.

Laurila, K. (1997). Noise robust speech recognition with state duration constraints. In *Proc. ICASSP*.

Leppänen, J., and Kiss, I. (2005). Comparison of low footprint acoustic modeling techniques for embedded ASR systems. In *Proc. Interspeech*.

Leppänen, J., and Kiss, I. (2006). Gaussian selection with non-overlapping clusters for ASR in embedded devices. In *Proc. ICASSP*.

McCulloch, N., Bedworth, M., and Bridle, J. (1987). NETspeak, a re-implementation of NETtalk. *Computer Speech and Language*, no. 2, pp. 289–301.

Ning, B., Garudadri, H., Chienchung, C., DeJaco, A., Yingyong, Q., Malayath, N., and Huang, W. (2002). A robust speech recognition system embedded in CDMA cellular phone chipsets. In *Proc. ICASSP*.

Olsen, J., and Oria, D. (2006). Profile-based compression of N-gram language models. In *Proc. ICASSP*.

Oria D., and Olsen, J. (2006). Statistical language modeling with semantic classes for large vocabulary speech recognition in embedded devices. CI 2006 Special Session on NLP for Real Life Applications.

Quinlan, J.R. (1993). *C4.5: Programs for Machine Learning.* Morgan Kaufmann Publishers Inc., San Mateo, CA.

Ramírez, J., Segura, J.C., Benítez, C., de la Torre, Á., and Rubio, A. (2004). Efficient voice activity detection algorithms using long-term speech information. *Speech Comm.*, vol. 42, no. 3–4, pp. 271–287.

Sarikaya, R., Gravano, A., and Yuqing, G. (2005). Rapid language model development using external resources for new spoken dialog domains. In *Proc. ICASSP.*

Schultz, T., and Waibel, A. (2001). Language-independent and language-adaptive acoustic modeling for speech recognition. *Speech Comm.*, vol. 35, no. 1–2, pp. 31–51.

Schultz, T., and Waibel, A. (2000). Language portability in acoustic modeling. In *Proc. IEEE Workshop Multilingual Speech Comm.*, pp. 59–64.

Sejnowski, J.T., and Rosenberg, C.R. (1987). Parallel networks that learn to pronounce English text, *Complex Systems*, vol. 1, no. 1, pp. 145–168.

Sethy, A., Georgiou, P., and Narayanan, S. (2006). Text data acquisition for domain-specific language models. In *Proc. EMNLP.*

Sethy, A., Ramabhadran, B., and Narayanan, S. (2007). Data driven approach for language model adaptation using stepwise relative entropy minimization. In *Proc. ICASSP.*

Sivadas, S. (2006). Additive noise reduction for speech recognition by filtering in modulation spectral domain. Technical Report, Nokia Research Center.

SpeechDat (2000). http://www.speechdat.org

Stolcke, A. (1998). Entropy-based pruning of backoff language models. In *Proc. DARPA Broadcast News Trans. Understanding Workshop*, pp. 270–274.

Varga, I., Aalburg, S., Andrassy, B., Astrov, S., Bauer, J.G., Beaugeant, Ch., Geissler, Ch., and Höge, H. (2002). ASR in mobile phones—an industrial approach. *IEEE Trans. Speech Audio Process.*, vol. 10, no. 8, pp. 562–569.

Vasilache, M. (2000). Speech recognition using HMMs with quantized parameters. In *Proc. ICSLP*, vol. 1, pp. 441–443.

Vasilache, M., and Viikki, O. (2001). Speaker adaptation of quantized parameter HMMs. In *Proc. Eurospeech*, vol. 2, pp. 1265–1268.

Viikki, O., Bye, D., and Laurila, K. (1998). A recursive feature vector normalization approach for robust speech recognition in noise. In *Proc. ICASSP.*

Virag, N. (1999). Single channel speech enhancement based on masking properties of the human auditory system. *IEEE Trans. Speech Audio Process.*, vol. 7, no. 2, pp.126–137.

Westphal, M. (1997). The use of cepstral means in conversational speech recognition. In *Proc. Eur. Conf. Speech Comm. Technol.* (Eurospeech).

Young, S., Kershaw, D., Odell, J., Ollason, D., Valtchev, V., and Woodland, P. (2002). *The HTK Book* (for HTK Version 3.1).

Young, S.J., Russel, N.H., and Thornton, J.H.S. (1989). Token passing: a conceptual model for connected speech recognition systems. Technical Report CUED/F-INFENG/TR.38, Cambridge University.

15

Handheld Speech to Speech Translation System

Yuqing Gao, Bowen Zhou, Weizhong Zhu and Wei Zhang

Abstract. Recent Advances in the processing capabilities of handheld devices (PDAs or mobile phones) have provided the opportunity for enablement of speech recognition system, and even end-to-end speech translation system on these devices. However, *two-way free-form* speech-to-speech translation (as opposite to fixed phrase translation) is a highly complex task. A large amount of computation is involved to achieve reliable transformation performance. Resource limitations are not just CPU speed, but also the memory and storage requirements, and the audio input and output requirements all tax current systems to their limits. When the resource demand exceeds the computational capability of available state-of-the-art hand-held devices, a common technique for mobile speech-to-speech translation system is to use a client-server approach, where the handheld device (a mobile phone or PDA) is treated simply as a system client. While we will briefly describe the client/server approach, we will mainly focus on the approach that the end-to-end speech-to-speech translation system is completely hosted on the handheld devices. We will describe the challenges and algorithm and code optimization solutions we developed for the handheld MASTOR systems (Multilingual Automatic Speech-to-Speech Translator) for between English and Mandarin Chinese, and between English and Arabic on embedded Linux and Windows CE operating systems. The system includes an HMM-based large vocabulary continuous speech recognizer using statistical n-grams, a translation module, and a multi-language speech synthesis system.

15.1 Introduction

In recent years, there have been significant efforts to develop reliable and satisfactory automatic speech-to-speech translation systems, which are typically available on more powerful platforms such as desktop servers or laptop computers. However, because such devices are not compact, they are not convenient for mobile applications. This limits the usefulness of this form of translation technology. Many circumstances where translation is required can only be effectively aided by truly *mobile devices* such as a *Personal Digital Assistant* (PDA).

On the one hand, automatic *speech-to-speech translation* is a highly complex task. A large amount of computation is required to achieve reliable translation performance. Memory and storage requirements, and the audio input and output requirements all tax current systems to their limits. Therefore, when the resource demand exceeds the computational capability of available state-of-the-art hand-held devices, a common technique for mobile speech-to-speech translation system is to use a client-server approach. Here, the hand-held device (either a mobile phone or PDA) is treated simply as a system client, and the speech input is compressed and transmitted from this client to a back-end server that is much more powerful, either over a wireless telephone network or a wireless LAN connection such as Wi-Fi (IEEE 802.11b). The entire end-to-end speech translation task is conducted at the server. Finally, the spoken utterance in the target language is sent back to the hand-held device, thus providing the user audio output on location.

Obviously, there are several disadvantages of the client-server based approach. First, the service area is limited to locations where wireless connections are available. Second, large vocabulary speech recognition over conventional telephone channels, especially unreliable wireless channels, will degrade the quality of the translation. Third, this approach limits the flexibility of the user's control over the overall translation system, making highly customized applications much more difficult to design and deploy.

On the other hand, the development of increasingly powerful mobile devices is reaching a level that is comparable to the power of desktop systems of only a few years ago. In order to bridge the gap between the requirements of contemporary translation systems and the current mobile computing platforms, we have employed a number of optimizations, significantly enhancing the accessibility of our automatic speech translation technology. We have developed our speech-to-speech translation systems on PDA with an embedded Linux platform as well as the popular Window-CE platform.

Our PDA-based system achieves comparable translation performance and speed to that found in our MASTOR desktop system. Numerous optimizations were employed to improve translation speed and to reduce resource demands. However, the system still maintains a large vocabulary continuous speech recognizer that operates in real time, or near-real time. The typical response time for an end-to-end translation is under 5 s.

The organization of this chapter is as follows. Section 15.2 describes the system overview. Section 15.3 covers each major components of the system. Section 15.4 explains the experiments results and has some discussions. And we give the conclusion in the final section.

15.2 System Overview

15.2.1 System Architecture

Our system employs the same architecture as its desktop counterpart, the MASTOR system (Gao et al. 2002). Specifically, the system consists of a Large Vocabulary Continuous Speech Recognizer (LVCSR) that operates in real-time or near-real time to recognize input utterances, a fast translation module to translate the recognized text from the source language into text in the target language, and a multi-language speech synthesizer to convert the translated text into audio output in the target language. The system GUI is designed to let the user check both recognition and translation results, and to allow the user to re-play the output. Logging of results from the recognition and translation modules, as well as system configuration is also implemented in the system.

Figure 15.1a, b show the architectures of our speech translation systems when two different translation approaches are used respectively. Figure 1a shows the *concept based translation* approach is used. The input speech is recognized using an automatic speech recognizer (ASR) and then parsed by a statistical natural language understanding (NLU) module. An information extraction component is responsible for analyzing the semantic tree obtained from the NLU. This component is responsible for representing the (recognized) spoken sentence information in a language independent "interlingua" representation (Gu et al. 2006). This is combined with the canonical representation of "named entities" such as numbers and other attributes detected by our semantic model. The resulting representations are sent to a natural language generation (NLG) engine to render in the target language. The two types of information are translated using distinct models, with the specific attributes of items, such as times and dates, using conventional techniques familiar to the machine translation community. The *interlingua* translation, however, uses statistical techniques and can perform considerable surface changes when required for the target language. Finally, when a textual representation of the utterance in the target language is complete, a text-to-speech synthesizer is used to produce spoken output.

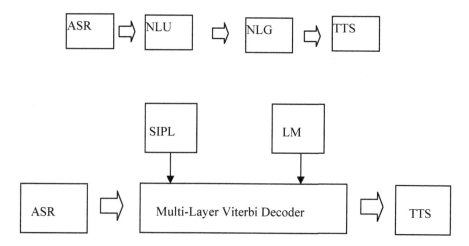

Fig. 15.1 System architecture of a speech-to-speech translation system on a handheld device: **a** Statistical NLU-NLG based concept translation approach. **b** statistical finite-state transducer based phrase translation approach

Figure 15.1b shows the architecture when statistical phrase based translation approach is used. The target translation can be obtained from a multi-layer Viterbi search, given the Statistical Integrated Phrase Lattices (SIPL) and a statistical language model, which is a novel framework for performing phrase-based statistical machine translation. IBM internal finite-state transducer toolkit is used during the development. In this work, we propose a novel framework for performing phrase-based *statistical machine translation* using weighted finite-state transducers (WFST's) that is significantly faster than existing frameworks while still being memory-efficient. In particular, we represent the entire translation model with a single WFST that is statically optimized, in contrast to previous work that represents the translation model as multiple WFST's that must be composed on the fly. While the language model must still be dynamically combined with the translation model, we describe a new decoding algorithm (Zhou et al. 2006) that can be viewed as an optimized implementation of dynamic composition.

15.2.2 Hardware and OS Specifications

To demonstrate the feasibility of building a system with our speech translation architecture on a standard *PDA*, we have built our system on three target *hardware platforms*. The first one (Zhou et al. 2004) is an iPaq PDA model H3800. It is equipped with an Intel's String ARM CPU with 206 MHz. The system has 64 MB of RAM and 32 MB of flash ROM. The original of this iPaq is shipped with Microsoft's Pocket PC 2002 and it is replaced with Familiar, a full featured Lunix distribution for mobile devices, based on the embedded Linux kernel. This PDA is also equipped with either an IBM Compact-Flash Micro drive, or a MultiMediaCard. We use it to store n-gram language models, translation models and dictionaries.

The second one (Zhu et al. 2006) is a customized PDA (referred to as the P2 below). It is equipped with an Intel 400 MHz XScale PXA 255 processor. The system has 256 MB of RAM and an SD (Secure Digital) card for additional storage. The P2 has been ruggedized for outdoor use. The original P2 was equipped with software for one-way, fixed phrase translation.

The third one is the HP iPaq PDA model H2750, a popular and commercially available device. The processor shipped with this product is Intel's XScale PXA 270 running at a frequency of 624 MHz. The system has 128 MB of RAM and an 82 MB iPAQ File Store system. It is also equipped with an SD card for additional storage. We use SD card to store the weighted finite state transducer based translation models and TTS voice files for the concatenated embedded TTS. All systems have a built-in microphone for speech input and an integrated speaker for audio output. The buttons on the PDA are used as push-to-talk for speech input, and user can also use the stylus to start or stop speech input.

15.2.3 Interface

Figure 15.2 shows the MASTOR user interfaces for several the handheld devices. On the left is on a Linux-based iPaq, on the middle is on a custom designed PDA using Microsoft Windows *WinCE.NET* 4.2 operation system, on the right is on HP iPAQ H2750 using Microsoft Pocket PC 2003 OS system. Usually, on the top of the screen, there are two radio buttons or two Start/Stop buttons for each direction showing the status of the microphone. There are also buttons showing the current speech translation direction. In the middle, there are two edit controls, one for displaying recognition results, the other for displaying translation results. The translation button

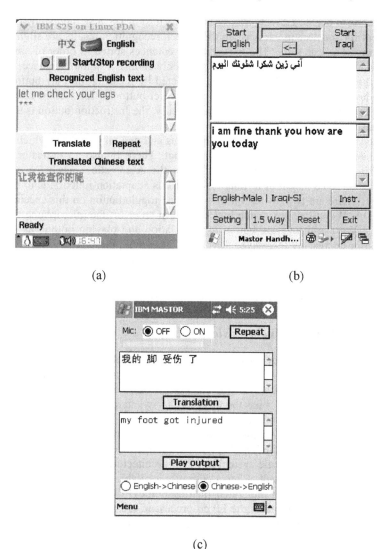

(a)   (b)

(c)

Fig. 15.2 Screenshots of Mastor system on several platforms **a** iPAQ H3800 using Embedded Linux; **b** Customized PDA using WinCE.net 2.4.0; **c** iPAQ H2750 using Pocket PC 2003

is for initiating the translation. The play output button is for playing or re-playing the audio output.

On the custom designed ruggedized PDA system. There are five buttons on the bottom of the screen: The Setting button is for setting the system parameters; 1.5

way button is to enable 1.5-way translation mode. Using 1.5 Way mode may be more convenient for the user in cases when the speech recognition results are slightly different from what was said. The list contains sentences sorted in order of similarity to what was spoken and allows the user to quickly choose a similar sentence. A sentence in the list can be clicked and then translated by pressing the Play button. 1.5 Way mode displays common, short phrases. Reset button is for system reset; Exit button is for exiting the Mastor system; and finally, the Instruction button is used for playing instruction of the system.

On Pocket PC 2003 system, there is one plus and one minus button which let the user change the volume level of speech output. There is also an indicator which shows the state of the adaptation function in the Automatic Speech Recognition (ASR) engines. The action of turning on or off this adaptation function is in the system menu. There are two switch buttons and one toggle button on this specially designed PDA. We utilize the switch buttons to turn on or off the microphone for both languages. We use the toggle button for translation and playing output, as well as increasing or decreasing the playback volume. Therefore, the system on this PDA is stylus free, making it possible to operate with one hand.

15.3 System Components and Optimization

15.3.1 LVCSR on Handheld Devices

The recognition module developed for our mobile speech translation system is an HMM-based *LVCSR* engine using statistical n-grams. Unlike most grammar-based embedded speech recognition systems, our system has the advantage of large vocabulary coverage. Moreover, it has the flexibility to switch to new application domains, which is typically only found on desktop-based systems. To accomplish this, IBM's large vocabulary speech recognition engine, as featured in the popular ViaVoice dictation product, was ported to the XScale processor architecture.

On porting this large scale system to ARM architecture, it is first noted that the StrongARM platform (as well as most currently available handheld devices), unlike the Intel x86 series, has no integrated floating point (FP) hardware. It depends entirely on software that emulates the FP co-processor. Despite much of the IBM recognizer being developed to use mostly integer computations, our initial profiling experiments showed that substantial amounts of time were consumed by FP calculations. Therefore, significant efforts were made to integerize the most of the signal processing front-end and search components of this system. This includes a fixed point math implementation of the following major recognizer components: the Mel-cepstrum feature extraction, the Gaussian likelihood computation of the context dependent phone models, as well as the procedures of fast match and detailed match during the decoding process. Particularly, at the feature extraction front-end computation modules such as high-pass filtering, discrete cosine transformation, Fast Fourier transform, LDA, pitch calculation and silence detection have been mostly integerized.

The English acoustic model uses an alphabet of 52 phones. Each phone is modeled with a 3-state left-to-right hidden Markov model (HMM). This system has approximately 3,500 context-dependent states modeled using 42 K Gaussian distributions and trained using 40 dimensional features. The context-dependent states are generated using a decision-tree classifier. The Chinese acoustic model uses 162 phones, including some phones that are tone-dependent. Each phone is also modeled with a 3-state left-to-right HMM. It has about 3,000 context-dependent states modeled using 40 K Gaussian distributions. The colloquial Arabic acoustic model uses about 30 grapheme phones that essentially correspond to letters in the Arabic alphabet, not including any diacritics such as short vowels. The colloquial Arabic HMM structure is the same as that of the English model. The colloquial Arabic acoustic model is also built using 40 dimensional features. It has 28 K Gaussian distributions. All models are trained using discriminative training (Povey and Woodland 2002).

A statistical trigram language model is built for English, Chinese and colloquial Arabic languages recognized in this speech translation system. The English language model is built using a corpus of 6.4 million words. This corpus is split as training and holdout sets with 5.7 million and 0.64 million words, respectively. The vocabulary size is about 30 K. The language model is smoothed using a deleted interpolation technique. Due to memory limitations of the P2 device, the language model is further pruned with bigram and trigram thresholds of 1 and 2, respectively. The size of the English language model is about 7 MB. In Chinese, the language model is built using a corpus of 2 million words, with a vocabulary size of 10 K. The size of Chinese language is about 5 million.

In Arabic, words can take *prefixes and suffixes* to generate new words that are semantically related to the root form of the word (stem). As a result, the vocabulary size in modern standard Arabic as well as dialectal Arabic can become very large even for specific domains. For colloquial Arabic, we used a corpus of 3.3 million words that is again split as training and holdout sets with 2.9 million and 0.32 million words, respectively. The vocabulary size for this corpus is about 98 K, which is too large to be used on the P2 system due to the CPU and memory limitations on the device. Aggressive pruning of both the vocabulary and the counts of the language model would be required. Instead, we built the language model on morphologically tokenized data. Applying the *morphological analysis*, we split some of the words into prefix + stem + suffix, prefix + stem, or stem + suffix forms. We refer the reader to (Afify et al. 2006) to learn more about the morphological tokenization algorithm. Morphological analysis reduced the vocabulary size to 58 K without sacrificing coverage. Nevertheless even this was too large to be used in the P2 device. Next, we eliminated singletons from the vocabulary, which reduced the vocabulary further down to 37 K, and applied cutoff thresholds to the bi-gram and tri-gram counts in the language model. The size of the final language model is about 9MB.

Adaptation to a new speaker or environment is becoming very important in embedded speech recognition as these systems are deployed in unpredictable real world situations. *Feature space Maximum Likelihood Regression* (fMLLR) has proven to be especially effective for this purpose, particularly when used for *incremental unsupervised adaptation* (Li et al. 2002). Unfortunately the standard implementation used by most authors requires unacceptable CPU power for embedded speech recognition systems. The CPU requirements can, to a degree, be lowered by using the block diagonal transformation matrix, but we will show that there are other problematic issues with the standard approach later.

We have decided to use the *stochastic gradient descent* approach. It has been successfully implemented in IBM's Embedded ViaVoice (EVV), where we face the fundamental problems of embedded systems: limited CPU performance, slow and small memory, no floating point unit. Adaptation is implemented through a feature space transform of the form $O' = AO + B$, where O are the speech frames, A is the transformation, and B is the bias. The total amount of parameters to estimate is only $n(n + 1)$, where n is the dimension of the feature vector. The adaptation is thus effective even with just a few seconds of data. One of the main challenges we face when deploying fMLLR on embedded platforms is *integerization*. The classical approach used in (Li et al. 2002) requires the need to compute the inverse of A. The inverse is usually performed using the Choleski decomposition algorithm. The implementation in integer arithmetic is fast, but unfortunately very sensitive to numerical errors (due to the necessary scaling and rounding), and can end up with completely wrong eigenvalues. This Choleski decomposition break down can be detected and an extra fail-safe mechanism usually takes care of resetting the transform in this case. In our solution (Zhu et al. 2006) we use the stochastic gradient descent approach which avoids the computation of the inverse (Balakrishnan 2003) and thus the eigenvalue related problems do not exist.

15.3.2 Natural Language Understanding and Generation Based Translation

We have developed two approaches for translation module. One is composed of a statistical natural language understanding (NLU) and a statistical natural language generation (NLG) module. The other is Weight Finite State Transducer based approach. We address the NLU/NLG approach in this section.

The NLU/NLG based concept translation approach is similar to the interlingua approach, which have been explored within C-STAR project by CMU (Lavie et al. 1997; Levin et al. 2000), ATR (Yamamoto 2000), ITC-IRST (Lazzari 2000), CAS (Zhou et al. 2004) and CLIPS (Blanchon and Boitet 2000), etc.

The NLU module is based on the statistical parser employed in IBM telephony natural language dialog systems. This component utilizes statistical decision-tree models to determine the meaning and structure of the input utterance, which is achieved by assigning a hierarchical tree structure to the recognized sentence as predicted by the statistical model. The *semantic parser* examines the class-tagged sentence and determines the meaning of the sentence by evaluating a large set of potential parse tree in a bottom-up left-to-right fashion. The parse hypothesis that scores the highest based on the statistical models is returned as the best parse hypothesis.

Current English and Chinese corpora include 10,000 sentences for each language in the domain of security and emergency medical care. 68 distinct labels and 144 distinct tags are used to capture the semantic information. An example of an annotated English sentence is illustrated in Fig. 15.3. In this parse tree, "FOOD" is a semantic concept represented by one or a group of words, while "food" is a tag that refers to a semantic concept represented by only one word. The concepts and tags in Fig. 15.3 are not designed to exclusively represent semantic meanings and may represent syntactic information as well, such as those shown in the label of "SUBJECT" and the tag of "query". While the semantic information remains the main annotation target, syntactic-related labels and tags are used to group the semantically less

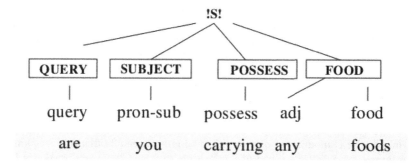

Fig. 15.3 Example of (concept-based) semantic parse tree

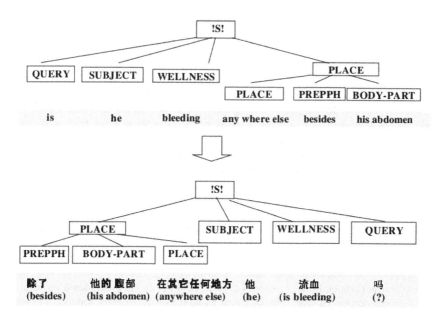

Fig. 15.4 Example of concept-based English-to-Chineses translation

important words into classes, which were found very useful in the NLG procedure to deal with the serious data sparseness problem.

An example of English-to-Chinese translation using the statistical interlingual approach is illustrated in Fig. 15.4. The source English sentence and the corresponding Chinese translation are represented by a set of concepts—{PLACE, SUBJECT, WELLNESS, QUERY, PREPPH, BODY-PART}. Some of the concepts (such as PLACE, WELLNESS and BODY-PART) are semantic representations while some of the concepts (such as PREPPH) are syntactic representations. There are also concepts (such as SUBJECT and QUERY) that represent both semantic and syntactic information. Note that although the source-language and target-language sentences share the same set of concepts, their tree structures could be significantly different because of the distinct nature of these two languages (i.e., English and Chinese). Therefore, in our approach, a natural language generation (NLG) algorithm, and in particular, a natural concept generation (NCG) algorithm, is required to transform the tree structures in the source language into appropriate tree structures in the target language, so that the source language sentences can be reliably translated into the target language sentences.

While the NLU module is not a significant computational bottleneck, it is important to improve the runtime speed of this module to lower the overall response time of the system. An effort was made to reduce the runtime memory requirements and to improve the parsing speed.

As we briefly mentioned in previous paragraph, very little work has been done using a statistical learning approach to produce natural language text directly form a semantic representation. Such as in our case, Ratnaparkhi (2000) introduced a statistic method to generate noun phrase from a simple semantic representation, attribute-value pairs, which is a special subclass of the semantic representation we want to deal with. We have developed our NLG component using a similar approach. The high-level semantic translation is accomplished by NLG in the target language from the semantic representation. More specifically, statistical NLG is used to discover the preferred concept ordering and to assign the lexical form of a grammatical sentence in the target language. The statistical models are directly learned from a training corpus, using no manually designed grammars or knowledge bases. In our speech translation system, the statistical NLG component has three kinds of input: a set tree-structured language-independent semantic variables, as shown in Fig. 15.3; a set of unordered translation attributes in the target language; a probability model for language generation.

During the translation, the source sentence is parsed, yielding the constituent structure of the semantic tree that is kept, while the concept ordering information is discarded. The word generation probability model is a maximum likelihood prediction based on *maximum entry modeling*.

On porting this component to ARM platform, this NLG module is re-implemented to fit with low computational resources available on PDA. This includes a more efficient implementation of search procedures, as well as significantly reduced I/O routines.

15.3.3 Weighted Finite State Transducer Based Translation

There is another approach based on a statistical MT methodology, originally proposed for written-text translation by an IBM group (Brown et al. 1993). It was then
applied to spoken language translation by the RWTH group (Ney et al. 2000; Ney
2003) and used in VerbMobil project (Wahlster 2000). More recently, finite state
methods have been widely applied in various speech and language processing applications (Mohri et al. 2002). Of particular interest are the recent efforts in approaching the task of machine translation using Weighted Finite State Transducers (WFST).
Various translation methods have been implemented using WFST in the literature.
Among them, Knight and Al-Onaizan (1998) described a system based on word-to-
word statistical translation models in the light of Brown et al. (1993). Bangalore and
Riccardi (2001) propose to apply WFST to select and reorder lexical items, and
Kumar et al. (2005) implemented the alignment template translation models using
WFST. One of the reasons why WFST-based approaches are favored is because of
the availability of mature and efficient algorithms for general purpose decoding and
optimization. For the task of speech-to-speech translation where our ultimate goal is
obtain a direct translation from source speech to target language, the WFST framework is even more attractive as it provides the additional advantages of integrating
speech recognition and machine translation more coherently. In addition, the nature
of WFST that combines cascaded models together as compositions offers an elegant
framework that is able to incorporate heterogeneous statistical knowledge from multiple sources. This should be particularly valuable when the translation task is more
complicated by the presence of conversational disfluent speech and recognition
errors. On the other hand, compared with word level SMT (Brown et al. 1993),
phrase-based methods explicitly take the word contexts into consideration to build
translation models. Koehn et al. (2003) compared several schemes proposed by various researchers as how to establish phrase-level correspondences and they showed
that all of these methods achieved consistently better performance over word-based
approaches.

We use *Weighted Finite State Transducer* to build the entire translation model. A
Viterbi decoder is used to combine the translation model and language model FST's
with input lattice efficiently.

The phrase-based translation task can be framed as finding the best path in the
following FSM, S = I o H, where, the "o" denotes the composition operation, I represents the source sentence with possible reordering, and,

$$H = P \circ T \circ W \circ L \tag{15.1}$$

here P, T, W, and L refer to the transducers of source language segmentation, the
phrase translation, the target language phrase-to-word, and the target language model,
respectively.

To minimize the amount of computation required at translation time, it is desirable to perform as many composition operations in Eq. 15.1 as possible, ahead of
time. The ideal situation is to compute H offline. At translation time, one need only
compute the best path of S = I o H. However, it can be very difficult to construct H

Fig. 15.5 A portion of source sentence segmentation transducer P. Each arc is labeled using the notation "input:output". The token "<epsilon>" (ε) denotes the empty string and '#' is used as a separator in multi-word labels. For simplicity, costs associated with arcs are not displayed in the graph

given practical memory constraints. While this has been done in the past for word-level and constrained phrase-level systems (Zhou et al. 2005), this has yet to be done for unconstrained phrase-based systems. In Zhou et al. (2006), this issue is tackled as the following.

First, we note that the source language segmentation transducer P explores all "acceptable" phrase sequences for any given source sentence. It is crucial that this transducer to be deterministic because this can radically affect translation speed and memory usage. In Zhou et al. (2006), we introduce an auxiliary symbol, denoted EOP, marking the end of each distinct source phrase. By adding these artificial phrase boundary markers, each input sequence corresponds to a single segmented output sequence and the transducer becomes determinizable.

Secondly, while it may not be feasible to compute H in its entirety as a single FSM, we separate H into two pieces: the language model L and the translation model M:

$$M = Min(Min(Det(P) \text{ o } T) \text{ o } W)$$ (15.2)

where *Det* and *Min* denotes the *determinization and minimization* operation respectively. In spite of the fact that T and W in (2) are not deterministic, and that minimization is formally defined on deterministic machines (Mohri et al. 2002), in practice, we often find that minimization can help reduce the number of states of non-deterministic machines. It should also be noted that due to the determinizability of P, M (the SIPL) in the above equation can be computed offline using a moderate amount of memory. See Fig. 15.5 for a sample portion of the resulting transducer.

In this approach, translation has been defined as finding the best path in I o M o L. To address the problem of efficient computation, Zhou et al. (2006) have developed a multilayer search algorithm. Specifically, as shown in Fig. 15.6, we have one layer for each of the input FSM's: I , L, and M. At each layer, the search process is performed via a state traversal procedure starting from the start state, and consuming an input word in each step in a left-to-right manner. This can be viewed as an optimized version of on-the-fly or dynamic composition integrated with a Viterbi search procedure. However, this specialized decoding algorithm has the advantage of not only significant memory efficiency and being possibly many times faster than general composition implementations found in FSM toolkits, but it can also incorporate information sources that cannot be easily or compactly represented using WFST's. For example, the decoder can allow us to apply the translation length penalties and phrase penalties to score the partial translation candidates during search.

We represent each state $\hat{S}$ in the search space using the following 7-tuple: $(S_I, S_M, S_L, C_M, C_L, \hat{h}, \hat{S}_{prev})$, where S_I, S_M, and S_L record the current state in each input FSM; C_M and C_L record the accumulated cost in M and L in the best path up to this point; $\hat{h}$ records the target word sequence labeling the best path up to this point; and $\hat{S}_{prev}$ records the best previous state. The initial search state $\hat{S}_0$ corresponds to being located at the start state of each input FSM with no accumulated costs. To reduce the search space, two active search states are merged whenever they have identical S_I, S_M and S_L values; the remaining state components are inherited from the state with lower cost. In addition, two pruning methods, histogram pruning and threshold or

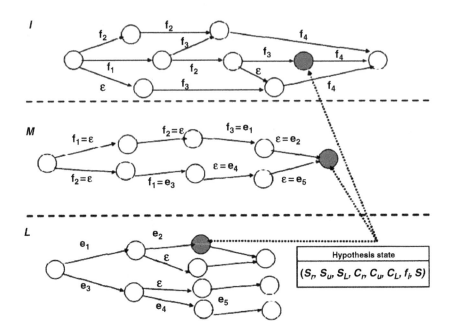

Fig. 15.6 Search state of the multi-layer search

beam pruning, are used to achieve the desired balance between translation accuracy and speed. The search algorithm is implemented using fixed-point arithmetic for deployment on PDA devices that lack a floating point processor. This results in translation speeds of hundreds of words per second on a PDA device, while using less than 20 MB runtime memory.

15.3.4 Embedded Speech Synthesis

Once an utterance is translated by the translation model, it is sent to the screen for display and to a *text-to-speech* (TTS) engine. Considering the limited resources available in a mobile device, formant TTS could be a reasonable choice. But high quality synthesized speech is vital for speech to speech communication. Therefore, compact concatenate voices have been developed both English and foreign languages.

Our synthesis system uses a set of speaker-dependent decision-tree state-clustered hidden Markov models to automatically generate a leaf level segmentation of a large signal-speaker continuous-read-speech database. During synthesis, the phone sequence to be synthesized is converted to an acoustic leaf sequence by descending the

HMM decision tree. Duration, energy and pitch values are predicted using separate trainable models. To determine the segment sequence to concatenate, a dynamic programming search is performed overall the waveform segments aligned to each leaf in training. The dynamic programming attempts to ensure that the selected segments join each other spectrally, and have durations, energies and pitches such that the amount of degradation introduced by the subsequence of using TD-PSOLA is minimized. More detail about IBM *concatenate TTS* is in paper (Donovan and Eide 1998). Due to the limited memory space available in a typical PDA device, the actual TTS voice segment fonts are stored in SD card while the search algorithm runs in the memory. The typical time for synthesis a sentence with 10 words is about one second.

15.4 Experiments and Discussions

15.4.1 Speech Recognition Experiments

Speech recognition experiments are designed to measure the effectiveness of our integrization algorithms and the proposed fast incremental adaptive method. Experiments were conducted on 3 different speech data sets. Tables 15.1 and 15.2 show the results on read speech recorded in quiet and noisy conditions. The first data set (shown in Table 15.1) is read speech from 5 speakers, each read 150 English sentences in quiet office. The second data set (shown in Table 15.2) was recorded in noisy condition from 3 speakers, each read 100 English sentences. These noisy speeches were recorded by using Andrea NC-65 microphone. The speakers' lips are about 2–3 in. away from the microphones. The vehicle noise source is about 5–6 ft away from the microphones, and the noise level of 70–75 dbA measured next to the microphones. The third test data set is spontaneous speech (results shown in Table 15.3). The spontaneous speech data was recorded in two different conditions. Test 1, Test 2 and Test 3 were recorded from S2S system mediated cross-lingual conversations. Test 4 was also from cross-lingual conversations but it was mediated by human interpreters, therefore the speech is much more casual and includes a lot of disfluencies.

Speech recognition results measuring in word error rates (WERs) with and without the adaptation method are shown also in Tables 15.1–15.3. As for the comparison, the results from PC version of ASR code are also shown in the same tables. In Table 15.1, clean speech, the average of WER of integerized ASR engine is 5.21% while WER on laptop version ASR engine is 4.60%. The degradation is reasonable and within our expectation. With *fMLLR* adaptation, the average WER goes down to 4.28% which is 22.09% relative improvement and is very close to results of laptop version (4.25%). In Table 15.2, for noisy speech, on average, the difference between integerized engine code and PC version is less than 1%. Here we also see the significant gain by using the adaptation algorithm. In Table 15.3, the degradation of integerization varies within different test, and the range is between 1% and 3%. With fMLLR, we see a significant gain on Test 1 data set, but not for Test 2 data set. Since the speech in Test4 data set was from conversations between people mediated human interpreters, therefore they are highly spontaneous, the WERs are particularly high. We noticed that the adaptation algorithm does not work well when the WERs are high.

Overall, these results indicate that the performance degradation of the integerization compared with the float-point engine is within expectation and significant gain is achieved by using proposed fast adaptation method.

Table 15.1 Read speech in quiet condition

Different ASR code	WER (%)
PDA (integerized code)	5.21
With fMLLR	4.28
PC (floating point code)	4.60
With fMLLR	4.25

Table 15.2 Read speech in noisy condition

Different ASR code	WER (%)
PDA (integerized code)	15.82
With fMLLR	11.14
PC (floating point code)	14.69
With fMLLR	12.46

Table 15.3 Spontaneous speech recognition

English	Test 1 (%)	Test 2 (%)	Test 3 (%)	Test 4 (%)
PDA (integerized code)	21.84	17.54	15.23	46.18
with fMLLR	15.57	18.76	13.80	46.29
PC (floating point code)	18.07	16.79	14.24	39.86
with fMLLR	13.19	14.93	12.00	35.83
Arabic PC (floating point code) with fMLLR	25.80	24.56	23.00	29.58

We also measured CPU usage with adaptation enabled and disabled. On average, the proposed stochastic gradient descent method only increases CPU usage by about 15%. Fortunately, most of this extra usage occurs at the end of recognition, after the user has been presented with recognition results, so it does not affect real time recognition performance.

15.4.2 Translation Experiments

The English–Mandarin recognition and translation experiments were done on the DARPA CAST Aug'04 offline evaluation data, which has an English script of 130 sentences and a Chinese script of 73 sentences for medical domain. Each script was read by 4 speakers. The recognition word error rate for English is 11.06%, while the character error rate for Mandarin is 13.60%, both are run on speaker-independent models. The translation experiments are done on both clean text and the ASR decoded scripts. The 4-g *Bleu score* results measured using 8 human translations as references are shown in Table 15.4. The oracle scores show that if one can combine the translation results from these two different approaches, the accuracy can be further improved. Currently, we present two alternate translations to users in the real-time system and give them more information for communication purpose. It is very useful to notice that the translation results generated by our two approaches are always consistent in meaning.

English–Arabic experiments are done on several S2S system mediated cross-lingual conversations (a subset of DARPA development set). In each dialog, an English speaker and an Arabic speaker were talking to each other via a speech-to-speech translation device. We extracted 395 English utterances and 200 colloquial Arabic utterances from the dialogs. Three human translation references are created for measuring the BLEU score purpose. The results are shown in Table 15.5. Since the data is spontaneous conversational speech, the recognition WERs for both English and Arabic are not as high as those observed in human interpreter mediated conversation. The BLEU scores of English-to-Arabic is slightly lower than that of Arabic-to-English. One possible reason is that spelling of words in colloquial Arabic dialect is not standardized (more variations for the same word), which can lead to a low BLEU score. Another observation is that the ASR errors degrade the BLEU score more significantly for English-to-Arabic. Although the ASR WERs look similar for English and Arabic, we notice that the WER of English content words is higher than that of Arabic. A possible reason is that the English acoustic model is not trained from spontaneous speech, while the Arabic acoustic model is trained with more conversational style speech mainly from in-domain data.

Table 15.4 BLUE score of English–Mandarin translation

Input	En-to-Cn		Cn-to-En	
	Clean	ASR	Clean	ASR
NLU/NLG	0.578	0.513	0.276	0.245
WFST	0.572	0.504	0.276	0.246
NLU/NLG+WFST(Oracle)	0.691	0.606	0.365	0.342

Table 15.5 BLUE score of English–Arabic translation

	ASR WER	BLUE (Clean)	BLUE (ASR)
En-to-Ar	15.9%	0.388	0.202
Ar-to-En	25.8%	0.596	0.416

We described two different statistical approaches for speech-to-speech translation. The concept based approach focuses on understanding and re-generating the meaning of the speech input, while the finite-state transducer based approach emphasizes both system development and search speed and memory efficiency. The former approaches usually involve large amount of human effort in linguistic information annotation, although the amount of annotated data needed is not very large. The latter approach, weighted FST, may exploit un-annotated parallel corpora at the cost of potential meaning loss and the requirement of large amount of parallel text data.

Both approaches have shown comparable results. The oracle scores show that if one can combine the translation results from these two different approaches, the accuracy can be further improved significantly. Currently we present two alternate translations to users in the real-time system to enhance the communications. It is very useful to notice that the translation results generated by our two approaches are always consistent in meaning.

In cases where under studied languages (low resource) are involved in speech translation, the task would be more complex due to a number of reasons. Here are two particular ones in our concern. (1) lack of large amount of speech data which represent the oral language spoken by the right target native speakers, consequently traditional statistical translation approach is not applicable, the speech recognition error rate is much higher than popular languages, such as English or Chinese; (2) lack of linguistic knowledge realization in annotated corpus. Therefore neither linguistic knowledge based approaches (such as our concept-based approaches) nor pure statistical approaches (such as IBM model 1–5 and FST-based methods) are suitable for rapid development of applicable systems.

We believe that integration of the two research paradigms into a unified framework, e.g., in a unified FST composition, should be the way to go. A shallow semantic/syntactic parser is designed and implemented to enable statistical speech translation using knowledge-based shallow semantic/syntactic structures. This information is further utilized to process inevitable speech recognition errors and disfluencies in the colloquial speech. While the shallow-structure parser is initiated upon lightly annotated linguistic corpora and trained using statistical model, it can be greatly enhanced and expanded by applying machine learning algorithms on un-annotated parallel corpora. The integration of the two approaches should increase the system end-to-end performance, and reduces the amount of parallel text data required by the statistical algorithm.

15.5 Conclusion

We present our recent effort to develop two-way free-form speech-to-speech translation systems on a PDA using embedded Linux and Window CE or general Pocket

PC platform. Due to the limited resources in both computational capability and memory and storage constrain in a typical handheld device, building an entire end-to end system on such devices is a highly complex task.

We developed the handheld MASTOR systems (Multilingual Automatic Speech-to-Speech Translator) for between English and Mandarin Chinese and between English and Arabic on embedded Linux and Windows CE operating systems. The system includes an HMM-based large vocabulary continuous speech recognizer using statistical n-grams, a translation module, and a multi-language speech synthesis system.

References

Afify, M., Sarikaya, R., Kuo, J., Besacier, L., and Gao, Y. (2006). On the use of morphological analysis for dialectal Arabic speech recognition. In *Proceedings of Inter-Speech*.

Balakrishnan, S.V. (2003). Fast incremental adaptation using maximum likelihood regression and stochastic gradient descent. In *Proceedings of EUROSPEECH*.

Bangalore, S., and Riccardi, G. (2001). A finite-state approach to machine translation. In Proceedings of North American Chapter of the Association for Computational Linguistics (NAACL).

Brown, P., Della Pietra, S.A., Della Pietra, V.J., and Mercer, R.L. (1993). The Mathematics of statistical machine translation: Parameter estimation. *Computational Linguistics*, vol. 19(2), pp. 263–311.

Donovan, R.E., and Eide, E.M. (1998). The IBM trainable speech synthesis system. In *Proceedings of ICSLP*, Sydney, Australia.

Gao, Y., Zhou, B., Diao, Z., Sorensen, J., and Picheny, M. (2002). MARS: A statistical Semantic Parsing and Generation-Based Multilingual Automatic Translation System. *Machine Translation*, vol. 17, pp. 185–212.

Gao, Y., Zhou, B., Gu, L., Sarikaya, R., Kuo, H-K., Rosti, A-V.I., Afify, M., and Zhu, W. (2006). IBM MASTOR: Multilingual Automatic Speech-to-Speech Translator. In *Proceedings of ICASSP*.

Gu, L., Gao, Y., Liu, F., and Picheny, M. (2006). Concept-based speech-to-speech translation using maximum entropy models for statistical natural concept generation. *IEEE Transactions on Speech and Audio Processing*, vol. 14, no. 2, pp. 377–392.

Knight, K., and Al-Onaizan, Y. (1998). Translation with finite-state devices. In *Proceedings of 4th Conference of the Association for Machine Translation in the Americas*, pp. 421–437.

Koehn, P., Och, F., and Marcu, D. (2003). Statistical phrase-based translation. In Proceedings of North American Chapter of the Association for Computational Linguistics/Human Language Technologies.

Kumar, S., Deng, Y., and Byrne, W. (2005). A weighted finite state transducer translation template model for statistical machine translation. *Journal of Natural Language Engineering*, vol. 11, no. 3.

Lavie, A., Waibel, A., Levin, L., Finke, M., Gates, D., Gavalda, M., Zeppenfeld, T., and Zhan P. (1997). JANUS-III: Speech-to-Speech Translation in Multiple Languages. In *Proceedings of ICASSP*, Munich, Germany, vol. 1, pp. 99–102.

Levin, L., Lavie, A., Woszczyna, M., Gates, D., Gavalda, M., Koll, D., and Waibel, A. (2000). The Janus-III Translation System: Speech-to-Speech Translation in Multiple Domains. *Machine Translation*, vol. 15, pp. 3–25.

Lazzari, G. (2000). Spoken Translation: Challenges and Opportunities. In *Proceedings of ICSLP*, Beijing.

Li, Y., Erdogan, H., Gao, Y., and Marcheret, E. (2002). Incremental on-line feature space MLLR adaptation for telephony speech recognition. In *Proceedings of ICSLP*.

Mohri, M., Pereira, F., and Riley, M. (2002). Weighted finite-state transducers in speech recognition. *Computer Speech and Language*, vol. 16, no. 1, pp. 69–88.

Ney, H., Niessen, S., Och, F.J., Sawaf, H., Tillmann, C., and Vogel, S. (2000). Algorithms for statistical translation for spoken language. *IEEE Transactions on Speech and Audio Processing*, vol. 8, no. 1, pp. 24–36.

Ney, H. (2003). The statistical approach to machine translation and a roadmap for speech translation. In *Proceedings of Eurospeech*.

Povey, D., and Woodland, and P.C. (2002). Minimum phone error and I-smoothing for improved discriminative training. In *Proceedings of ICASSP*.

Ratnaparkhi, A. (2002). Trainable method for surface natural language generation. In *Proceedings of 1st Meeting of North American Chapter of ACL*.

Tillmann, C., Vogel, S., Ney, H., and Sawaf, H. (2000). Statistical Translation of Text and Speech: First Results with the RWTH System. *Machine Translation*, vol. 15, pp. 43–74.

Wahlster, W. (ed.) (2000). Verbmobil: Foundations of Speech-to-Speech Translation. Springer, Berlin.

Yamamoto, S. (2000). Toward speech communications beyond language barrier— Research of spoken language translation technologies at ATR. In *Proceedings of ICSLP*, Beijing.

Zhou, B., Dechelotte, D., and Gao, Y. (2004). Two-way Speech-to-Speech Translation on Handheld Devices. In *Proceedings of ICSLP*.

Zhou, B., Chen, S., and Gao, Y. (2005). Constrained phrase-based translation using weighted finite-state transducers. In *Proceedings of ICASSP*.

Zhou, B., Chen, S., and Gao, Y. (2006). Folsom: A Fast and memory-efficient phrase-based approach to statistical machine translation. In *Proceedings of IEEE/ACL 2006 Workshop on Spoken Language Technology*.

Zhou, Y., Zong, C., and Xu, B. (2004). Bilingual chunk alignment in statistical machine translation. In *Proceedings IEEE International Conference on Systems, Man and Cybernetics*, vol. 2, pp. 1401–1406.

Zhu, W., Zhou, B., Prosser, C., Krbec, P., and Gao, Y. (2006). Recent advances of IBM's handheld speech translation system. In *Proceedings of Interspeech*.

16

Automotive Speech Recognition

Harald Höge, Sascha Hohenner, Bernhard Kämmerer, Niels Kunstmann,
Stefanie Schachtl, Martin Schönle, and Panji Setiawan

Abstract. In the coming years speech recognition will be a commodity feature in car. Control of communication systems integrated in the car infotainment system including telephony, audio devices and destination inputs for navigation can be done via voice. Concerning speech recognition technology biggest the challenge is the recognition of large vocabularies in noisy environments using cost sensitive hardware platforms. Further intuitive dialog design coupled with natural sounding text to speech systems has to be provided to achieve a smooth man-machine interaction. This chapter describes commercial driven activities to develop and produce speech technology components for various automotive applications including the used speech recognition, speaker characterization, speech synthesis and dialog technology, the used platforms, and a methodology for the evaluation of recognition performance.

16.1 Introduction

Man Machine Interaction in car is a typical application demanding a "hands-free," "eyes-free" operation mode. Speech recognition is well suited to fulfill these demands. Yet the specific acoustic environment and specific platforms found in cars are challenging:

— Noisy environment with a signal-noise ratio in the range of 20 dB till −5 dB
— Low cost microphones mounted 30–100 cm from the speaker
— Embedded platforms with restricted computing power and memory

First applications were focused on command and control functions as name dialing to handle the telephone integrated in car. Nowadays destination input for navigation containing more than 100 000 street and city names is the most challenging task for speech recognition technology.

In the following recent advances and activities performed in Siemens Corporate Technology by the group "Siemens Speech Processing" are described.

16.2 Siemens Speech Processing—From Research to Products

Siemens Corporate Technology was founded in the late 70th to support Siemens Operating Groups to secure a forefront position. Speech processing was one of the first topics of Corporate Technology, therefore now looking back on a history of more than 25 years. While the first decade—the 80th—was mainly research oriented, there was already demand for large vocabulary recognition (at that time 1000–2000 isolated words in a speaker-dependent mode). Requests came, among others, from the Medical Group to support doctors for report generation from Computer Tomography images.

The second decade—the 90th—saw increased utilization of the speaker-independent recognition in communications. Large switching systems were enhanced with generic recognition units for later application integration—implemented on integer DSPs—as well as with dedicated speech-enabled functions e.g., for voice-mail control.

The late 90th and the beginning of the new century finally brought speech technology on embedded platforms which became cheap enough to serve vast consumer markets like mobile phones and car *infotainment*.

While command and control like recognition allows for basic voice access to services, more user convenience can be realized with natural voice dialogs. Dialog systems control the flow of user input and system output to collect all parameters as dates or money amounts needed for an application query. For applications in communications/telephony, dialog systems control speech in and speech out, but future systems are foreseen to provide multi-modal interaction combining voice, graphics, and haptics in a synchronized and consistent way.

Speech signals carry more than just words and sentences: there is implicit information about the speaker—gender, age, language, and mood or stress—which is of value for many applications. In order to make this information accessible, Siemens Speech Processing developed components for *speaker recognition* and *speaker characterization*. While speaker recognition has to be trained on the person to be recognized (*enrollment*) speaker characterization derives age/gender or language decisions speaker-independently.

Finally, universal voice feedback to users needs a flexible *text-to-speech* synthesis system which is optimized for the chosen application domain.

16.2.1 Development for Performance and Quality

Speech technology serves as the connector between users and systems, requesting safe and reliable operation. Development at Siemens Speech Processing therefore follows a well defined process from requirements analysis (driven by market pull, product component demands, and research/technology push), development frameworks, and acceptance tests. Tools for versioning, workflow management, and defect/change tracking help to maintain a high level of software and functionality quality. Apart from bug-free software and functional completeness the (recognition) performance of speech components is crucial for the final deployment. Therefore procedures and measures have to de developed that serve both the supplier and the customer to gain confidence and trust in the operation of the components. As an example, a systematic approach for SNR-based measurements of recognizer performance has been developed which is described later on.

Siemens Speech Processing is organized into "Innovation," "Technology," "Products," and "*Natural Language Understanding*," implementing a chain of valueads that narrows down from the broad spectrum of science and research over proven technology pieces to products that match market requirements. The focus on Natural Language Understanding combined with dialog capabilities—often found in separate companies—opens the way to natural interaction with optimized functionality.

The following will give a short overview of components developed and delivered by Siemens Speech Processing.

16.2.2 High-Performance Recognizer

With the event of cellular phones, processing power became cheap enough to bring speech recognition on mobile devices. For that purpose a dedicated recognizer product was developed that offers various benefits. The Siemens *Recognizer Embedded* is targeted for mobile phones, car *infotainment* and *navigation*, PDA/PNA deployment, and dedicated *embedded systems* in hearing aids, medical devices, or industrial panels and comes with selected European, US and Asian languages (see Fig. 16.1).

Fig. 16.1 User interface for voice driven applications in car

This Siemens *Recognizer Embedded* is complemented by the Siemens *Recognizer Server* that offers standard interfaces and protocols like *MRCP* and *RTP*, multi-port and multi-threading with load-balancing, and optimizations for Windows and Linux. Acoustic models for selected languages of Europe, US, China, and India are provided to serve these important markets. The Siemens Recognizer Server is targeted for call-center automation, auto-attendant solutions, and industrial applications.

16.2.3 Ultra-Compact Text-to-Speech Synthesizer

The generation of artificial voices has to meet several requirements—the speech must be intelligible and natural while the footprint must match given hardware limits. Siemens Speech Processing decided to concentrate on a solution for extremely small footprints. A state-of-the-art diphone based technology is used where short segments from real speech are concatenated, adjusted at the boundaries, and modulated by the prosody contour. For advanced applications a dedicated text pre-processing module resolves e.g., abbreviations and numbers.

The Siemens *Text-to-Speech* Synthesizer *Embedded* starts from just around 250 kB for tasks like caller name announcement in mobile phones and reaches 1.5 MB for email or SMS reading (one language). The system is available for European languages, US English and Mandarin Chinese and is targeted for low-footprint, low-cost devices like mobile phones e.g., for developing countries or mobile industrial devices.

Fig. 16.2 Carefully designed dialog machines are needed for a smooth man-machine interaction

16.2.4 Natural Voice Dialog

Mixed-initiative conversational voice dialog systems offer a maximum of convenience to users. These "how can I help you" systems were first implemented in telephony voice portals and advanced *IVR* systems. Siemens Speech Processing gained considerable experience from these areas before converting the technology to embedded systems. By that, convenient voice access to complex information sources like web-services or *manuals* can now be experienced for example in cars that are "*always-on*" (see Sect. 16.3).

Traditional dialog systems operate in a state-based manner on VoiceXML scripts that are hand-crafted to implement certain functionality. While this approach showed to be widely accepted, it is very time consuming for a first implementation. Siemens Speech Processing therefore developed a slot-filling solution which allows for a descriptive dialog design. This *dialog engine* knows about the task and associated parameters, allows for multi-parameter input and over-answering and performs an automatic dialog in case of unclear or missing entries.

Additional state-based procedures and a complementary communication with the platform over dynamic VoiceXML pages combine the best of both worlds.

Especially in *automotive* applications there is a need for coherence between the GUI and a voice dialog. The Siemens dialog engine is perfectly prepared to achieve this, relieving the UI developer of a heavy burden and securing a consistent user experience (see Fig. 16.2).

16.2.5 Speaker Characterization and Recognition

Imagine a system that adapts to a driver without knowing him: *speaker characterization* determines automatically age, gender, and language of the current user as basis for adaptive dialogs. Spoken language identification allows a user to interact in his language, even if it differs from the system language (e.g., in rental cars). With age classes "child," "teenager (f/m)," "adult (f/m)," and "senior," dedicated dialog styles can be chosen—from uncouth to serious. And knowing the gender the recognition performance can be optimized as well as, e.g., the content of services re-ordered.

The Siemens Speaker Characterization adds age/gender recognition and language identification to the Siemens *Recognizer Server* while the Siemens *Speaker Recognition* performs a biometric recognition of individuals after *enrollment*—text dependent as well as text-independent.

16.3 Example Automotive Voice Applications: Infotainment, Navigation, Manuals, and Internet

High performance recognition opens new opportunities for a more natural interaction by voice in cars. When vocabularies are no longer restricted to few commands or names but extend to several thousand words, and when those recognizers are combined with an appropriate dialog engine and *Text-to-Speech* synthesizer, especially in the automotive scenario new speech applications become reality that will significantly

enhance usability. Siemens Speech Processing explores various applications that benefit from speech user interfaces, already today or in the near future. The following gives
a short survey on the results and solutions obtained so far.

16.3.1 Radio Station Selection

A typical use-case for speech recognition is the control of entertainment sources of car
infotainment systems. Available radios already display the name of the tuned station,
provided as the "Program Service Name" by the radio data system RDS. A serious
issue for voice-based selection of radio stations is the limitation to at most eight characters for the Program Service Name. Due to this it is a common to transmit abbreviations or short forms for the station names like, e.g., "CEREDIGN" for Radio Ceredigion
in the UK. This raises the question how an inexperienced user knows what in order to
switch to the corresponding station. One solution here is to provide an exception list for
the recognizer which contains multiple *pronunciation*s for each (known) RDS name.
An automatic conversion of text to phonemes is then utilized only for RDS names not
contained in this list.

16.3.2 MP3 Title Selection

There is an increasing demand to consume audio and video media wherever they are.
The development of effective compression techniques for audio like *MP3*-coding and
the availability of portable players, even integrate in various recent cell phones accelerated this trend. The use of speech control for the administration of large amounts of
audio files, playback control, and the selection of titles and artist becomes a desirable
feature, especially for the case of limited interaction possibilities of portable players or
car infotainment systems. What makes the task of voice control for portable players so
specific?

First of all, with genres, interprets, albums, and titles there is a large amount of
partly structured data to be operated on. The phoneme strings for genres are normally
provided by the supplier and fit to the chosen speech recognition language, while the
names of interprets, titles, or albums need not to stem from the current speech recognition language. Although there are activities going on to supply possible pronunciations
for interprets and titles (see e.g., www.gracenote.com), suitable information in this area
is still sparse. This situation leads to the third particularity of the *MP3* selection, the
need to deduce a phonetic representation for arbitrary song names where the originat-
ing language is not known beforehand. Unfortunately, usually provided language information in subjective property frames of an ID3 tag is not reliable enough.

The task of language identification is made even more ambitious when the titles do
not contain valid words, or when there are purposeful spelling "errors" like for the
album "Konvicted" by Akon. Furthermore, song titles are often short and consist of
only a few words not necessarily typical for the chosen language. A related issue arises
from phrases which contain words from different languages, e.g., in the title "Femme
Like U" of the album "La Good Life" by K-Maro.

Usually employed techniques for language identification employ n-gram statistics, decision trees, or neural networks and operate on phrase level or on word level. On the word level, useful cues are language-specific letters or letter sequences, e.g., the German "ß," or "th" for English. On phrase level the approaches try to combine the indicative features from single words in an intelligent way or one use the sequence of words. Presumably, the best solution for the language identification task in the given situation will be a weighted combination of different approaches.

Once the system has decided on the language of a phrase and of the underlying words, the grapheme-to-phoneme conversion deduces the phoneme string corresponding to a possible pronunciation.

Even if the language was correctly detected there arises another issue that has great impact on the recognition performance: in many situations the language of the speech engine is different from the language of the title to be spoken. As a consequence, e.g., a non-French who wants to select "Je ne regrette rien" by Edith Piaf will produce a more or less strong foreign accent. Whether a sophisticated phoneme substitution technique can handle such cases in a satisfying way remains to be clarified when voice control for portable players becomes more widespread.

At the moment most systems operate with an English speech recognizer as the majority of titles and albums are in English. As a consequence, various issues of multilinguality are still the most prominent challenges for the *MP3 title selection* use case.

16.3.3 Navigation Destination Entry

A challenging task for speech recognition is the input of a destination into *navigation* system. Depending on what is regarded as a city there are between 50 k and 100 k entities in Germany, comprising up to 12 k streets. Due to this large number of active words combined with the adverse sound conditions in a moving car the recognition result is typically provided as an n-best-list on the display of the head unit for final selection by the user.

Since the pronunciation of city and street names is often non-regular, the corresponding phoneme strings are normally contained in the navigation database. In case of a destination entry, the phonetic data for all city names (or all street names respectively) has to be provided to the recognition engine. This poses strong requirements on either data transfer rates in case of separated units for navigation and recognition or on memory space to cache all information in advance, since the response time of the recognizer increases considerably when adding the time for data transfer to the recognition time itself.

It can be observed that there a different cities with the same name as well as identical phonetic representations for names with different writings. As an example, the phonetic pronunciation "*Snalt*" for a German city corresponds to 5 different orthographic forms of 11 cities:

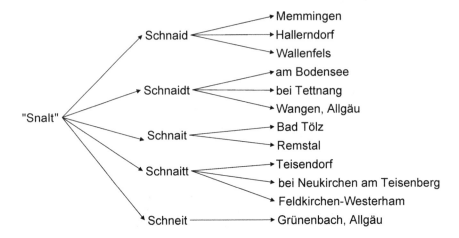

The task here is to provide a multi-modal HMI for recognition in combination with an appropriate approach to resolve such ambiguities. As a solution the list of different written forms is displayed and the final destination is selected in a consecutive dialog step.

16.3.4 Manuals and Help Systems

"Where is the gas cap located?" When approaching a gas station with a rental car this question is not unlikely to arise, and a spoken dialog *help system* in that car would then be very welcome. However, to provide an appropriate interaction the dialog has to perform much more than simple *command and control*. Such a system must be able to understand questions in natural language, and it has to understand as many variations of this question as possible.

User:	How do I start a conference call?
System:	In order to start a conference call, call the first participant. Then press the call back button, ...
User:	Which one is the call back button?
System	The call back button is found on ....

Fig. 16.3 Example: spoken dialog help system for a PABX

Siemens Speech Processing explored this task by generating an interactive help system for a PABX telephone system (see Block et al. 2004), for which about 200 different help topics have been modeled. For each topic a key grammar for possible phrasings of the question (all in all about 1150 words) as well as an answer prompt with the required help was developed. In addition, prompts for each topic were formulated that are used for a clarification sub-dialog when the user input could not be interpreted straight forward. These elements then constitute the parameters for the dialog engine which combines them into a spoken dialog system (cf. Fig. 16.3). While no further tuning was performed, a usability test showed already 85% task completion.

It seems obvious that—while driving—a user would prefer to access the car *manual* by voice instead of filing through a (even electronic) booklet, making this approach a major step towards improved overall usability in cars.

16.3.5 Access to Structured Web Content

The *internet* offers a lot of information that might be of value for drivers. Again the question arises how a driver can access this information (which was originally designed for graphical interaction) in an eyes-free/hands-free mode when on the road. As participants in the research project SmartWeb[1] Siemens Speech Processing explored the automatic generation of spoken dialogs from structured *Web content*. These dialog applications then allow the driver to gather relevant information in a free voice dialog without too much distraction, leaving his/her hands on the steering wheel.

The chosen approach focuses on Web content that is represented as HTML tables. By structuring its information in headers, columns and lines, a table can be perceived by humans at a glance. Another important characteristic of tables is that they allow for easy comparison of values, cf. example below:

Table 16.1 Gas prices for Bonn and surroundings

Station	Fuel	City	Code	Address	Price/l
Name 1	Normal	Bonn	53115	Street 1	1,189
Name 4	Normal	Bonn	53115	Street 4	1,199
Name 2	Normal	Siegburg	53721	Street 2	1,199
Name 3	Normal	Rheinbach	53359	Street 3	1,179
Name 1	Super	Bonn	53115	Street 1	1,239
Name 4	Super	Bonn	53115	Street 4	1,249

These characteristics are now used to automatically generate speech dialog applications. Tables are collected by a web crawler, sorted according to their usefulness, and normalized. After these steps, the linguistic content of a table is parsed and transformed into three units: introduction, key grammars, and answer prompts.

The introduction is necessary to tell the user what the new dialog application is about, in order to prevent out-of-vocabulary and out-of-domain questions by the user. From the Table 16.1 in Example2 the following introduction would be generated: "*Gas*

---

[1]This work was partially funded by the German Ministry of Education and Research BMBF in the framework of the SmartWeb project under grant 01IMD01K. See Wahlster (2004) and www.smartweb-projekt.de for more information on the SmartWeb project.

Prices in Bonn and surroundings, with information on station, fuel, city, code, address and price per liter. You can ask me for example, what do you know about Name 1?".

In order to model possible questions a user might ask after such an introduction— e.g., *"Where do I get the cheapest super in Bonn?"* or *"How much is super in Bonn?"*—several mechanisms are deployed. First, the system tries to identify the type of a column automatically. If this was successful, it assigns standard grammars that belong to the type determined. For example, the heading *"Price per liter"* followed by many numerical values is likely to be of type "price" and would be assigned the standard grammars provided for querying the price of something, i.e., phrases like *"How much is,"* *"What's the price of"* etc. If no type can be identified, grammars are generated from the values themselves by putting a *"Which"* in front of the column headline, e.g., *"which city,"* and enumerating the values, e.g., *"Bonn"* etc. Through this mechanism a question like *"Where do I find a Shell station in Bonn?"* is captured. Finally, typical phrases for comparison are incorporated wherever a numeric value is found, adding *"cheapest,"* *"cheaper than"* etc. to the vocabulary.

For the automatic generation of an answer prompt it is important, that the system tells the user first what was understood before the answer is given, e.g.: *"As answer to your question about address, super, Bonn, the cheapest price, I found: Hauptstraße 14."*

The content of the table is stored in a serialized file to feed the answers, and some algorithms are added in order to allow cross comparison of the numeric values triggered by the comparative grammars. By this, all relevant information is contained in the system once it is generated from the *internet*, which means that the application does not need to be online while in use. Only when the contents of the table change the system has to be updated. See Berton et al. (2006) for a description of the transport of internet applications into the car and their integration in a multimodal *infotainment* HMI of the research prototype.

16.3.6 Access to Web Services

Another way to access *web content* is given by web services, which offer standardized methods for accessing enclosed data. A web service can be seen as a database or a collection of databases for which a description of fields and access methods are provided in the WSDL format. An increasing number of those services appear in the web, offering access to all kinds of information, e.g., to yellow pages (e.g., dialo.de), event calendars (e.g., eventful.com), or weather reports (wetteronline.de). The information provided in these services is always up to date while the interface remains stable in comparison to normal web pages.

If these services are to be accessed by speech, necessary parameter values have to be gathered in a dialog. This is depicted in Fig. 16.4 for voice access to an event web service, where recognition results are (indirectly) confirmed and missing parameters are asked from the user before the query to the web service is started.

An interface layer for the dialog application was implemented that transforms the question of the user into a correct WSDL query and transmits it, e.g., via UMTS, to the database.

User:	Are there any concerts in Munich?
System:	Concerts in Munich, on which day?
User:	Tomorrow.
System:	Tomorrow, Tuesday, 10th of April?
User:	Yes.
System:	I found seven concerts. Alban Berg Quartett-Haydn, Schönberg, Beethoven. Herkulessaal der Residenz. Start time 8 p.m.
User:	–
System:	More than Soul. Nightclub Bayerischer Hof. Start time 8 p.m.
…	…

Fig. 16.4 Spoken dialog access to web services

The answer sent by the web service—in this case again via UMTS—usually consists of a long table consisting of all entries matching the query. A pre-processing unit phrases the answer so that it can be well read out by the Text-to-Speech and understood by the driver. Two approaches for this pre-processing step were studied. In the first procedure, the items on the answer list are read one after the other and the user can barge in anywhere with commands like "details," in order to hear more information on this specific item, or "drive me there" in order to enter an intermediate destination into the navigation system. The second approach implements an additional dialog built on the answers from the web service. This dialog allows the user to ask for items which otherwise would have been out of vocabulary, e.g.: *"Are there any concerts with Alfred Brendel next week?"*

In contrast to the processed HTML tables described in the section above—for which a one way communication such as broadcast suffices—web service information must always be accessed online. But this also implies the potential to provide the driver with up-to-the-minute information where necessary, e.g., when looking for a free parking lot at the destination—and this can pay in terms of fuel and time savings.

16.4 Automotive Platform Issues and Challenges

The environment for the implementation of speech recognition on *automotive* platforms differs from the one on a mobile phone for a couple of reasons. Before going

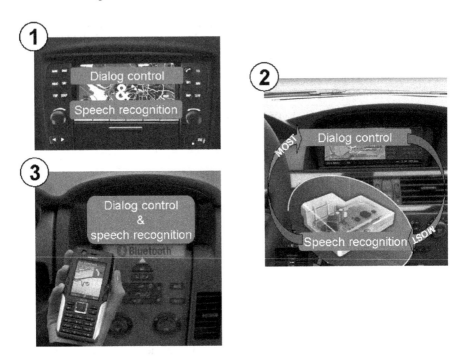

Fig. 16.5 Distribution of dialog control and speech recognition between different processing units in cars: Both in the head unit (a), in different units connected by an automotive bus system (b), in dedicated telephone unit connected to the mobile phone via Bluetooth (c)

into detail here we take a very coarse look at the situation. There is more than one control unit in the car where the actual recognition engine can reside on (cf. Fig. 16.5).

16.4.1 Hardware Constraints

The most salient difference of an *automotive* environment when compared to a mobile device is the amount of electric energy. By this some of the restrictions known from mobile devices vanish and higher clock rates for processors (therefore processing power) and busses (throughput) become possible.

But concerning hardware there are other important topics for automotive platforms to be considered. The quality requirements regarding temperature range for electronic components, data retention time for non-volatile memory and operational life time as formulated in the quasi-standard "AEC-Q100" (Automotive Electronic Council 2003) are much higher than in the ordinary consumer electronics industry. Table 16.2 shows a comparison of typical processors used in automotive infotainment systems to a processor as currently used in office PCs (rightmost column). Apart from the most prominent difference—the clock speed of the processor—there are quite a couple of issues to be mentioned in this context. For systems like the blackfin BF533 (which is a derivate from a DSP development line) it is still important to support an integer implementation of the relevant signal processing parts of speech recognition. Another topic with sometimes underestimated importance is an efficient access to code and data in the memory. Although modern *embedded systems* already feature L1-caches for code and data, L2-caches are often missing and the clock speed for the memory bus is way below the one for current desktop processors. Future developments, e.g., the so-called CarPCs might reduce the gap between automotive hardware to desktop systems, but the special requirements on robustness prohibit the catch up for basic performance numbers.

Table 16.2 Comparison of typical processors for automotive infotainment

	Analog devices ADSP BF533	Freescale i.MX31	Renesas SH7785	AMD Athlon 64 FX
Clock speed	594 MHz	532 MHz	600 MHz	3000 MHz
Native data Type	16/32 Bit	32 Bit	32 Bit	32/64 Bit
L1 I–Cache	16 k	16 k	32 k	64 k (per core)
L1 D–Cache	32 k	16 k	32 k	64 k (per core)
System memory clock	120 MHz	133 MHz	300 MHz	2000 MHz
Floating point	SW	HW	HW	HW

16.4.2 Software Constraints

Not only the hardware, but also the software in automotive environments has to obey certain qualification criteria. The so-called *MISRA* guidelines (The Motor Industry Software Reliability Association 2004) together with an approved development process (The SPICE User Group 2005) set the touchstones for any development of automotive software.

Furthermore, possible complexity of use cases makes it necessary to employ an advanced real time operating system. Even in the case of challenging tasks like the concurrent streaming of two *MP3*-streams for rear seat entertainment with additional recalculation of the route, voice control of, e.g., the radio for station selection should not be deferred on the head unit.

The typical real time operation systems in automotive environments are VxWorks and QNX, with Windows CE becoming a constantly maturing alternative.

The above mentioned software conditions for the deployment of speech recognition as well as the demands on the hardware in automotive environment can be partly contributed to the product life cycle for cars which—being around 6 years—is much higher than for any mobile consumer electronics device.

16.4.3 User Constraints

In the car voice control is part of a multimodal man machine interface. Switching of the input modality between a touch screen, other haptic input, and voice control—which is commonly activated by pressing a push-to-talk key—should be possible at any time. Hence, at system design time there has to be a clear decision on the functionality and use cases which should be operable by voice input.

If the chosen approach attaches variable voice commands to the wording of any possible screen on the display, quite a large number of different recognizer configurations may emerge. This inherently complicates verification and testing of dialog flows.

A crucial point for the acceptance of an automotive control device is the start-up time of the entire system. People expecting the vehicle engine to start at turn-key are unlikely to accept waiting for a long time until the car infotainment system is operable.

16.4.4 Acoustic Channel

The acoustic environment constitutes one of the greatest challenges for automotive speech recognition. The degree of noise to be handled by the recognizer depends on the position of the *microphone*(s) with respect to the speaker and on the quality of the microphone(s) themselves. The microphone is typically mounted in the roof of the car somewhere near the sunshields or next to the rear-view mirror. Hence there is no close-talk situation and the direction of ventilation might be right towards the microphone, e.g., in the defrost operation mode.

The combination of multiple microphones for microphone array processing with beam-forming can be used for the reduction of ambient noise and the masking of people talking on the co-driver's seat. However, since an amount of four microphones quadruples the price on the bill-of-material and additional space and mounting is needed as compared to a single microphone, this option is not chosen very often by the car manufacturers, even though DSPs for microphone array signal processing are available.

16.5 Noise Robust Recognition Technology

As described in Sect. 16.4.4 the reduction of noise captured by the acoustic channel is a challenging task for speech recognition. Various noise sources contribute to a noise mixture that can often reach or exceed the level of the desired speech signal. Stationary noises produced by tires, airstreams, and fan noise sum up with non-stationary noises from the engine, exterior traffic or the indicator.

In speech recognition two basic approaches are used in parallel to handle the problem of noise:

Use of *noise reduction* algorithms
Use of "environment-matched" HMMs.

The purpose of noise reduction is to deliver features as MFCCs with minimal disturbances compared to clean speech. The feature extraction including the noise reduction is performed in the front-end of the Siemens Recognizer Embedded. Currently this front-end is optimized with the goal to exceed the performance of the standardized "advanced front-end" (Ramabadran et al. 2004). Environment-matched HMMs are used in the back-end of the recognizer. They are trained with speech databases which were recorded in an acoustic environment as expected in the applications (Höge 2000). This is motivated by the fact that environment-matched HMMs are theoretically providing the best recognition performance. As already mentioned in Chap. 14 (Speech Recognition in Mobile Phones) databases dedicated to car environment have been produced for many languages by EC-funded projects.

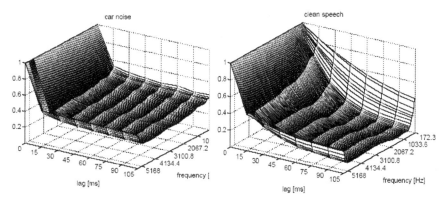

Fig. 16.6 *Autocorrelation function of power spectra* of car noise (left) and of clean speech (right)

As shown in Sect. 16.5.5 noise reduction algorithms improve recognition also in the case when "environment-matched" HMMs are used. The reason for these findings can be seen in the imperfectness of current HMM technology which assumes statistical independencies of features across frames and do not model the strong temporal correlation of spectral features. As shown in Fig. 16.6 and 16.7, the autocorrelation of power spectra extracted per FFT-bin, which are the basis for the MFCC features, show high correlation in time, where the noise reaches after ca. 20 ms its offset (average value of power spectrum) and speech beyond 100 ms. This behavior is quite uniform over the different frequency bins.

As shown in Fig. 16.6, the offset of *car noise* is quite high, which shows the slow change of the level of noise. Another property of the car noise is shown in Fig. 16.8, which shows that the power spectra from car noise and clean speech are quite different (e.g., in contrast to babble noise). This property explains further why recognition in car works quite successfully under noisy condition.

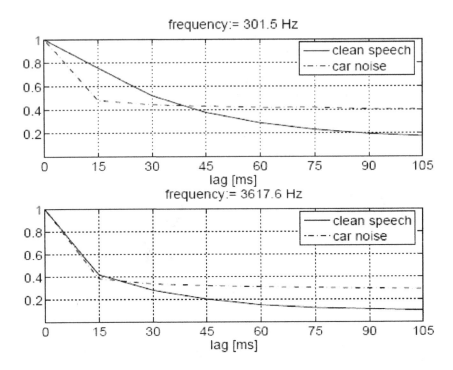

Fig. 16.7 Autocorrelation of power spectra of clean speech and car noise for two selected bins (301,5 Hz; 3617,6 Hz)

In the following we describe shortly the Siemens Recognizer Embedded, afterwards the noise reduction algorithms with related recognition results.

16.5.1 ASR Front-End

The ASR front-end (Varga et al. 2002) is based on the MFCC feature extraction method. Sampling frequency is 11.025 kHz, the length of the audio frames is 23 ms with a 15 ms frame-shift. Per frame a 256 bin FFT is performed and the power spectrum per bin is calculated. This power spectrum is used for noise reduction delivering noise reduced power spectra. After Mel-filtering and logarithmic compression of the power spectra 12 cepstral coefficients and one energy coefficient are computed per frame. To remove spectral bias a maximum likelihood based channel compensation technique is used. From these coefficients delta and delta-delta coefficients are calculated and a "super-vector" containing the coefficients of two consecutive frames is built. The super-vector is reduced to dimension 24 using LDA.

The back-end is based on HMM technology. In the experiments presented below we use a phoneme modeling approach (Bauer 1997).

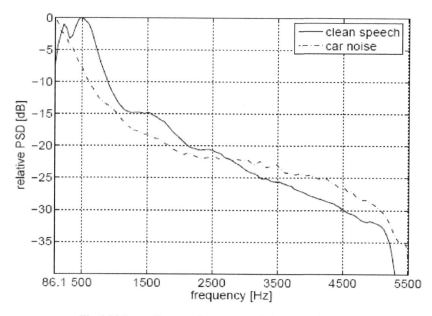

Fig. 16.8 Power Spectra of *car noise* and clean speech

16.5.2 Minimum Mean Square Weighting Rules

Most noise reduction algorithms are computed in the frequency domain, where an individual gain factor is applied for every frequency bin. Short-time spectral analysis based on overlapping speech frames is usually employed as a frequency domain transformation. Many weighting rules are derived in a way, that the mean square error between the original speech signal and the resulting estimated speech signal is minimized, the most famous being the Wiener Filter

$$G_k^W(l) = \frac{SNR_k(l)}{SNR_k(l)+1},$$
(16.1)

where l denotes the frame index, k the frequency index, and $SNR_k(l)$ the Signal-to-Noise-Ratio of the actual frame. As the original speech signal $S_k(l)$ is usually not available $SNR_k(l)$ has to be estimated. An algorithm which represents the state-of-the-art in speech enhancement, the so-called a priori SNR Wiener Filter (Scalart and Filho 1996), is based on the a-priori Signal-to-Noise-Ratio SNR_{prio}. SNR_{prio} is estimated by a decision-directed approach which was first described in Ephraim and Malah (1984). The estimated values of the previous frame for the clean speech spectrum $\hat{S}_k(l-1)$ and noise variance $\hat{\lambda}_{N_k}(l-1) = E\{|N_k(l-1)|^2\}$ are considered recursively for the computation of the actual SNR value:

$$SNR_{k,prio}(l) = (1-\varepsilon) \cdot \max\{SNR_{k,post}(l)-1, 0\} + \varepsilon \frac{|\hat{S}_k(l-1)|^2}{\hat{\lambda}_{N_k}(l-1)} \ , \qquad (16.2)$$

ε being the smoothing constant. The a-posteriori Signal-to-Noise-Ratio $SNR_{k,post}(l)$ is computed from the noisy speech spectrum $Y_k(l)$ and the estimated noise variance of the actual frame:

$$SNR_{k,post}(l) = \frac{|Y_k(l)|^2}{\hat{\lambda}_{N_k}(l)} \ . \qquad (16.3)$$

16.5.3 Recursive Least Squares Weighting Rules

Only the actual frame is considered in the derivation of weighting rules based on the minimum mean square error cost function. However, the key signal components used in the weighting rules, estimations of speech and noise power spectral densities, are computed recursively incorporating the frame(s) before. A good example is the computation of SNR_{prio} in Eq. 16.2.

To avoid this contradiction a family of weighting rules based on the Recursive Least Squares (RLS) criterion can be derived. The dependency of the actual frame on the previous ones is explicitly considered in the corresponding cost function

$$J_{LS}(M) = \sum_{l=0}^{M} w(l) |E_k(l)|^2 \ , \qquad (16.4)$$

where $w(l)$ is a weighting coefficient and M denotes the actual frame. The error $E_k(l)$ is defined as

$$E_k(l) = S_k(l) - G_k(l) \, Y_k(l) \, , \qquad (16.5)$$

$S_k(l)$ representing the clean speech spectrum and $\hat{S}_k(l) = G_k(l) \, Y_k(l)$ its estimate, where $G_k(l)$ denotes the desired gain factor and $Y_k(l)$ the noisy speech signal.

Using Eq. 16.5 in Eq. 16.4 and minimizing the result with respect to $G_k(l)$ leads to the well-known recursive least squares weighting rule

$$G_k^{LS}(M) = \frac{\sum_{l=0}^{M} w(l) Y_k^*(l) S_k(l)}{\sum_{l=0}^{M} w(l) |Y_k(l)|^2} \ . \qquad (16.6)$$

From this generic weighting rule various implementations can be derived, mainly differing in the way of estimating the clean speech signal.

16.5.4 Implementations of RLS Weighting Rules

For the following implementations an exponential weighting coefficient $w(l)$ is used, taking into account all frames from the past,

$$w(l) = \rho^{M-l}, \quad 0 \le l \le M, \quad 0 < \rho < 1. \tag{16.7}$$

1. Recursive Least Squares (RLS) Algorithm

For the derivation of the Recursive Least Squares algorithm described in Beaugeant et al. (2002) the noisy speech signal in the nominator of (16.6) is expressed as $Y_k(l) = S_k(l) + N_k(l)$ and the assumption is made that on average speech and noise are uncorrelated

$$\sum_{l=0}^{M} \rho^{M-l} S_k^*(l) N_k(l) = 0 . \tag{16.8}$$

This results in

$$G_k^{LS}(M) = \frac{E_{S_k}(M)}{E_{S_k}(M) + E_{N_k}(M)} \quad \text{with} \quad E_{U_k}(M) = \sum_{l=0}^{M} \rho^{M-l} |U_k(l)|^2 \quad \text{and} \quad U \in \{S, N\}. \tag{16.9}$$

Subsequently $S_k(l)$ is approximated by the noisy observation $Y_k(l)$. Using different weighting coefficients for noise and speech power spectral densities and introducing an overestimation factor α for the noise term the following weighting rule is obtained:

$$G_k^{RLS}(M) = \frac{E_{Y_k}(M)}{E_{Y_k}(M) + \alpha E_{N_k}(M)} \quad \text{with} \quad E_{U_k}(M) = \sum_{l=0}^{M} \rho_U^{M-l} |U_k(l)|^2 \tag{16.10}$$

$$\text{and } U \in \{Y, N\}.$$

The exponential forgetting factor ρ_U smoothes the involved signals. Usually different values are applied for ρ_Y and ρ_N.

2. Recursive Gain Least Squares (RGLS)

A new weighting rule is obtained by directly applying Eq. 16.6. A recursive formulation for the weighting factors $G_k(l)$ can be stated (Setiawan 2005a):

$$G_k^{RGLS}(M) = G_k^{RGLS}(M-1) + K_k^M r_k^M , \tag{16.11}$$

where K_k^M is a gain factor and r_k^M is called the residual with

$$r_k^M = S_k(M) - Y_k(M) G_k^{RGLS}(M-1) . \tag{16.12}$$

The mathematical derivation is given in Setiawan (2005a). The clean speech signal $S_k(M)$ is approximated by

$$\hat{S}_k(M) = \sqrt{E_{Y_k}(M) - \alpha E_{N_k}(M)}\; e^{j\Theta_{Y_k}(M)}\,, \tag{16.13}$$

$e^{j\Theta_{Y_k}(M)}$ being the phase component of the noisy signal.

3. RLS with Modified Spectral Subtraction

Another method for estimating the clean speech signal is given in Setiawan (2005a): The same derivation as for the RLS algorithm is used, but the average power spectral density of speech in the nominator of Eq. 16.6 is now computed from the power spectral densities of noisy signal and noise signal averaged over all past frames:

$$\sum_{l=0}^{M} \rho^{M-l} |S_k(l)|^2 = \sum_{l=0}^{M} \rho_Y^{M-l} |Y_k(l)|^2 - \alpha \sum_{l=0}^{M} \rho_N^{M-l} |N_k(l)|^2\,. \tag{16.14}$$

Using the same notations as above results in the weighting rule

$$G_k^{RLS-sub}(M) = \frac{E_{Y_k}(M) - \alpha E_{N_k}(M)}{E_{Y_k}(M)}\,. \tag{16.15}$$

16.5.5 Recognition Results

Table 16.3 shows some recognition results achieved with the new algorithmic approaches. The recognition tests have been carried out on the Spanish versions of the *SpeechDat Car*, *SPEECON* Car, and SPEECON Adult databases (SpeechDat 2000).

The channels given in the table represent signals recorded with far-talk microphones at a medium distance of 0.5m–1m. For the SpeechDat Car recordings the microphones have been mounted at typical positions at the car ceiling (channel 2: A-pillar, channel 3: sun visor in front of driver, channel 4: rear mirror). Different isolated word recognition tasks have been examined. The back-end of the ASR, a phoneme-based HMM recognizer trained with 20000 Gaussian densities has been used (Bauer 1997). As a baseline system serves the RLS implementation (Beaugeant et al. 2002). Results using this kind of noise reduction have also been published in the context of the ETSI Aurora front-end evaluation (Andrassy et al. 2001). For comparison purposes the results achieved with the state-of-the art Wiener approach using a-priori SNR estimation (Scalart and Filho 1996) are given as well. It can be seen from the table that a relative improvement of the word error rate up to 18 % can be achieved on the average, the best results are achieved with the modified spectral subtraction approach. We were able to improve these results up to 22 % relative word error rate improvement by combining the new weighting rules with root compression algorithms (Setiawan 2005b).

Table 16.3 Recognition results with SpeechDat Car ES and SPEECON Adult ES—Word Accuracies (ACC) as defined in Eq. 16.16; WER=1—ACC

Algorithm →			RLS Eq. 16.10	A-priori SNR Wiener Eq. 16.1	RGLS Eq. 16.11	RLS with mod. spectral subtr. Eq.16.15
Database ↓			Word accuracies			
SpeechDat Car ES(306 speakers)	858 commands	Channel 2	93.0	92.8	92.0	93.0
		Channel 3	92.0	92.5	92.4	92.3
		Channel 4	88.9	88.6	89.9	90.3
	150 city names	Channel 2	87.8	88.0	90.6	90.0
		Channel 3	88.4	89.2	90.7	91.1
		Channel 4	87.6	88.8	89.2	89.0
Speecon Car ES(625 speakers)	208 application specific words	Channel 2	85.4	88.0	88.2	88.8
		Channel 3	87.2	90.2	89.6	90.0
	100 city names	Channel 2	79.8	82.7	82.3	82.7
		Channel 3	81.8	85.6	85.8	85.8
Speecon Adult ES(561 speakers)	208 application specific words	Channel 2	92.2	95.0	94.2	94.3
		Channel 3	89.5	92.3	92.6	92.6
		Mean word accuracy	87.8	89.5	89.8	90.0
		Rel. WER improvement	0.0	13.9	16.4	18.0

16.6 Methodology for Evaluation of Automotive Recognizers Quality Measurement Using SNR Curves

The evaluation of speech recognition systems is an essential issue for the customer acceptance of a product. However, evaluation results of individual customer-specific tests are hardly comparable due to different test setups, while database tests often do not reflect the real-life environment for the final product.

Therefore, we have developed a well-defined procedure for an independent evaluation of speech recognizer products in real-life car-environments. The procedure has been exemplified and validated in comprehensive in-car tests by deploying the Siemens Recognizer Embedded based on the RLS Algorithm shown in Eq. 16.10 of

Sect. 16.5 allowing a comprehensive, objective and comparable assessment of recognizer performances under real-life conditions.

16.6.1 Common Evaluation Procedures

The common evaluation procedures for speech recognition products can roughly be subdivided into three different approaches: customer-specific tests, database tests, and a hybrid approach of both.

With customer-specific tests the fulfillment of different customer requirements can be verified very well. However, there is nearly no possibility to compare evaluation results from different customer-specific test setups.

With tests on common databases it is possible to produce comparable test results. However, such database tests often do not reflect the real-life environment for the final product: for example, the audio path of the final product is not taken into account, and the recordings were in most cases performed in cars different to the target car. Finally, common databases often do not contain all commands of the final product, and therefore allow to verify only a subset of the overall commands.

The following approach can be seen as a hybrid of a customer-specific test and common database tests: clean speech from databases is mixed with noise recorded with the final target in the final environment. However, this approach has also some disadvantages. First, as for database tests already stated, often not all commands of the final product are included in a database. And second, the mixing of clean speech and noise is not the same as speech recorded in real noise, as e.g., due to the Lombard-effect speech characteristics often change under noise (Junqua 1993).

16.6.2 Proposed SNR-Approach

In order to overcome the constraints of the previously described approaches, we propose in the following an objective and practical evaluation procedure especially for automotive environments based on SNR (Signal-To-Noise Ratio) values. For our evaluation procedure, recordings are taken on a normalized roundtrip with typical traffic and road situations. For each utterance a specific signal-to-noise ratio is calculated to assign the utterance to an SNR-bin of the main car noise range. In a defined evaluation procedure, the SNR-bins are compiled into normalized SNR recognition curves. This SNR-based approach has the advantage of better comparability in opposition to conventional tests with fixed driving speed (e.g., 0/50/130 km/h), as environmental conditions like weather, tires or road type are implicitly considered. Furthermore an SNR-based approach takes the speaker loudness correctly into account.

16.6.3 Data Recording

To provide a comparable and comprehensive assessment of the recognizer performance, recordings from 12 test speakers (6 male and 6 female from target group) are taken on a normalized roundtrip. All recordings are performed with the final target. The audio signals are recorded as provided to the recognizer engine as well as the recognition results. The normalized roundtrip should contain most of normal traffic situations and road types like town traffic, country roads and highways. To diversify the in-car situation, recordings are taken with opened as well as with closed windows and comprise different settings of the air conditioning. All test speakers get a list of the same test utterances. This list should contain all commands, that have to be tested, and every command should occur in the same quantity (preferably at least five times each).

16.6.4 Evaluation

After the calculation of the signal-to-noise ratio with a well-defined algorithm (Höge and Andrassy 2006) for every recorded utterance, all recordings are grouped into SNR-bins from 2 to 16 with a step of 2, where every SNR-bin X includes all recordings with a signal-to-noise ration higher or equal $X - 2$ and lower $X + 2$. The idea of this grouping is that every utterance within the same SNR-bin has the same "level of challenge" for the recognizer, as the challenge notably depends on the ratio between the intensity of speech signal and environmental noise. Furthermore the grouping summarizes the utterances in few SNR-bins improving the statistical relevance of recognition results.

After this grouping, the word accuracy ACC_{mean} for every SNR-bin S is calculated:

$$ACC_{mean}(S) = 1 - \frac{subst(S) + del(S) + ins(S)}{utt(S)} \qquad (16.16)$$

where $utt(S)$ is the total number of recorded utterances in the SNR-bin S, $subst(S)$ is the number of false recognized (substituted) utterances, $del(S)$ is the number of not recognized (deleted) utterances, and $ins(S)$ is the number of additionally recognized, but not spoken (inserted) utterances in the SNR-bin S.

To avoid a strong influence of very good or very bad recognized speakers on the word accuracy, the best and the worst speaker are removed from all SNR-bins. For this purpose, an individual word accuracy ACC_n over all SNR-bins S is calculated for every speaker n as follows

$$ACC_n = \frac{\sum_{S=min}^{max}[ACC_n(S) - ACC_{mean}(S)] * utt_n(S)}{\sum_{S=min}^{max} utt_n(S)} \qquad (16.17)$$

where $ACC_{mean}(S)$ is the mean word accuracy over all speakers for the corresponding SNR S from equation (1), and $utt_n(S)$ is the number of utterances of speaker n in the SNR-bin S.

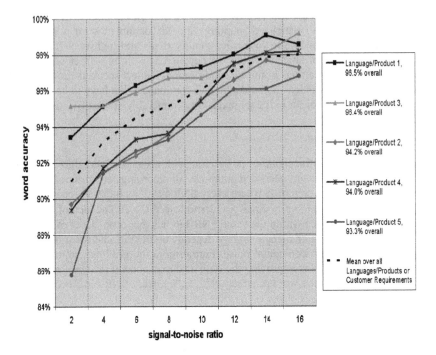

Fig. 16.9 SNR-curves for different languages/products

With this approach the decision on best and worst speaker is based on their relative performance to the other speakers per SNR-bin rather than on an overall word accuracy, which heavily depends on the particular noise conditions per speaker.

After removing all utterances from the best and worst speakers, the final word accuracy for every SNR-bin is calculated receiving SNR-curves like shown in the Fig. 16.9. These SNR-curves shown in Fig. 16.9 give a comparative overview about the performance of a recognition product under different aspects. First, the distribution of a SNR-curve shows the characteristics of the recognizer under different noise levels. Second, all recognition curves (e.g., for different languages or products) are directly comparable, as every SNR-curve has been created under the same conditions and subdivided into the SNR-bins by the same algorithm. For example, if two recognizers with a similar performance have been recorded under different weather conditions (e.g., sunny vs. rainy), the overall word accuracy will normally differ due to the different noise levels. However, with our approach, SNR-curve of both recognizers can be compared directly, as every SNR-bin reflects a similar noise level for the corresponding utterances.

The SNR-curves can finally be translated into normalized SNR-curves as shown in Fig. 16.10, taking the mean over all curves within every SNR-bin (or e.g., customer requirements) as baseline

$$ACC_{relative}(S) = \frac{ACC(S) - ACC_{baseline}(S)}{ACC_{baseline}(S)} \tag{16.18}$$

With such a normalized representation it is now very easy to compare the recognition performances visually.

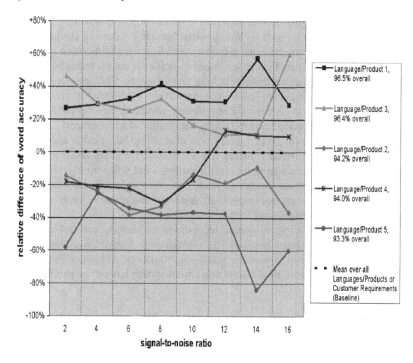

Fig. 16.10 Normalized SNR-curves for different languages/products

16.6.5 Best Practice

We experienced some issues to be considered to get a good coverage of the utterances over the whole SNR-range and to avoid unnatural accentuation of the test utterances. First of all, the sequence of the test utterances should be varied a little bit for every speaker to avoid dependences between certain utterances in the sequence and certain traffic situations, as often the same circuit will be driven. Second, the sequence of utterances should not contain a series of same commands. Otherwise the test speakers sometimes start to play with the accentuation of this command. The same applies for unnatural digit-sequences (e.g., like 0102030405). Finally, the round trip should contain as much different driving situations as possible to get a good distribution of the utterances over the whole SNR-range for all speakers and therefore to get a comprehensive assessment of the recognizers performance.

16.7 Conclusion

Various voice driven automotive applications will be installed in car in the coming years. The most challenging task is destination input for navigation due to the large vocabulary of city and street names. The Siemens Speech Processing Group delivers a set of speech processing components, which are suited for realizing these applications on the automotive platforms. Recent advances in noise reduction technology are presented which lead to further improvement in recognition rate. The proposed methodology for evaluation of recognizers has been exhaustively field-tested in comprehensive in-car tests by deploying the Siemens Recognizer Embedded. Several car manufacturers request the proposed procedure for judging the quality of a recognizer.

Future developments will go in the direction of multimodal, speaker adapted dialog technology, where also infrared cameras will be involved. Fusing facial and acoustic features will improve the recognition rate (lip reading) and will improve speaker characterization parameters (stress, uncertainty, etc.) to allow speaker state adapted dialog steering. For the next years we expect advances in microphone array technology (hardware and software) improving further recognition.

Still some "hard" problems in speech recognition technology will not be solved in the near future and need basic research. Substantial improvement in recognition performance on phoneme level is needed to achieve human performance. Further variations in pronunciation caused by dialectal and casual speech have to be handled.

References

Andrassy, B., Hilger, F. and Beaugeant, C. (2001) Investigations on the combination of four algorithms to increase the noise robustness of a DSR front-end for real world car data. In *Proceedings of Automatic Speech Recognition and Understanding Workshop.*

Automotive Electronic Council (2003) Stress Test Qualification for Integrated Circuits, AEC—Q100—Rev-F.2, 2003-07-18, Automotive Electronics Council, Component Technical Committee.

Bauer, J.G. (1997) Enhanced control and estimation of parameters for a telephone based isolated digit recognizer. In *Proceedings of IEEE International Conference of Acoustics, Speech, and Signal Processing (ICASSP)*, pp. 1531–1534.

Beaugeant, C., Gilg, V., Schönle, M., Jax, P. and Martin, R. (2002) Computationally efficient speech enhancement using RLS and psycho-acoustic motivated algorithm. In *Proceedings of World Multi-Conference on Systemics, Cybernetics and Informatics.*

Berton, A., Regel-Brietzmann, P., Block, H.U. and Schachtl, S. (2006) Integration of Scalable Dialog Systems in Cars. In *Proceedings of ESSV*, Freiberg.

Block, H.-U., Caspari, R. and Schachtl, S. (2004) Callable Manuals – Access to Product Documentation via Voice. *"it" Information Technology*, Vol. 46, Oldenburg Verlag, München, pp. 299–305.

Ephraim, Y. and Malah, D. (1984) Speech enhancement using a minimum mean-square error short-time spectral amplitude estimator. *IEEE Transaction on Acoustics, Speech and Signal Processing*, Vol. 32, no. 6, pp. 1109–1121.

Höge, H. (2000) Speech database technology for commercially used recognizers-status and future issues. In *Proceedings of Workshop XLDB on LREC 2000*, Athens.

Höge, H. and Andrassy, B. (2006) Human and machine recognition as a function of SNR. In *LREC 2006 ELRA, Genoa, Italy*, pp. 2060–2063.

Junqua, J.C. (1993) The Lombard reflex and its role on human listeners and automatic speech recognizers. *Journal Of the Acoustical Society of America*, Vol. 93, pp. 510–524.

Ramabadran, T., Sorin, A., McLaughlin, M., Chanzan, D., Pearce, D. and Hoory, R. (2004) The ETSI extended distributed speech recognition (DSR) standards. In *Proceedings of IEEE ICASSP*, Vol. I, pp. 53–56.

Scalart, P. and Filho, J., (1996) Speech enhancement based on a priori signal to noise estimation. In *Proceedings of ICASSP*, pp. 629–632.

Setiawan, P., Beaugeant, C., Stan, S. and Fingscheidt, T. (2005a) Least-squares weighting rule formulations in the frequency domain. In *Proceedings of Electronic Speech Signal Processing Conference* (ESSP), September 2005.

Setiawan, P., Suhadi S., Fingscheidt, T. and Stan, S. (2005b) Robust speech recognition for mobile devices in car noise. In *Proceedings of European Conference on Speech Communication and Technology (EUROSPEECH)*.

SpeechDat (2000) http://www.speechdat.org.

The Motor Industry Software Reliability Association (2004) MISRA-C: 2004—Guidelines for the use of the C language in critical systems, *MIRA Ltd., Warwickshire*.

The SPICE User Group (2005) Automotive SPICE Process Assessment Model, Version 2.2, 2005-08-21 (see www.automotivespice.com)

Varga, I., Aalburg, S., Andrassy, B., Astrov, S., Bauer, J.G., Beaugeant, Ch., Geissler, Ch. and Höge, H. (2002) ASR in Mobile Phones—An Industrial Approach. *IEEE Trans. Speech and Audio Processing*, Vol. 10, no. 8, pp. 562–569.

Wahlster, W. (2004) SmartWeb—Mobile applications of the semantic web. In P. Dadam and M. Reichert (eds.), Springer *GI Jahrestagung 2004*.

17

Energy Aware Speech Recognition for Mobile Devices

Brian Delaney

Abstract. As portable electronic devices move to smaller form-factors with more features, one challenge is managing and optimizing battery lifetime. Unfortunately, battery technology has not kept up with the rapid pace of semiconductor and wireless technology improvements over the years. In this chapter, we present a study of speech recognition with respect to energy consumption. Our analysis considers distributed speech recognition on hardware platforms with PDA-like functionality. We investigate quality of service and energy trade-offs in this context. We present software optimizations on a speech recognition front-end that can reduce the energy consumption by over 80% compared to the original implementation. A power on/off scheduling algorithm for the wireless interface is presented. This scheduling of the wireless interface can increase the *battery lifetime* by an order of magnitude. We study the effects of wireless networking and fading channel characteristics on distributed speech recognition using Bluetooth and IEEE 802.11b networks. When viewed as a whole, the optimized distributed speech recognition system can reduce the total energy consumption by over 95% compared to a software client-side ASR implementation. Error concealment techniques can be used to provide further energy savings in low channel SNR conditions.

17.1 Introduction

In this chapter we present software and hardware optimizations to reduce the energy consumption for distributed speech recognition on portable hardware with PDA-like functionality. We concentrate specifically on general-purpose hardware including StrongARM processors, IEEE 802.11, and Bluetooth networks. We explore the energy design space with respect to delay, wireless channel characteristics, and local processing capability. We begin with a brief overview of battery technology followed by a review of energy-aware design principles. Next, we present energy optimization results for the speech recognition front-end on the HP Labs Smartbadge hardware platform. Finally, energy tradeoffs with respect to the wireless network interface are explored for both IEEE 802.11 and Bluetooth networks.

17.1.1 Battery Technology

In the past 30 years, processor speeds and memory sizes have increased at a staggering rate, while battery technology has only increase by a factor of two to three.

New battery technologies are being developed to minimize this gap, but the fact remains that battery technology has traditionally lagged behind advances of processor and memory technology. With the proliferation of portable electronic devices, this emphasizes the need to use battery resources efficiently.

A battery technology can be rated according to several factors (Green and Wilson 2001):

Energy density The amount of energy stored per unit volume (Wh/l^3)

Specific energy The energy per unit weight of a battery (Wh/kg)

Nominal Voltage The average rated voltage output throughout the discharge cycle (V)

Rated Capacity The amount of current the battery can deliver over a specified period of time (milliamp-hours).

The energy density and specific energy are used to rate the amount of energy with respect to the size and weight of the battery. A battery with a rated capacity of 1000 mAh will be able to deliver current of 1000 mA for one hour, 500 mA for 2 h, or 2000 mA for half an hour. Given the rated capacity and nominal voltage, one can find the total *battery energy* by multiplying the two values.

In the area of rechargeable battery technology, there have been several types over the years. The first, nickel-cadmium (NiCd) technology is virtually non-existent in the marketplace today. This battery technology suffered from low energy densities and a memory effect that reduced the capacity after relatively few charge/discharge cycles. Nickel-metal hydride (NiMH) technology alleviates some of the memory effect of NiCd with increased energy densities, but the total lifetime of the battery is reduced. The most common technology is lithium-ion (Li-ion). Li-ion batteries have a much longer life cycle with increased energy densities, but the charging process requires more sophisticated electronics, which drives up cost. The newest battery technology, lithium-polymer, has even greater energy density than Li-ion but with increased cost. Despite the cost, lithium-polymer has found its way into smaller devices where weight and size are critical. While new technologies are being developed, such as miniature generators and fuel cells, there is an ever-increasing demand for improved battery technology with today's power hungry portable devices.

17.1.2 Energy Aware Design Principles

Given a fixed amount of battery energy, there has been an emphasis on *energy-aware* design principles in the literature. The goals of energy aware design are to put hooks or knobs into the hardware, software, or applications that allow scalability in quality vs. energy. This is different from *low-power* design, which often does not seek to allow scalability. The result of energy-aware design is a system that can adapt to changing conditions and modify its energy usage accordingly. For example, a hand-held video streaming application might opt to send and decode video of decreased quality to extend battery lifetime. In another situation, the user might demand high-quality video, even if only for a short time.

Energy-aware design and scalability must take place at all levels, from the device level to the application layer. Many CPUs already allow energy scalable operation through techniques such as dynamic voltage and frequency scaling. Running a particular application at its lowest frequency and voltage setting that still provides acceptable performance will save energy. Dynamic application of this technique can be difficult since the operating system must have knowledge of the operating requirements of various applications running on the system. For applications such as speech recognition, this information may be difficult to predict far in advance. Operating systems are becoming increasingly aware of energy considerations, but fine grained control requires the assistance from the application layer. Memory subsystems can be designed such that entire banks of memory are shut off when not needed, but, once again, the operating system must maintain control of these adaptations.

Software optimization techniques can also help to reduce energy consumption. By writing software that will run efficiently on a particular platform, the program can use fewer resources, including battery energy. Compiler optimizations only offer marginal improvements in energy consumption. Any significant gains will require optimizations that address bottlenecks with respect to the particular architecture studied. This may include limiting the mathematical precision (i.e. fixed point arithmetic), efficient data structure organization to reduce cache misses, and the use of approximate algorithms when hardware accelerated versions are not available, such as square root, logarithmic, or trigonometric functions.

Table 17.1 Power dissipation for major subsystems of the HP Labs Smartbadge IV

1. Subsystem	2. Power (mW)	3. Percentage
4. CPU	5. 694	6. 21
7. Memory	8. 1115	9. 34
10. Wireless	11. 1500	12. 45
13. Total	14. 3309	15. 100

The wireless network can use significant amounts of power in an embedded system. Table 17.1 shows the power dissipation of various components of the HP Labs Smartbadge IV embedded system. These are average power measurements during some moderate CPU processing and wireless network activity. The 802.11b network interface used almost half of the power of the total system; therefore wireless network optimization is an important consideration.

17.1.3 Related Work

The wireless network power optimization problem has been addressed at different abstraction layers, starting from the semiconductor device level to the system and application level. Energy efficient channel coding and traffic shaping to exploit battery lifetime of portable devices were proposed in Chiasserini et al. (2002). A physical layer aware scheduling algorithm aimed at efficient management of sleep modes in sensor network nodes is illustrated in Shih et al. (2002). Energy efficiency can be improved at the data link layer by performing adaptive packet length and error control (Lettieri et al. (1999). At the protocol level, there have been attempts to improve the efficiency of the standard 802.11b, and proposals for new protocols (Jones et al. 1999). Packet scheduling strategies also can be used to reduce the energy consumption of transmit power. A server-driven scheduling methodology aimed at reducing power consumption for streaming MPEG4 video was introduced in Acquaviva et al. (2003). Savings of as much as 50% in the wireless local area network (WLAN) power consumption, relative to just using 802.11b power management, were reported.

Traditional system-level power management techniques are divided into those aimed at shutting down components and policies that dynamically scale down processing voltage and frequency (Simunic et al. 2001). Energy-performance tradeoffs based on application needs have been addressed (Kravets and Krishnan 2000). A different approach is to perform transcoding and traffic smoothing at the server side by exploiting estimation of energy budget at the clients (Shenoy and Radkov 2003). A new communication system, consisting of a server, clients and proxies, which reduce the energy consumption of 802.11b compliant portable devices by exploiting a secondary low-power channel is presented in Shih et al. (2001). Since multimedia applications are often most demanding of system resources, a few researchers studied the cooperation between such applications and the OS to save energy.

There have been several studies of power consumption with respect to speech recognition. Analog signal processing techniques were used in Smith and Hasler (2002) to build both the signal processing front-end and HMM acoustic modeling. In analog signal processing, DSP algorithms are realized with analog CMOS circuits. Power consumption estimates for the front-end were less than 100 microwatts. Reducing the computation and memory access can also reduce energy consumption. By using a subset of available features for likelihood computation Li and Bilmes (2005) report a 27%–43% reduction in power consumption using a cycle accurate energy consumption simulator. Other authors have considered custom chip architectures designed specifically for speech recognition. By exploiting parallelism in the speech recognition process Krishna et al. (2003) were able to increase battery lifetime by about 25% while improving recognition speed on a custom XScale-based architecture. In Nedevschi et al. (2005) a custom hardware architecture for low-power speech recognition was introduced. Power consumption was found to be about 12 times lower than that of a software-based ARM processor solution. In Mathew et al. (2003) a hardware accelerator for Gaussian evaluation was built alongside a general purpose processor. The resulting system used 100 times less energy than a Pentium 4 system when running the CMU Sphinx recognition system. Figure 17.1 shows the power consumption estimates of several published speech recognition applications on general purpose hardware, DSP chips, and custom ASIC designs.

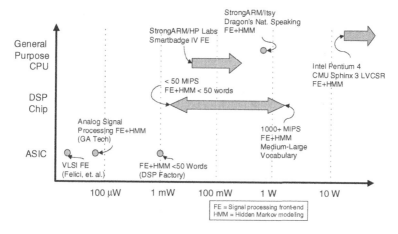

Fig. 17.1 Power consumption figures for various ASR hardware/software configurations

17.2 Case Study of Distributed Speech Recognition Using the HP Labs Smartbadge System

As we have seen, implementing high quality speech recognition on an embedded system, such as a cellular phone, PDA, or other device is a difficult challenge. In this section, we discuss some of these challenges in detail and present some solutions. First, we present a software based front-end feature extraction for a distributed speech recognition system that is designed for minimal power consumption. Through algorithmic, architectural optimizations, and dynamic voltage scaling, we are able to reduce the energy consumption of the signal processing algorithm on a general purpose processor by 89%. Next, we model and analyze the energy required to transmit speech features across a network using IEEE 802.11 and Bluetooth networks.

17.2.1 Signal Processing Front-End

This section describes the optimization of a signal processing *front-end feature extraction* for a distributed speech recognition system. The baseline system used in the experiments is version 0.3 of the open-source Sphinx II speech recognizer from Carnegie Mellon University. The optimization methods used for the algorithm substantially decrease the power usage while increasing speed (measured in processor cycle counts). Estimates of total power usage are performed using a cycle-accurate energy consumption simulator (Simunic et al. 2001a).

The architecture of the *embedded system* simulated in the experiments mimics that of the Smartbadge IV system developed at the Appliance Platform department of Hewlett-Packard Laboratories (Maguire et al. 2004). It is based on a *Strong-ARM* processor running a lightweight Linux O/S. In addition to performing energy consumption simulations to evaluate the quality of source code optimizations, we also implemented and ran the optimized version of the front-end on Smartbadge IV hardware. We found that real-time signal processing of speech is possible at eleven discrete CPU frequency and voltage settings, thus enabling further power savings.

Since the front-end feature extraction step is relatively low in complexity, it is desirable to perform this step on the embedded device and to send compressed features across the network. The signal processing itself consists of a pre-emphasis filter, an FFT, filter-bank computation, a DCT, and a logarithm. It has been shown that mel-frequency cepstral coefficients can be compressed with little effect on the error rate of the speech recognizer (Zhu and Alwan 2001). The ETSI standard for distributed speech recognition describes algorithms to compute, compress, and transmit these speech features (Pearce 2001). We consider several bit rates and quantization levels, including one that is similar to the ETSI standard.

The source code optimizations can be grouped into two categories. The first category, architectural optimizations, aims to reduce power consumption while increasing speed by using optimization methods targeted to a particular processor or platform (e.g. an embedded system with no floating-point hardware). Ideally, many of these optimizations should be done by a compiler. However, currently available compilers for most embedded systems do not have these optimizations built-in. In addition, measurements presented in Simunic et al. (2001) show that the improvements that can be gained using standard compiler optimizations are marginal compared to writing energy efficient source code. The second category of source code optimizations is more general and involves changes in the algorithmic implementation of the source code with the goal of faster performance with less power consumption.

The final optimization presented in this work, *dynamic voltage scaling* (*DVS*), is the most general since it can be applied at run-time without any changes to the source code. Dynamic voltage scaling algorithms reduce energy consumption by changing processor speed and voltage at run-time depending on the needs of the applications running. The maximum power savings obtained with *DVS* are proportional to the savings in frequency and to the square of voltage.

Profiling of the original source code under a *StrongARM* simulator revealed that most of the execution time was spent in the computation of the DFT (which is implemented as an *FFT*). Since speech is a real-valued signal, an N-point complex FFT can be reduced to an N/2-point real FFT. Some further processing of the output is required to get the desired result, but this overhead is minimal compared to the reduction in computation. Additional savings can be obtained when the trigonometric functions used in the computation of the FFT are pre-computed and stored in a lookup table, thus eliminating multiple function calls in the FFT loop. Algorithmically, the source code is now ready for optimizations specific to the StrongARM architecture.

Further profiling of the source code on a StrongARM simulator revealed that over 90% of the time was spent in floating-point emulation. The StrongARM has no on-chip floating-point processor, so all floating-point operations must be emulated in software. Simply changing from double- to single-precision floats improved the performance considerably. However, additional profiling showed that 80% of the time was still being spent in floating point emulation. Any further gains require *fixed-point* arithmetic. Implementing a pre-emphasis filter and Hamming window using fixed-point arithmetic is straight-forward. Fixed-point *FFT*s are well studied and have often been implemented on digital signal processor chips. Careful attention must be paid to the location of the decimal point to avoid overflow while maintaining precision.

After passing the input frame through the *FFT*, the mel filter bank must be applied. The filter bank amplitudes are calculated using the squared magnitude. This presents some challenges since this squared number multiplied by the filter coefficients, $H_i[k]$, can easily overflow the 32-bit registers. A 64-bit result can be obtained from the StrongARM multiplier using assembly language, but overflow can be avoided simply by rewriting the filter bank equation to use just the magnitude:

$$Y[i] = \sum_{k=0}^{N/2} \left(|X[k]| \sqrt{H_i[k]} \right)^2 \qquad (17.1)$$

This avoids overflow since $H_i[k] < 1$, therefore the result of each multiplication is small. The coefficients, $\sqrt{H_i[k]}$, are stored in a lookup table. The one drawback to this method is that computing the magnitude requires a square root operation. Fast integer square root algorithms exist, but they must be used on each output from the *FFT*, which is costly. Fortunately, the magnitude can be estimated as a linear combination of the real and imaginary parts using the following equation (Frerking 1994):

$$|x| \approx \alpha \max(|\Re\{x\}|, |\Im\{x\}|) + \beta \min(|\Re\{x\}|, |\Im\{x\}|) \qquad (17.2)$$

where α and β can be chosen to minimize the mean squared error, and $\Re\{x\}$ and $\Im\{x\}$ represent the real and imaginary parts of the complex number x.

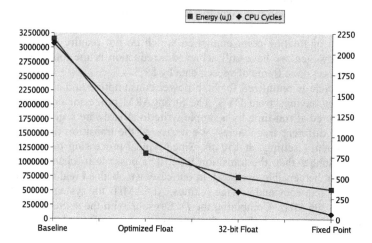

Fig. 17.2 Cycle count (left axis) and energy consumption (right axis) per frame of speech (Delaney et al. 2002, © 2002 IEEE)

Computing the first 13 coefficients of the *DCT* is relatively easy to do in *fixed-point* arithmetic, but taking the natural *logarithm* is a more difficult task. One possible option is to perform a floating-point logarithm, but profiling showed that the logarithm itself as well as the transition to and from fixed-point is costly. A fixed-point logarithm using a polynomial expansion requires some divides, which are slow on the StrongARM. However, we can approximate the logarithm in base 2 using simple bit manipulation (Crenshaw 2000). A shift and scaling of the result is used to obtain the natural logarithm for a fixed point number x, whose decimal is located at the nth bit.

$$\ln\left(\frac{x}{2^n}\right) = \left[\log_2(x) - n\right]\ln(2) \qquad (17.3)$$

Three main criteria are considered in order to evaluate the effectiveness of a particular optimization: performance (in terms of processor cycle count), energy consumption, and accuracy or word error rate (WER). Simulation results for processing one frame (25 ms) of speech on the Smartbadge IV architecture running at 202.4 MHz are shown in Fig 17.2. The x-axis shows the source code in various stages of optimization. The "baseline" source code contains no software optimizations. The "optimized float" code contains the algorithmic optimizations as well as some additional source code optimizations. Double-precision floating-point numbers were changed to single-precision 32-bit floats in the "32-bit float" version of the code. Finally, the "fixed-point" implementation contains all of the source code optimizations described in this chapter. For each version of the code, we report the performance (in CPU cycles) and the total *battery energy* consumed (in μ Joules). The simulation results are computed by the cycle-accurate energy simulator, and include processor core and level 1 cache energy, interconnect and pin energy, energy used by the memory, losses from the DC/DC converter, and battery inefficiency. The reduction in energy consumption is not as dramatic as the reduction in cycle count for the fixed-point version due to an increase in memory references per unit of time. In fixed-point code, basic math operations are reduced to a few cycles as opposed to long iterations of floating-point emulation which do not require as many memory references. However, we have still achieved a reduction in the total battery energy required to process one frame of speech data by 83.5%.

Once the code is optimized for both power consumption and speed, we investigate the energy savings from DVS. The StrongARM processor on Smartbadge IV can be configured at run-time by a simple write to a hardware register to execute at one of eleven different frequencies. We measured the transition time between two different frequency settings at 150 ms. Since typical processing time for the front-end is much longer than the transition time, it is possible to change the CPU frequency without perceivable overhead. In our case, we obtained real time performance at all possible frequency and voltage settings. At 59 MHz the system uses 34.7% less power than at 206 MHz. Combining the *DVS* results with the source code optimizations, we calculate the overall reduction in power consumption to be 89.2%.

Finally, we include the fixed-point vector quantization code in our profiling and consider different *bit rates* and *quantization* levels. Although some differing techniques have been proposed, the most common technique for compressing *Mel-frequency cepstral coefficients* (*MFCC*) is some form of *vector quantization*. For our system, we use an intra-frame product code vector quantization scheme presented in Pearce (2001). We train a set of codebooks using a K-means training algorithm with bit rates ranging from 1.2 kbps to 2.0 kbps. We include an additional bit allocation that is similar to the ETSI standard that will operate at 4.2 kbps. In general, we can expect increased WER at lower bit rates.

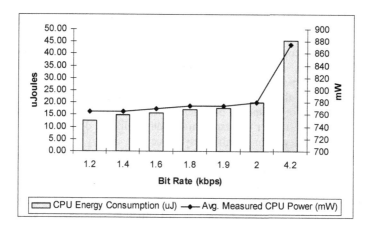

Fig. 17.3 Computational energy usage and measured average power for different quantization bit allocation schemes

Source code to perform the quantization of the MFCC data was written in *fixed-point* for the StrongARM processor and profiled using the energy consumption simulator. Figure 17.3 shows a comparison of energy consumption for various vector quantization bit allocation schemes. The bars represent the total energy consumption per frame of speech for the quantization step, and the line represents the measured CPU power dissipation at each bit rate. The measured values closely match the results from the energy consumption simulator. There is approximately a 14% increase in CPU power consumption but a greater than 50% reduction in WER between the highest and lowest bit rates. Even at the highest bit rate, the vector quantization is only 12% of the total energy usage. This suggests that speeding up the quantization process by using smaller codebooks would produce minimal reductions in energy consumption and would have a much greater impact on speech recognition accuracy.

In this section, we have outlined some optimization techniques to reduce the energy consumption of a particular signal processing algorithm. On embedded systems with no floating-point hardware, fixed-point arithmetic is an important step in lowering the power consumption of a program. However, careful attention must be paid to basic math functions (i.e. cosine, log, etc.) and overflow/underflow issues. Approximate algorithms perform well for certain applications and can result in savings in both time and power usage. By using software optimizations, we were able to achieve a reduction in energy usage by 83.5% compared to the non-optimized source code. We show that additional power savings are possible by scaling processor frequency and voltage at run time, while still meeting the performance requirements. At the lowest frequency/voltage setting, we calculate an overall reduction in power consumption by 89.2%. With the addition of vector quantization, the total energy required to process one frame of speech data is approximately 380 μ Joules.

17.2.2 Energy Consumption of DSR with IEEE 802.11 Wireless Networks

In this section we address the issue of energy consumption of the wireless interface for a *distributed speech recognition* system. As we have discussed earlier, the wireless interface can occupy almost half of the energy budget on many mobile wireless devices. We introduce techniques to minimize the energy consumption required to transmit speech parameters to an ASR server. We model the energy consumption of a DSR system using the IEEE 802.11b wireless interface. By employing synchronous burst transmission of speech parameters, we can maximize the amount of time spent in a low power state or off state while adding minimal delay to the application. Using this technique, we can significantly reduce the energy consumption required for transmission. We explore these tradeoffs with respect to latency, channel conditions, and energy consumption. These techniques can provide reductions in energy consumption of over 90% compared to a software based *client-side* ASR system.

Given the relatively low bit rates used in DSR, these networks will operate well below their maximum throughput range. In this situation, more energy saving opportunities will develop from exploiting moderate increases in application latency by transmitting more data less often. This allows the network interface to either be powered down or placed into a low-power state in between transmissions. Other wireless networks with throughput in the low kbps range, such as many cellular telephony networks, may require other techniques, such as better compression, to minimize energy consumption. However, we do not consider such wireless networks here.

In order to estimate the power consumption for wireless transmission, we directly measure the average current into the network interface. These measurements are performed under ideal conditions with no competing mobile hosts or excessive interference. Using these measurements as a baseline, we are able to tailor a simple energy consumption model to investigate the effects of increased application latency. By buffering compressed speech features, we maximize the amount of time spent in the low-power or off state. We introduce a power on/off scheduling algorithm for the 802.11b device that exploits this increased latency. Given the medium access control (MAC) scheme for both 802.11b and Bluetooth, we can incorporate the effects of channel errors into the energy model. We use these results to investigate which techniques should be used to maintain a minimum quality of service for the speech recognition task with respect to channel conditions.

The *802.11b* interface operates at a maximum bit rate of 11 Mbps with a maximum range of 100 m. The MAC protocol is based on a carrier sense multiple access/collision avoidance schemes, which includes a binary exponential back-off system to avoid collision. It uses an *automatic repeat request (ARQ)* system with CRC error detection to maintain data integrity. We used a PCMCIA 802.11b interface card and measured the average current going into the interface to get the power dissipation.

Our measurements indicate there is only a difference of a few mW in power consumption between the highest and lowest bit rates. This is expected since the bit rates are low, and the transmit times are very short. Also, the use of UDP/IP protocol stacks and 802.11b MAC layer protocols both add significant overhead for small packet sizes. The 11 Mbps WLAN interface is under-utilized with this type of low bit rate traffic. The other 802.11 standards, including 802.11g, have similar operation but with different modulation schemes which provide higher bit rates. However, we can obtain some improvement in power consumption by increasing the number of frames per packet. This increases the total delay of the system, but less battery energy is used since the various networking overhead is amortized across a larger packet size. However, due to the relatively high data rates provided by 802.11b, the WLAN interface spends most of its time waiting for the next packet to transmit. The 802.11b power management (PM) mode can provide some savings in energy consumption but this does not hold under heavy broadcast traffic conditions (Acquaviva et al. 2003), defined as a higher than average amount of broadcast packets. In addition, the PM mode is not available in the ad-hoc (as opposed to infrastructure) topography. We present an on/off scheduling algorithm to reduce the total energy consumption of the 802.11b device under these conditions. While operating in the 802.11b power management mode, a WLAN card goes into an idle state. Every 100 ms it wakes up and receives a traffic indication map, which is used to indicate when the base station will be transmitting data to this particular mobile host. With heavy broadcast traffic, the WLAN interface will rarely be in the idle state and it will consume power as if it were in the always-on mode. This is because the time required to analyze the broadcast packets is larger than the sleep interval. This increase in power consumption will happen even if there are no applications running on the mobile host.

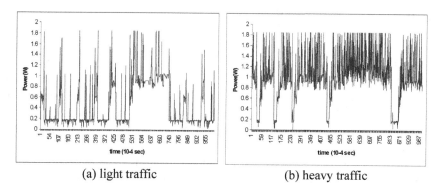

(a) light traffic   (b) heavy traffic

Fig. 17.4 WLAN power consumption in 802.11b PM mode in light and heavy traffic conditions

Figure 17.4 shows the *power consumption* of the WLAN card in the 802.11b power management mode in both heavy and light traffic conditions. Notice that in the left graph, under heavy traffic, the card is unable to transition to the *low-power* idle state very often. The average power approaches the always-on mode. Measurements in Acquaviva et al. (2003) indicate that even in less than average amounts of broadcast traffic, energy is wasted by the extra processing.

Since the energy consumption of PM mode on 802.11b networks breaks down in heavy traffic conditions, we consider an alternate technique. If we are only interested in transmitting speech recognition related traffic and not any other broadcast traffic, we can simply power off the WLAN card until we have buffered enough data to transmit. However, powering the card on and off has an energy-related cost that needs to be accounted for.

Figure 17.5 shows the timing of this scheduling algorithm. The period, T, is determined by the number of speech frames sent in one packet. The transmission is synchronous such that every T seconds we will send that amount of compressed speech features and stay in the off state for the remainder of the time. With larger values of T we can hope to amortize the cost of turning the WLAN card on and off at the expense of longer delay. Assuming that a speech recognizer server is able to process speech at or near real-time, the user will experience delay near the value of T. Depending on the type of application, a longer delay may or may not be acceptable to the end user.

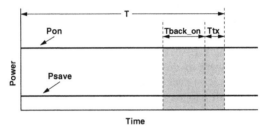

Fig. 17.5 The timing of the 802.11b scheduling algorithm (Delaney et al. 2005, © 2005 IEEE)

The two interesting parameters to consider are the power on time (T_{back_on}) and the number of speech frames transmitted at once, which dictates the total period T. Figure 17.6 shows the power on delay on the x-axis and estimated energy consumption on the y-axis. We fixed the value of T to 0.48 s, or 48 frames of speech data. The PM mode configuration in light traffic almost always outperforms the proposed scheduling algorithm except for very small values of T_{back_on}. Typical values may range from 100 ms to 300 ms, with newer hardware possibly using less time. However, in heavy traffic conditions, the PM mode approaches the always on power consumption (shown by the top line in the plot), so the scheduling algorithm can give better performance under these conditions. With T_{back_on} at 100 ms, the total energy consumption per packet is approximately 75 mJ for the scheduling algorithm and approximately 390 mJ for PM mode in heavy traffic conditions (from Fig. 17.6). This is a reduction in energy consumption by about 80%. However, this only holds true for heavy broadcast traffic conditions, so the mobile device will have to monitor

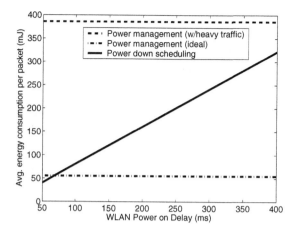

Fig. 17.6 Wireless LAN power on delay vs. energy consumption per packet (Delaney et al. 2005, © 2005 IEEE)

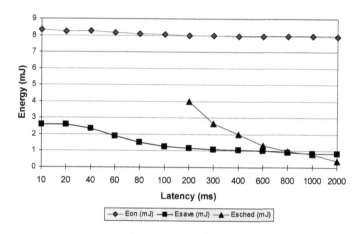

Fig. 17.7 Average energy consumption per 10 ms speech frame versus DSR latency for various 802.11b power management schemes. (WLAN power on delay is fixed at 100 ms.) (Delaney et al. 2005, © 2005 IEEE)

the broadcast traffic and decide between the standard 802.11b PM mode or the scheduling algorithm.

Finally, we consider increased delay or latency, T, in Fig. 17.7 with T_{back_on} fixed at 100ms. In this plot, the energy cost was determined using measured values of power consumption. The energy cost has been normalized to show the average energy required to transmit one frame of speech data. As the total number of frames approaches 80 ($T = 800ms$), we can see that the scheduling algorithm (E_{sched})

will be able to outperform the PM mode configuration (E_{save}) regardless of traffic conditions. This will result in less than one second of delay for a user interface application with speech recognition. Shorter power on (T_{back_on}) times can help move this crossover point to shorter delays. Longer delays of two seconds or more can further reduce energy consumption and are good candidates for applications requiring lower interactivity such as dictation.

Since the 802.11b MAC protocol uses an automatic-repeat-request (ARQ) protocol with CRC error detection to maintain data integrity, the energy consumption will be a function of channel signal to noise ratio (SNR). After the reception of a good packet, an ACK is sent across a robust control channel. For a given bit error rate and packet length, the probability of a packet error in the absence of any error correction coding techniques is:

$$P_r = 1 - (1 - BER)^L \qquad (17.4)$$

where L is the packet length, and BER is the bit error probability for the current channel conditions. For our analysis, we used the BER probability for 256-QAM modulation in a Rayleigh fading channel to approximate the 802.11b CCK modulation (Proakis 1995).

Given the probability of retransmission (P_r), the expected number of retransmissions (T_r) is given by (Wicker 1995):

$$T_r = \frac{1}{1 - P_r} \qquad (17.5)$$

Using these equations, an energy model can be constructed that incorporates the energy used in the MAC overhead as well as the energy required for repeated retransmissions, assuming the average SNR remains the same. Such an energy model is presented in Ebert et al. (2002) and is summarized here:

$$E_{tx}(BER, L) = E_{aq} + T_{ack} \times P_{rx} + \left(E_{aq} + T_{tx} \times P_{tx}\right) \times \frac{1}{(1 - BER)^L} \qquad (17.6)$$

where E_{aq} is the average energy required to acquire the channel, T_{ack} is the time required to receive the ACK packet, P_{rx} is the receive power for the robust control channel, and P_{tx} is the power used during transmission. Given this energy model, we can incorporate it into our scheduling shown in Fig. 17.5.

We use this expression in our final comparison to quantify the energy consumption of 802.11b vs. channel SNR. In particular, we show how larger packet sizes and lack of error correction techniques force 802.11b to operate in higher channel SNR. However, techniques such as packet fragmentation and error correction can be used to extend the lower SNR range of 802.11b.

17.2.3 Energy Consumption of DSR Using Bluetooth Networks

The *Bluetooth* personal area network provides a maximum bit rate of 1 Mbps, and a variety of different packet types are available to support different traffic requirements. It supports a range that is considerably less than 802.11b, on the order of 10 m. Bluetooth supports both data and voice traffic packets as well as a hybrid packet containing both voice and data. Media access is handled via a time-division duplex (TDD) scheme where each time slot lasts 625 μ s. Voice packets are given priority over data packets in scheduling. In this work, we consider only pure voice or pure data packets. Data packets are available in both high-rate and medium-rate packets. These are DHn or DMn packets for both high and medium data rate respectively, where n depicts the number of TDD slots the packet occupies: 1, 3, or 5. High-rate packets use a stop-and-wait automatic-repeat-request (*ARQ*) protocol with CRC error detection within the packet. Medium-rate packets use a 2/3 rate (15,10) shortened Hamming code in addition to the *ARQ* protocol. Voice packets, due to their time-sensitive nature, do not use an *ARQ* protocol. Voice packets are available in HV1, HV2, or HV3 types, where the number denotes the amount of *error correction* rather than slot length. All voice packets occupy one TDD slot with varying data payloads. HV3 packets use no error correction. HV2 packets use the (15,10) Hamming code, and HV1 packets use a 1/3 rate repetition code. Given the soft time deadlines with speech data intended for a machine listener, we can easily use either data packets or voice packets without consideration of packet jitter or delay characteristics.

First we use a model for the energy consumption of a single Bluetooth voice or data packet given in Delaney (2004). We then consider the use of Bluetooth power saving modes to reduce the energy consumption during the idle time, similar to the 802.11b scheduling algorithm. Finally, we investigate the implications of bit errors on both voice and data packets.

Using power measurements of a USB Bluetooth device attached to the Smart-Badge IV combined with our energy model, we are able to estimate the energy usage for our system. Figure 17.8 shows the energy required to transmit one frame of speech data at various DSR compression rates over a Bluetooth link. We consider the use of both high-speed and medium-speed data packets. We assume an error-free channel with no retransmissions. We can see in Fig. 17.8 that there is a higher energy cost for medium-rate packets due to the *forward error correction* (*FEC*) overhead. However, these packets will be a better choice for lower SNR conditions. Energy consumption approximately doubles between the 1.2 kbps and 4.2 kbps bit rates. However, these estimates do not consider idle time between packets that will consume energy as well.

We can incorporate the Bluetooth power saving modes into our model to account for the idle time in between packets. A node within a Bluetooth piconet can operate in a variety of different power management modes. These are connected/transmit, park, hold, and sniff. There is a fixed cost to transition from one mode to the next, and the power consumption of each mode can be measured directly. For our analysis, we will use the *park*, since it provides competitive transition times as well as the

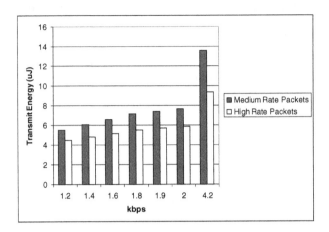

Fig. 17.8 Energy used to transmit one frame of speech with varying compression rates for Bluetooth radio

ability for a master node performing multiple speech recognition requests to support more nodes.

A Bluetooth node in park mode will wake up upon activity to transmit some data and then enter the park mode when finished. The energy consumption of this scenario is as follows:

$$E = P_{tx} \times T_{tx} + E_{transition} + P_{park} \times T_{park} \qquad (17.7)$$

where $E_{transition}$ is the total energy used to transition to/from the various operating states, and P_{park} and T_{park} are the power dissipation and times in the park mode respectively. The time spent in the park state is a function of the overall latency of the system and the amount of data being transmitted. We measure 0.18 watts in the transmit mode, and 0.077 watts in the park mode. Transition times to and from the park state are on the order of several milliseconds each.

Next, we investigate how the presence of bit errors on the wireless channel will affect both the energy consumption and, in the case of voice packets, speech recognition accuracy. We use this data to identify which types of packets can be used effectively in various channel conditions. The main difference between the two types of packets is that voice packets rely only on FEC and no ARQ, while data packets can use both FEC and ARQ. The energy consumption of Bluetooth voice packets is independent of channel conditions. Therefore, we can estimate the energy consumption using an equation similar to Eq. 7. The main difference in energy consumption per frame of speech will come from the reduced user payload due to FEC bits.

Bluetooth voice packets have energy consumption that is independent of SNR since no ARQ protocol is used. By using increased delay, as with the data packets, we can minimize the energy consumption by increasing the amount of time spent in the low-power park state. However, since ARQ is not used, bit errors can have an impact on speech recognition accuracy. Using a combined interleaving and error concealment approach (Delaney 2004), increased delay can improve the ASR accuracy as well as reduce energy consumption in low channel SNR conditions.

In Fig. 17.9, the energy consumption per frame of speech is plotted vs. the interleaving delay. This energy consumption includes the time between transmissions, including the low-power park state. For a given packet type the reduction in energy consumption with respect to increased delay levels off after 64 frames. This knee coincides with accuracy experiments in Delaney (2004), suggesting that delays beyond 0.64 s have little benefit in terms of improved accuracy and decreased energy consumption.

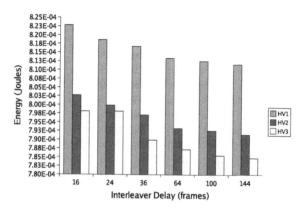

Fig. 17.9 Energy vs. interleaver delay for Bluetooth voice packet types

Conversely, data packets in the presence of bit errors will continue to be retransmitted until they are received correctly or a timeout occurs. For the purposes of this analysis, we calculate the BER using BFSK modulation under a Rayleigh fading channel. By accounting for the various modes of packet failure and error correction performance, the probability of a packet retransmission for a given BER can be derived theoretically (Valenti et al. 2002). Given this information, the energy can be estimated as:

$$E_{Dxn} = P_{tx} \times 625\,\mu s \times n \times \frac{1}{1 - P_r} \qquad (17.8)$$

where P_r is the probability of a retransmit for the appropriate packet type. By dividing the energy by the number of frames in a packet, which varies with packet length and coding technique, we can get the energy required to send one frame of speech.

17.2.4 Comparison of 802.11 and Bluetooth in DSR

We have provided energy models for both 802.11b and Bluetooth wireless networks for distributed speech recognition traffic. The two main variables of interest are the total delay, T, and the average channel SNR. Bluetooth networks do not generally benefit from increased delay, as the energy consumption spent in the park mode dominates the energy usage after about 100 ms. Powering off a Bluetooth node is not an option because the paging/inquiry process to rejoin a piconet can take in excess of 10 s.

For the purposes of this work, it is sufficient to describe the energy requirements for local ASR as the product of the average power dissipation of the processor and memory under load and the time required to perform the speech recognition task. For the Smartbadge IV, we have measured the average CPU and memory power dissipation as $P_{cpu} = 694$ mW and $P_{mem} = 1115$ mW when under full load. Given the real-time factor R for the speech recognition task, we can estimate the energy consumption to recognize one frame of speech as:

$$E_{local} = (P_{cpu} + P_{mem}) \times R \times \frac{1}{100} \qquad (17.9)$$

Therefore, for a speech recognition task that runs R = 2.5 times slower than real-time, we can expect to use approximately 45 mJ of *battery energy* to process one frame of speech. Similarly, we can estimate energy usage for speech recognition that occurs at or near real-time, R = 1, thus providing a range of realistic energy consumption estimates for a *client-side* implementation.

By using the client-side ASR energy model and the DSR energy model for both Bluetooth and 802.11b wireless networks, we can examine the energy tradeoffs with respect to channel quality, delay, and ASR accuracy. Higher bit rates have small increases in system level energy consumption due to the overhead of the power saving algorithms on the wireless device.

In Fig. 17.10, we plot the energy consumption per frame of speech for client-side ASR and DSR under both 802.11b and Bluetooth wireless networks with respect to channel quality. For DSR, we include the both the communication and computation (feature extraction/quantization) energy costs. For 802.11b, we consider the energy consumption of the power on/off scheduling algorithm with a latency of 240 ms, 480 ms, and 2 s and unlimited ARQ retransmissions. For the Bluetooth interface we show the energy consumption for both medium- and high-rate data packets as well as the three types of voice packets with latency of 480 ms. To the right of the Y-axis we have the approximate energy savings over client-side ASR operating 2.5 times slower than real-time. We can expect a scaled down speech recognition task (i.e. simpler acoustic and language models or smaller vocabulary) running at real-time to give 60% energy savings. However, this will come at a cost of reduced functionality for the user, perhaps going to a more constrained vocabulary and speaking style. We have not quantified the cost of reduced utility for the user in this work. However, for the various DSR scenarios in Fig. 17.10 we assume little to no reduction in quality for the end-user by maintaining sufficient data integrity through source coding techniques and/or ARQ retransmissions. Table 17.2 shows the percentages of computation and communication energy for a few different configurations as well as the expected battery lifetime with a 1400 mAh/3.6V lithium-ion cell. The 802.11b interface with long delays gives the lowest overall energy consumption and an almost even division between energy spent in computation and communication. DSR with Bluetooth uses a higher percentage of communication energy, and this amount does not decrease

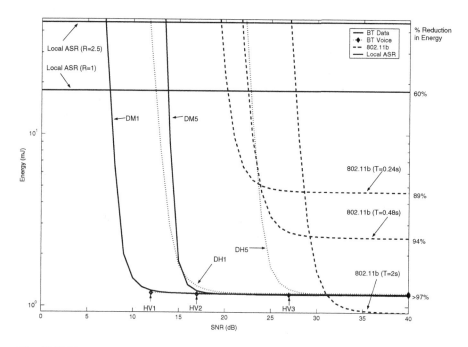

Fig. 17.10 The energy consumption of client-side ASR and DSR under Bluetooth and 802.11b vs. SNR. The Y-axis is log-scale (Delaney et al. 2005, © 2005 IEEE)

Table 17.2 Summary of speech recognition energy consumption in high channel SNR conditions (Delaney et al. 2005, © 2005 IEEE)

Type	Comp. (%)	Comm. (%)	Total/Frame (mJ)	Battery lifetime (h)
DSR w/Bluetooth (T = 0.48 s)	32	68	1.17	43.1
DSR w/802.11b (T = 0.48 s)	15	85	2.5	20.2
DSR w/802.11b (T = 2 s)	42	58	0.92	54.8
Local ASR (R = 2.5)	100	0	45	1.12

significantly with increased delay due to the overhead of the park mode. Even modest delays of less than half a second can yield significant battery lifetime with constant streaming of DSR data.

In a good channel with high SNR, Bluetooth allows system wide energy savings of over 95% compared with full client-side ASR. DH5 packets offer the lowest overhead and best energy savings, while DM1 packets offer the most robust operation down to around 10 dB with some minimal energy cost. The ARQ retransmission protocol causes rapid increases in energy consumption after some SNR threshold is reached. It is possible to operate in lower SNR through packet fragmentation, which will lower the probability of a packet being received in error. This is evident in Fig. 17.10 by comparing DH1 and DH5 data packets. The longer packet length in DH5 packets causes a sharp increase in retransmits and energy consumption at around 25 dB, whereas DH1 packets can operate down 15 dB before the number of retransmits becomes excessive. In addition, FEC bits can be used to lower the probability of a packet retransmit. The Hamming code in DM1 and DM5 packets allows operation down to around 10 and 16 dB respectively.

IEEE 802.11b networks allow system wide energy savings of approximately 89–94% with relatively small values of T. With larger values of T, such as one second or more, we can use less energy than Bluetooth. However, due to the larger packet overhead, larger maximum packet sizes, different modulation, techniques, and lack of error-correcting codes, the 802.11b network does not operate as well in lower SNR ranges. Packet fragmentation or a switch to a more robust modulation technique with lower maximum bit rate can extend the lower SNR range at the cost of increased energy consumption, but we have not considered these effects here. However, 802.11b does offer increased range and may be more appropriate in certain scenarios.

Table 17.3 Lower SNR bound for Bluetooth packets using server-side error concealment and interleaving

Packet type	SNR lower bound (dB)		Energy (μJ)
	ETSI	Error concealment	
HV3	27	17	23.4
HV2	17	10.5	37.5
HV1	12	5	70.3

In Table 17.3, the practical lower SNR bound for distributed speech recognition using Bluetooth voice packets with and without server-side error concealment and interleaving is shown. The table is derived from a series of experiments under various channel conditions from Delaney (2004). The ETSI bit-stream is corrupted by burst errors and speech recognition is performed. The WER is calculated in each case, and the error concealment technique provides more graceful degradation in the presence of bit errors. Error concealment and interleaving can reduce the energy consumption by allowing Bluetooth packets with higher data payloads to be used in lower SNR conditions. Between 27 and 17 dB a 37% reduction in transmit energy is possible since HV3 packets can be used instead of HV2 packets. A 46% reduction in transmit energy between 17 and 10.5 dB since HV2 packets can be used instead of HV1 packets. DSR can still be used down to 5 dB SNR, so the much more expensive client-side ASR does not need to be used.

17.3 Conclusion

In this chapter, we investigated the system-level energy consumption of distributed speech recognition on a portable wireless device. We considered energy usage from both computation and communication in our final analysis. Careful optimization of the signal processing front-end from an energy consumption perspective was performed. The advantages of DSR from an energy consumption perspective are clear. Client-side speech recognition in software can consume several orders of magnitude more energy than a DSR system. However, the use of low-power ASIC chips for speech recognition may help reduce the energy consumption of client-side ASR below that of off-the-shelf hardware.

In our analysis of DSR, we have considered both 802.11b and Bluetooth wireless networks. Given the relatively high bit rates these standards provide with respect to DSR traffic, we investigated the use of synchronous burst transmission of the data to maximize the amount of time spent in a low-power or off state. While this adds a small delay to the end-user, the energy savings can be significant. With 802.11b, we can reduce the energy consumption of the wireless interface by around 80% with modest application delays of just under half a second. Bluetooth offers lower energy consumption for smaller values of delay, T, but as delay increases, the Bluetooth energy consumption is dominated by the time spent in park mode. The 802.11b interface with on/off scheduling can operate with a lower energy consumption than Bluetooth when T exceeds 1.3 s. Through the use of error concealment and interleaving, we can operate Bluetooth voice packets in low SNR conditions with minimal impact on speech recognition and accuracy while still consuming small amounts of energy.

References

Acquaviva, A., Simunic, T., Deolalikar, V., and Roy, S. (2003). Remote power control of wireless network interfaces. In *Proceedings of PATMOS in Lecture Notes in Computer Science*, Springer-Verlag, Turin, September 2003.

Chiasserini, C., Nuggehalli, P., and Srinivasan, V. (2002). Energy-efficient communication protocols. In *Proceedings of DAC*.

Crenshaw, J. (2000). *Math Toolkit for Real-Time Programming*. CMP Books, Lawrence, Kansas.

Delaney, B. (2004). Reduced Energy Consumption and Improved Accuracy for Distributed Speech Recognition in Wireless Environments. Ph.D. Thesis, Georgia Institute of Technology.

Delaney, B., Jayant, N., Simunic, T. (2005). Energy-aware distributed speech recognition for wireless mobile devices. *IEEE Design and Test of Computers.* vol. 22, pp. 39–49.

Delaney, B. Jayant, N. Hans, M. Simunic, T. Acquaviva, A. (2002). A low-power, fixed-point, front-end feature extraction for a distributed speech recognition system. In *Proceedings of ICASSP*, pp. I-793–796.

Ebert, J.-P., Aier, S., Kofahl, G., Becker, A., Burns, B., and Wolisz, A. (2002). Measurement and simulation of the energy consumption of a WLAN interface. Tech. Rep. TKN-02-010, Technical University of Berlin, Telecommunication Networks Group.

Frerking, M. E. (1994). Digital Signal Processing in Communications Systems. Van Nostrand Reinhold.

Green, K. and Wilson, J. C. (2001). Future power sources for mobile communications. *Electronics and Communication Engineering Journal*.

Jones, C., Sivalingam, K., Agrawal, P., and Chen, J. (1999). A survey of energy efficient network protocols for wireless networks. In *Proceedings of DATE*, pp. 77–81.

Kravets, R. and Krishnan, P. (2000). Application-driven power management for mobile communication. *Wireless Networks*, vol. 6, no. 4, pp. 263–277.

Krishna, R., Mahlke, S., and Austin, T. (2003). Architectural optimizations for low-power, real-time speech recognition. In *Proceedings of CASES*.

Lettieri, P., Schurgers, C., and Srivastava, M. (1999). Adaptive link layer strategies for energy efficient wireless networking. *Wireless Networks*, vol. 5, pp. 339–355.

Li, X. and Bilmes, J. (2005). Feature pruning for low-power ASR systems in clean and noisy environments. *IEEE Signal Processing Letters*, vol. 12, no. 7, pp. 489–492.

Maguire, G. Q., Smith, M., and Beadle, H. W. P. (1998). Smartbadges: A wearable computer and communication system. 6th International Workshop on Hardware/Software Codesign, Invited Talk.

Mathew, B., Davis, A., and Fang, Z. (2003). A Low-power accelerator for the SPHINX 3 speech recognition system. In *Proceedings of CASES*, 2003.

Nedevschi, S., Patra, R., and Brewer, E. (2005). Hardware speech recognition for user interfaces in low cost, low power devices. In *Proceedings of DAC*.

Pearce, D. (2001). Developing the ETSI Aurora advanced distributed speech recognition front-end and what next?. In *Proceedings of ASRU*, pp. 131–134.

Proakis, J. G. (1995). *Digital Communications*. McGraw-Hill, 3rd edition.

Shenoy, P. and Radkov, P. (2003) Proxy-assisted power-friendly streaming to mobile devices. In *Proceedings of MMCN*.

Shih, E., Cho, S., Ickes, N., Min, R., Sinha, A., Wang, A., and Chandrakasan, A. (2001). Physical layer driven protocol and algorithm design for energy efficient wireless sensor networks. In *Proceedings of 7th Annual International Conference on Mobile Computing Network*.: SIGMOBILE.

Simunic, T., Benini, L., and Micheli, G. D. (2001a). Energy-efficient design of battery-powered embedded systems. *IEEE TVLSI* (*Special Issue*), pp. 18–28.

Simunic, T., Benini, L., Glynn, P., and Micheli, G. D. (2001b). Event-driven power management. *IEEE Transactions on CAD*.

Smith, P. and Hasler, P. (2002). Analog speech recognition project. In *Proceedings of ICASSP*, pp. IV-3988–IV-3991.

Valenti, M., Robert, M., and Reed, J. (2002). On the throughput of Bluetooth data transmissions. *IEEE Wireless Communications and Networking Conference*, vol. 1, pp. 119–123.

Wicker, S. B. (1995). *Error Control Systems for Digital Communication and Storage*. Simon and Schuster.

Zhu, Q. and Alwan, A. (2001). An efficient and scalable 2D DCT-based feature coding scheme for remote speech recognition. In *Proceedings of ICASSP*.

Index